Don C. Wallace

W6AM

Amateur Radio's Pioneer

IN MEMORIAM

"What am I in this world for? I hope to be loved and respected by my community whether my `community' be five, five thousand, or five million. I hope to leave this world a better place to live in, some of this improvement being due to my efforts. In my own way and in my own chosen field, things are going to be helped along, to a happier, more pleasant status for all."

My Creed D. C. Wallace July 3, 1924

Don C. Wallace

W6AM

Amateur Radio's Pioneer

Jan David Perkins

N6AW

The Vestal Press. Ltd.
Vestal NY 13851-0097

Library of Congress Cataloging-in-Publication Data

Perkins, Jan David, 1948-
Don C. Wallace : W6AM, amateur radio's pioneer / Jan David Perkins.
p. cm.
Includes bibliographical references and index.
ISBN 0-911572-99-6 : $29.95
1. Wallace, Donald Clare, 1898- . 2. Radio operators — United States —Biography. 3. Amateur radio stations — History. I. Title.
TK6545.W25P47 1991
621.3841'6'092—dc20
[B] 91-8094
CIP

First printing 1991
Printed in the United States of America

Contents

Acknowledgments

There are a number of people whose contribututions made this book possible. Don's heirs gave me unrestricted access to all of his personal papers and documents. Bud Held, the stepson of Charles Stuart, entrusted me with his father's archives, and Bruce Kelley, W2ICE, the curator of the Antique Wireless Association Museum, provided me with a detailed explanation, as well as a demonstration, of the operation of all of the various types of early spark transmitters. Bill Wallace, W6TCG, Ludwell Sibley, KB3EVN, and Ray Furlong, W6QIL, edited portions of the manuscript and Harvey Roehl proofed it. My bride of just a few months, Jill, was very patient while I spent countless hours in front of the computer. To these people, and to the many others who unselfishly gave of their time and effort, my heartfelt thanks.

A tremendous amount of work was required to dismantle Don's 25-acre antenna farm after he became a silent key in 1985. It was a combined effort of numerous people and organizations in the area.

A total of four months, working on a daily basis, was necessary to complete the task of dismantling the rhombic farm, at least the outside work. The final tally came to about 1,200 man hours. The W6AM team put out a tremendous effort. The consistent support of Don Barnett, N6EKX, Bob Kliman, KB6CIO, Jim Stevenson, KM6B, Tom Alford, W6KP, Jerry Hagen, N6AV, Wayne Gingerich, W6EUF, and members of the Southern California DX Club, San Diego DX Club and the Northern California DX Foundation made this formidable task achievable. Don Wallace, Jr. requested that these people and organizations be given a token of appreciation, and special plaques for them were made up at Trophies, Inc. in Long Beach.

Several of Don's grandchildren came out to the ranch and pitched in. Jonie Wallace Swanson, Alex Wallace, and Anne Wallace Hedley were always available to lend a hand and were a tremendous asset to the effort.

The task of removing and disseminating the outside hardware in an organized fashion was possible due to the cooperation of the new owners of the property, a group headed by Ron Florence, of Carriage Realty. They modified their anticipated construction schedule in order for the W6AM team to have adequate time to dismantle the ranch safely. The author is indebted to the heirs of the estate, Bill Wallace, Don Wallace, Jr., and Betty Jean Green for the opportunity to spend "Summer Camp" at the ranch.

The Wallace family donated one acre of the property to the city, a strip bordering the eastern edge, which was developed into a jogging trail. A total of 83 homes were being constructed on the remaining 24 acres at the site. In June 1987 the "Hundred-Foot-Long-ham Shack" was dismantled.

In order to verify the exact time that certain events occurred in the first two decades of the book, *200 Meters and Down* provided an invaluable reference source.

Preface

This book traces the life of Don Wallace and with it the early history of amateur radio. Don's experience with wireless in 1909 preceded regulations or any form of licensing. Ships at sea, commercial and government stations, and wireless experimenters all shared the same wavelengths.

Born in 1898, Don took an interest in wireless at the age of ten when the industry was in its infancy. The timing was right for a bright and dedicated man to come along and make a lasting imprint. He would be the radio operator for a President, build the best amateur radio station in the country, write a shortwave book, set the standards of operating achievement for many amateurs, share the podium with the most prominent amateurs of the era, build the largest and most powerful amateur radio station in the world, and be the role model for many of the young men entering the hobby. It was the latter that he relished most. Don held an annual open house and would go to great lengths to encourage newcomers to radio.

Although always a hobby to Don, many of the leaders of the radio industry came to know and respect him. The single word that was always used to describe Don and his attitude toward the wireless hobby was "enthusiasm."

About the Author

I grew up in North Beach and first met Don Wallace, W6AM, in 1965 when I was in high school. I belonged to an electronics-oriented Explorer Scout Post and went on a tour of the famous W6AM rhombic farm. When we arrived at Don's rhombic farm on top of Palos Verdes peninsula there were almost too many sights to absorb at once. Don's property had numerous telephone poles of various sizes spread across the hillsides which supported a maze of wire. Twenty years later, when I dismantling the rhombic farm, I was overwhelmed with nostalgia.

I received my novice license three weeks after the visit to the ranch, WN6PNB, and then my general class license that summer, WB6PNB. Over the next year I had contacted the 100 countries required to qualify for the American Radio Relay League DX Century Club and I had joined the Southern California DX Club. I saw W6AM an the first SCDXC meeting I attended and we chatted at length. Don remembered my visit the year before. After the meeting he showed me the mobile radio in his car. Don had a Swan transceiver next to the steering wheel and a kilowatt amplifier in the trunk. He sent with a speed key that he had next to him on the car seat.

In 1968 I was drafted into the United States Army, just one week after I had passed the amateur extra-class examination, and after a year of electronics and radio school was sent to Viet Nam. One evening shortly after I arrived in Pleiku I listened on 20 meters — the loudest signal on the band was W6AM. It was at that time I began to fully realize what a potent signal that Don put out with those rhombic antennas.

When I returned stateside in 1971 I saw Don at the SCDXC meetings just about every month over the next few years and we often chatted about DX. I was working full-time at TRW and going to night school at the California State University at Long Beach, but still found time to work DX contests and do some DXing. My current call sign, N6AW, was received in 1977.

By 1979, about the time I made the ARRL DXCC Honor Roll, Don was having numerous troubles with his equipment at the ranch. He hadn't had anyone to help him work on it for quite a while. I had graduated with a BSEE a year earlier and went to work at Hughes Aircraft Company as an RF design engineer, and had quite a bit of experience with radio equipment from my previous military and aerospace background. I went up to the ranch one evening and worked on an amplifier with a pair of 833s in the final and got it working properly. After that I was invited back and I used to go up to Don's station every couple of weeks. We would have dinner together and then work on the equipment for the rest of the evening. While I was there I noticed a lot of old certificates and awards an the walls. I asked Don about them and he began to relate the history of each one and his early experiences in amateur radio.

It was at that time that I realized that Don's early days in amateur radio started in the spark era, well before World War I, and had been continuous for over 70 years. I began to bring a tape recorder along to the ranch and I started recording his stories. I decided that these experiences would not be lost. Somehow I was going to put this information into a book and preserve this history of radio, as experienced by Don Wallace. Considering that I didn't type, this was to be a bigger challenge than getting on the Honor Roll.

The Dawn of Wireless

Don Wallace persevered. During his life he faced many formidable challenges and was dealt many setbacks. What made him different from most men, aside from being bright and hardworking, was that he wouldn't take no for an answer. Born Donald Clare Wallace on July 10, 1898, he spent his first few years in Bellevue, Minnesota, near Minneapolis. In 1904 the family moved to Long Beach in Southern California. The city, then only 16 years old, had a population of 6,000 people. At the time, according to Don, perhaps 50 people in Long Beach owned an automobile, and just 200 households had telephones. Except for a brief period during the Great War, World War I, and immediately afterward, a total of 10 years, Don would spend the rest of his life in Long Beach.

His father, William Wallace, established the Exchange National Bank at Third Street and Pine Avenue. The next year he purchased a 1905 Model C Ford for $800. Once the Wallace family settled into their new residence at 1431 Linden Avenue, young Don found a box of his father's college textbooks. Being inquisitive by nature, he started reading the physics books. He found the part about electricity fascinating. There were descriptions of bells, buzzers, transformers, and a variety of other gadgets. Don salvaged wire from old doorbells and other sources to build the projects described in the books. The possibilities were endless.

In 1909, guided by a description in *Electrician and Mechanic*, Don constructed a primitive receiving set, making use of a large tuning coil and a coherer-decoherer. Nearly everything was homemade. The coil was wound with enameled wire on an oatmeal box. In his first attempt he used a rolling pin as the coil form, he said, but his mother soon reclaimed it. One end of the oatmeal box coil was connected to the antenna. A wire trailed out the window, up to the eaves of Don's house, then across the fence to the neighbor's house next door. The other end of the coil went to a cold water pipe outside his bedroom window for the ground connection. A tap slid along one side of the coil, where the enameled wire was scraped bare, to tune in the desired station.

The coherer-decoherer served as the detector. It was a glass tube from a chemistry set partially filled with iron filings. Inserted into each end was a brass plug. The magnetic field from the received RF pulse of the desired station would align (cohere) the filings and create a low resistance, which triggered a relay. The sound of the relay closing, sort of a "thunk", would be interpreted as a dot or dash in the code. As the relay closed, its armature tapped on the tube and repositioned (decohered) the filings after the pulse had passed. The filings now presented a high resistance and released the relay for the next pulse of RF to be received. Due to the mechanical limitations of the relay the maximum code speed that could be deciphered was just ten words per minute. Voice signals, which were a rare occurrence even with government stations, could not be received with this apparatus.

The result of this effort was being able to hear perhaps a half-dozen stations up to 5 or 6 miles away. Broadcast stations would not come into existence for another eleven years. Those stations that transmitted at this time consisted primarily of land-based Naval stations, commercial stations, ships at sea, and the first wireless experimenters. They all shared the same portion of the radio spectrum. Theodore Roosevelt was President of the United States, and the first regulations governing radio licensing and operation were still three years away. These experimenters were later to become the first amateur radio operators, of which there were perhaps five to six hundred in the entire country. Most of them were concentrated in New York City and the northeastern vicinity. The western states and territories were sparsely populated so far — there were only two other experimenters in Long Beach.

The term "amateur" really isn't descriptive of the early pioneers of radio. The members of this group contained the whole spectrum of knowledge, from one end to the other. Youngsters like Don were largely self-taught. There were no wireless clubs or electronics classes yet in existence. An individual's own motivation to read and experiment was the only available means to learn this new development that came to be called radio. At the other end of the spectrum were those physicists and experimenters who were rewriting the books and changing history.

About fifteen years before Don was born, Edison had noted that when a metal plate was placed in one of his light bulbs and a battery connected to it, an electron stream flowed from the filament to the plate, across the ether (vacuum). He modestly described it as the Edison effect but didn't realize its potential. Some years later, in 1904, Fleming found that the filament and plate would function as a detector, and that it was a superior means of rectifying the received signal in a receiver, replacing the coherer-decoherer. Fleming was granted a patent on his "vacuum valve."

Another of the noted "amateurs" was Lee de Forest. He had been experimenting with adding additional elements to the Fleming vacuum valve. By placing a fine S-shaped wire between the filament and plate the first amplification was achieved. In 1907 de Forest was granted a patent on his audion. This, of course, would be known in later years as a triode (for three electrodes). It would be another five years before these "spherical" audions were commercially available in small numbers. Even then the quantities were quite limited for the next decade and the cost prohibitive for many people.

Now that Don had a basic receiver with which he could hear a handful of stations, he had another task to complete: learning the telegraph code. Because of its efficiency the majority of stations used code to communicate with. Voice modulation techniques were inefficient, the quality was poor, and even the large government stations had a severely limited range with voice transmission.

Don, now in junior high school, could venture some distance from home on weekends as long as he was back by supper time and his homework was done. The United Wireless commercial station PJ was just six miles away in San Pedro, a half-hour ride on his bicycle. The radio operators there were quite friendly, and soon Don was a frequent visitor on Saturdays. They drew up a list of the characters of the Morse code for him to learn, and after his homework was completed each day he studied the code (See appendix 1).

The Morse code was developed for sending messages on telegraph lines between land stations. It had been devised by Samuel F.B. Morse in 1835, nine years before land telegraphy lines between cities were first installed. American ships and wireless shore stations continued to use this code until the *Titanic* disaster in April 1912. Shortly after that they adapted the Continental or International Morse code for use when transmitting with radio apparatus. The International Morse code was relatively new, as Marconi had performed the first wireless demonstrations just a decade earlier.

Once Don had learned the basic code characters and could receive at a moderate speed, the PJ operators showed him how to build a telegraph key from some strips of brass by soldering them together with one of their big soldering irons. Soon he was spending much of his spare time practicing the code with a homemade buzzer. Through all of this activity Don managed to get good grades. His parents required that his school work not suffer as a result of this hobby.

The visits to PJ continued for some time and Don became quite proficient with the code. Often he would listen to the messages that were coming in from the ships and check his copy against that of the regular operator. The way they had their coherer-decoherer rigged, with a tobacco can pressed against the relay frame, enhanced the sound of the relay's "thunk" and made copying it easier. A friend of Don's from school who lived a couple of blocks away, John Cook, often accompanied him on these outings.

They strung a wire the three blocks between their houses and set up buzzers at each end. As telephones were available only to the affluent, this was a handy form of communication as well as code practice. It did have its pitfalls, however. One day while the boys were playing in the yard a railroad man came over, rolling up their telegraph wire as he approached. In the route between their houses it had crossed over the railroad tracks and when the railroad man discovered it he had cut it down. So, from time to time, Don and John had to splice and reroute their private telegraph wire.

Don yearned to build a transmitter but simply lacked the resources. The extra income he accumulated by mowing yards and doing odd jobs around the neighborhood was not substantial enough to purchase the required items. Several of the electrical magazines of the day had wireless columns and the PJ operators were very knowledgeable, so Don had a good idea of what components were required for a transmitter. On a summer day in 1910 his luck took a turn for the better.

One Saturday the Wallace family traveled out to Orange County to visit some of their relatives, the Beebe family. Their destination was Anaheim, near the area that would later become Disneyland. The distance, 30 miles, was a substantial trip over the dirt roads that traversed Orange County in 1910. The Wallaces started out for Anaheim early in the morning. A trip of that length usually meant having to repair one or two flat tires along the way, and the journey would take several hours to complete. The roads, unimproved, were along the borders of the one-mile square sections of land. One road they traveled on went past an orange farmer's house. He had two daughters, named Kate and Ella, and the Orange County portion of the road later became known as Katella Avenue.

The Beebe family had a fine home in Anaheim. Don's grand uncle, Albert J. Wallace, had been the lieutenant governor of California some years earlier when Hiram Johnson was governor. Don's uncle, Jay Beebe, was a doctor with a successful family practice. A short time after they arrived, Don and his cousins went outside to play. After throwing a ball around for a while, they went exploring. Don had never been this far out in the country before. There were orange trees as far as you could see. And there was a gas station down the street at the crossroads. Don and his cousins were playing hide-and-seek near the junkpile behind the station. It was there they discovered a gold mine, at least in Don's eyes. A pile of old car batteries had been discarded. Don asked the station owner if he could take some. The man replied that Don could have all he wanted, that Don would be doing him a favor. The batteries were not usable in cars any longer and it was a long trip to the nearest dump. At that time the "hot-shot" 6-volt dry-cell batteries were used only to provide voltage to an automobile's ignition system, as all cars were hand-cranked to start them. The first electric automobile starter would be introduced by Cadillac two years later, in 1912. With the help of his cousins, Don spent the rest of the afternoon carrying the batteries back to his uncle's house, and loading them into his Dad's car. A discarded Ford spark coil was also retrieved from the dump, as well as an old stove pipe which he strapped across the back of the car. The next morning when the Wallace family was preparing to return to Long Beach, his father noticed not only the extra baggage, but that their Ford was riding a bit low. Don 'fessed up to the fact that he had found some components for a wireless transmitter and wanted to take them back home. Little did Don's parents know the impact it would have on their neighborhood when they agreed to bring the items back to Long Beach.

Radio stores were not in existence yet, at least not on the West Coast. A few major components could be mail-ordered from sources such as the Electro Importing Company in New York, but everything else was home-built. Only a few magazines, such as *Modern Electrics*, were available for construction ideas. The state-of-the-art transmitter then was the spark gap. It consisted of a pair of electrodes which connected to a high voltage transformer. The energy developed by the spark across the electrodes when the key was pressed was coupled to the antenna and radiated as a radio signal.

Upon their return from the wilds of Orange County, Don got busy making a transmitter with his new treasures. The batteries were quite weak, but most had full open-circuit voltage across their terminals. By putting all of them in series, their voltage would add. Each one was six volts, and the total of all of them connected together was nearly 100 volts. By connecting them to a transformer, and rewiring an interrupter from a Ford spark coil, Don could generate a high voltage out of the

A spark key used at a commercial station.

transformer. The key would act as a switch in series, and the moment it was pressed there would be full voltage for an instant. Because the batteries were weak, the voltage would rapidly drop off. That was acceptable, as the principle upon which the spark gap worked full voltage was only needed for the instant the electrodes arced across. Although Don didn't know it at the time, the principles of plasma physics maintained the arc at a lower voltage as long as the key was pressed.

Next, he started to work on the high-voltage transformer. Commercial units were available by mail order, but the cost was prohibitive. Here is where the stove pipe came in. His father had a pair of sheep shears in the barn, and with some effort he was able to cut the stove pipe into strips. By interleaving waxed paper and pieces of stove pipe, Don fashioned a core for the transformer. A long piece of copper wire was needed. One with splices wouldn't do, as it would arc over to the adjacent turn of wire on the transformer core. This was the one item he purchased for his transmitter. Copper wire was expensive, and paying for it took all of his odd-job money.

The transformer input, or primary winding, was wound with large wire close to the core. The high voltage output, or secondary winding, was wound with several thousand turns of smaller wire on top of the primary winding. When keyed with the bank of batteries and the Ford interrupter, the result was in the vicinity of 10,000 volts being developed across the output of the transformer. Don had no way of measuring the voltage. He could estimate it by the space between the secondary leads and if the voltage would still arc over when the key was pressed. The transformer certainly drew a big, wicked-sounding arc when the secondary leads were brought within an inch of each other.

Don's next project was the electrodes for the arc. The large brass spheres pictured in the magazine's diagram were too expensive to fit into his budget. The fellows at PJ told him that for the electrodes a piece of carbon or zinc would work just as well. They had a zinc rod on hand and cut off two pieces for him. He drilled a hole in two larger brass rods that would hold the zinc electrodes and mounted the fixture on a wooden base.

All this activity did not go unnoticed by his parents. They were pleased that Don was so motivated by a hobby, but they were not especially happy about this one. The final straw came the first time Don fired up the spark transmitter in his bedroom. The noise of the high voltage arcing across the electrodes — similar to a firecracker exploding — brought everyone in the household running. The batteries, transformer, spark gap, and receiving appaatus were exiled to the barn. Nearly meeting a similar fate himself, Don was on good behavior for some time after that.

Actually, the move to the barn was for the better. Don put up a longer antenna, running from the roof of the barn to a 40-foot-high mast that he erected out in front of the house, and then to a palm tree down the street. The mast was made out of used 2 x 4 lumber and had cross pieces of wood bolted to it to make climbing easier. Don brought the antenna wire in by the transmitter and installed a ground stake a few feet away. After some experimentation he was able to get the transmitter operating properly with the antenna. Now he could use the rig without disturbing the household too much. A homemade hot-wire ammeter, built from a description in *Modern Electrics,* indicated RF current and showed that Don was getting on the order of 150 watts to the antenna. This was the fall of 1910, and with an apparatus such as this, distances of ten to twelve miles could be achieved at night. Don discovered that during the day the radio set was noisier and the maximum distance that could be achieved was somewhat less, perhaps five miles. The wireless experimenters and the government and commercial stations all shared the same wavelengths. To the casual observer this must have seemed like bedlam, but to those young enthusiasts like Don it was heaven. As there was no licensing authority established yet, Don selected the call sign WU. Many experimenters used their own initials for call signs. However, Don had always been impressed with Western Union so he borrowed theirs.

One should not underestimate the absolute determination of these early experimenters. Don's first attempt at putting up a mast was not a complete success. Seventy years later he recalled "*Well, I had bolted some 2 x 4's together to make the mast about thirty feet long. Several friends and neighbors had come out to help with its erection, and a few spectators gathered also. Once the mast was vertical I climbed up to about the 25-foot point and it started tilting. I guess that it wasn't guyed very well. I could feel it going over, so I loosened my grip and managed to climb and slide down a few steps and then just hung on. I was maybe 10 feet up when the mast went over. I was okay, except for a lot of splinters in my legs and arms and chest. So I laid on the ground for a while and all the girls gathered around with tweezers and got out most of them. Some of the splinters I had to take out myself, you know. It was great fun.*" Not easily deterred, Don added 10 feet to the mast and put it up the next week. The passage of time usually takes the rough edge off unpleasant experiences — undoubtedly Don's first attempt at putting up an antenna mast was quite painful.

The commercial wireless stations, which ran much more power than the experimenters and had tremendous antennas strung from high towers, usually overwhelmed anyone else on the air nearby. However, over the next couple of years the sheer volume of wireless experimenters coming on the air with transmitters was creating problems for the big stations. On August 13, 1912 the new radio law was signed by President Taft. Stations and operators would be licensed by the Department of Commerce. The best wavelengths, in the region of 600 to 2,000 meters (500 KHz to 150 KHz), were assigned to the government and commercial radio stations. The amateurs were restricted to what was considered useless wavelengths, 200 meters and down (1500 KHz and up). This corresponded to the region from the top of the modern AM broadcast band, up through the HF, VHF and UHF (television) bands, past the radar and satellite communications region, and on up to blue light.

As a result of the *Titanic* disaster on April 12 that year, American ships and shore radio stations began using the Continental, or International Morse, code (see Appendix 1). Morse code was still used for land telegraph communications. It was rather confusing for Don at first, because thirteen of the letter characters and nine of the numbers were different between the two codes, but with time and practice he became knowledgeable in both.

One day in 1913 Don was on the air when a station he contacted noted his two letter call sign, WU, and asked if he had sent in for a license yet. That was when Don learned about the new license requirements for wireless stations. There was no local radio club in the Long Beach area, and no amateur radio publication existed. Don went down to the Exchange National Bank, where his father was the manager, and got a statement notarized that he could send and receive the code at five words per minute. He sent this document to the Department of Commerce in Washington, DC, and was issued the call sign 6OC. His friend John Cook received 6OD. Their licenses arrived late in the year and their calls are first listed in supplement No. 3 (April 1, 1914) of the first edition of *Radio Stations of the United States* (July 1, 1913). For the next 19 years the Department of Commerce issued a call book annually, which was available from the U.S. Government Printing Office for 15 cents. By this time there were some 2,000 licensed amateur stations in the country. The actual assignment of calls was handled by the Bureau of Navigation of the Department of Commerce. Groups of states were placed in a district and assigned a number for the prefix of the call sign. There were a total of nine districts. The sixth district consisted of California, Nevada, Utah, the Territory of Arizona (it achieved statehood in 1912), and the United States possessions in the Pacific (see Appendix 2).

During this time Don was steadily improving his receiver and transmitter and experimenting with various methods of stringing an antenna. The coherer-decoherer had given way to a cat-whisker with a galena crystal mounted on the same base. The cat-whisker pressed against the crystal and was adjusted across its surface until an impurity was found in the crystal—that was the most sensitive spot to receive the desired signal. Instead of the thunk of a relay, a weak audio signal was generated by the rectification that occurred at the junction of the cat-whisker and the crystal. To listen to the signal Don made a single head-receiver (later known as headphones) which make code copying much easier. Faster speeds could also be copied because the receiving method no longer required a mechanical device. A loose coupler and variable condenser replaced the slide tuner.

Don climbing his antenna mast in 1912, thought to be the highest in Long Beach at that time.

This method of receiving gave improved selectivity, which was necessary with all of the new amateur stations coming on the air.

In 1914 Don's zinc spark gap transmitter gave way to a rotary spark gap (which had been a Christmas present). Brass contacts were mounted on an insulated bracket on the motor's armature and rotated past a fixed contact. When the high voltage was keyed the contacts on the armature arced across to the fixed contact as they rotated by it. This method of transmission gave a higher-pitched note, which was easier to copy at the receiving station. On a cat-whisker crystal set the received signal from a zinc spark gap had sounded somewhat like ignition noise. The disadvantage of the rotary spark gap was that it made a much louder noise when one transmitted. A large coil and condenser, to transfer the spark gap energy to the antenna and its feed line, had been added to the set. The condenser was fashioned by pasting pieces of tin foil on each side of a piece of plate (window) glass. Don wound a large coil out of some old wire used for house electrical wiring. After the insulation was stripped off it was suitable for the output inductor. This made the transmitted bandwidth of the signal narrower, in line with current wireless regulations.

Seventy years later Don received a letter from Laird (Jack) Jeffers, SOWP SGP 3372. "I was reading the Society of Wireless Pioneers Skippers Log, Issue 82-1. In the Morse telegraph story on page 28 the name Don Wallace seemed to jump at me. Back in 1912, when I was 12 years old, you were perhaps 14 and tall for your age. I lived at 1485 Linden Avenue, just up the street from you. We kids, about 4 of us, Edwin Merrill, Joseph Reardon, and two or three others all looked up to you and regarded you as a genius in mechanics and particularly in electricity. The resemblance in the photograph is still strong, allowing for the changes wrought by 70 years."

Don often chatted over the air with John Cook, who had also built a transmitter. Once it was operating their telegraph wire became obsolete. The last time it was cut down by the railroad men, Don and John added the wire to their antennas. Don fashioned a round wooden disc several inches across and 1/4-inch thick. He installed a sequence of brass screws and strips of brass along its edge which corresponded to the code letters 6OC. He then mounted it on a stand and wired it across his code key. At the end of a transmission Don would sign his call by reaching over and rotating the wheel.

The high-voltage transformer would get hot when he was transmitting, and occasionally smoke would pour out. Each time it developed a burnt odor he had to rewind it. Don also added more pieces of stove pipe to the core as it became available and got better at winding the wire on the core. It would now withstand being keyed while plugged into the 110 VAC line voltage and he estimated that it produced upward of 20,000 volts. The batteries were retired to a shed behind the barn and kept for emergency purposes. The past year Don had relocated his antenna to the palm trees out in front of his house. Long Beach city codes required that home owners plant palm trees in front of their homes along the street. The palm trees grew rather quickly, so Don's 3-block-long antenna was getting higher every year. His antenna twice crossed above the high-voltage lines of two Edison Company power poles. History has not recorded how Don managed to string the antenna wire safely, but his parents were very concerned over these forays of his.

A shortcoming of most antennas in those days were inefficient support insulators. Most commercially made insulators were very inefficient and would remain so for a number of years. Until the glazed porcelain type was developed, signal power would be absorbed by poor quality antenna insulators. Realizing this, Don supported his antenna by insulators made from the glass in house window panes. These were fashioned by taking an old piece of window glass, breaking it into long narrow strips, and drilling a hole in each end. A wooden spreader was placed on each end of the antenna to separate the individual wires and several antenna wires were supported in this fashion.

By that time many amateur stations were superior to the commercial stations. The commercial stations could legally use the loose-coupled receiving tuners only if they purchased Marconi equipment, which was quite expensive. Marconi controlled the famous "four sevens" patent on loose couplers. Many commercial stations used obsolete untuned or direct-coupled receiving equipment rather than paying the patent royalty. The amateur radio stations were not constrained by the patent, as their use of loose couplers was personal rather than commercial.

The power line frequency in Southern California was 50 cycles per second at that time. A straight carbon or zinc spark gap had a 50-cycle (Hertz) note, not unlike the sound of ignition noise in a modern AM receiver. The smoothness or roughness of the received note was determined by whether the spark gap electrodes were flat or filed to a point. The coupling circuit between the spark gap and the antenna determined if the signal faded fast or slowly trailed away when the transmitter was keyed. Tight coupling from the spark gap put more power to the antenna but gave the signal a fast fading effect. It also had another undesirable characteristic. The width of the transmitted signal was broadened and caused the signal to occupy two distinct places in the radio spectrum. Loose coupling resulted in a cleaner signal, but less of the power that was generated in the spark gap was coupled to the antenna. The signal from the non-synchronous rotary spark rigs had a higher pitched note at the receiving station than did the fixed spark gap. The number of times per second the brass contacts rotated past the fixed contact, causing a momentary spark between them, determined the pitch of the note. The motors that were available ran at varying speeds and different configurations of contacts mounted on the rotating armature gave variety of spark rates. In this manner, even though there were dozens of amateurs and experimenters on the air in Long Beach and the adjacent cities in the Los Angeles basin by 1913, each transmitter carried its own particular note, or signature. You could identify the owner before he gave his call sign.

Some 70 years later Don described his rotary spark rig:

"We had a barn for Nellie (his horse), and a corral. When I got interested in amateur radio — it was called wireless then — my dad had a Scandinavian fellow, who was a nice carpenter, build me a little room in part of this barn. Our buggy was there, and the horse. Here I would be operating five or six feet from the nose of my Nellie, who was eating her grain or whatever. I could shut the door and I had an oil heater in there to warm the room up. People didn't worry about oil heaters in those days.

"Long Beach, near the ocean, is consistently cold at night, especially in the winter. I liked to go out there and operate in the morning. Something happened in the morning at sunrise, even then. You could work a little further in the morning than you could at night.

"This little room had a floor on it, above the concrete of the stable, the carriage house. My Dad had built the barn so that the architecture looked like the house, and it was a nice-looking place. Well, the spark transmitter in this little radio room made so much racket that the neighbors mentioned to Dad that I made an awful lot of noise. I woke them up in the morning with my spark transmitter, which of course is about as loud as a cannon firecracker on the Fourth of July.

"So I built a box out of wood and I covered the spark gap with rug pieces, seven or eight layers deep. I had a lid that went over the whole thing with the same number of rug pieces on top. After that the spark was essentially shut up to the point that there were no more objections from the neighbors. How-

ever, that made the leads longer to the antenna inductance. I doubt if the transmitter was ever as efficient after that. Also, I would have to take it apart and put in new spark electrodes periodically because it didn't dissipate the heat that it generated. The heat of the spark, a kilowatt or so jumping across there like in the form of an arc, would heat up the electrodes. It was an early form of TVI (He laughed). "It was asserted over the air as sound waves rather than through a speaker.

"There was another thing that happened in those days. It was a peculiar effect, and the certain combination of inductance and transmission caused what we called a kickback. Some of the jolt in the power lines at a certain frequency would fire back into the line, and maybe light fixtures would blow up at our house, or somebody's a block away, or in between. There would be a big backfire. The sound would almost knock you out of the chair and you wouldn't know what happened. But it was an awful backfire, and this high-voltage spike would go through the whole system,up and down the block.

"We called it kickback. The best way to stop it was to put a carbon rod of enough ohmage so that it didn't short the 115-volt AC across the power line and ground the center tap. The surge would go across this big resistor at the point where the power went into the transmitter and be grounded instead of going back into the line. Eventually we all got these big carbon rods and the ohmage would be a few thousand ohms, so the power consumed would only be 30 or 40 watts when they were wired across the circuit.

"People talk about interference with the neighbors now, but we had it right off the bat. First I had the noise of the spark, which you could hear for a block or two. In the room it sounded like shooting off medium sized cannon firecrackers. That's what made it so much fun. Kids enjoyed all that racket. Well, it seems that the bigger racket you made, the more distance you achieved. So we all tried to get bigger, stronger signals, and bigger, longer aerials. I never heard of a one-wire aerial in those days. We always used big spreaders, I think I got up to about 30-foot spreaders at each end of the antenna, maybe six wires supported in between them. It seemed like the more wires you had the louder you would be with your transmitter, and the more stations you would hear.

"In the very early days all we had was a coherer. Those were very difficult to make and adjust. Later came the crystal detectors and it became quite a search by everyone to find a more sensitive piece of galena. We tried everything under the sun; we must have tried 30 or 40 different materials as the detector. You could hear signals on carbon. You could hear more signals on silicon. You could hear some on ferrite, pyrite. You could even have a couple of corroded wires put together and get some signals. For a detector, galena was the best material of the day. Then the question was, how to protect the crystal? Galena varied in sensitivity over a wide range, because the spots of impurity on the crystal varied. But if you transmitted nearby there was so much RF floating around that, you burnt off a spot on the crystal and you had less good spots. So we started putting a knife switch right next to the crystal, just like the power switch which started the transmitter. When I transmitted I would throw the switch and short the stray RF around the crystal.

"We didn't know any form of keying except to key the transformer. So there was quite a jerk on the line, from the surge through the transformer core, particularly if you happened to hit the reverse cycle

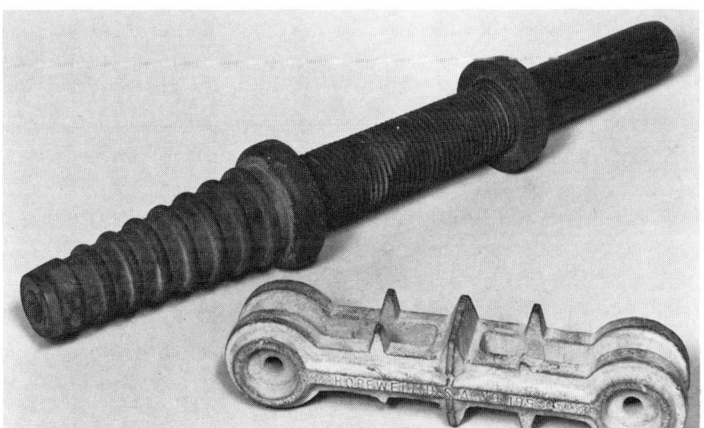

Early insulators were very lossey.

A state-of-the-art spark gap amateur radio station in 1914. (Tnx Bruce Kelley, W2ICE, and the Antique Wireless Association.)

of the AC line voltage that it left on the prior dot. So, periodically you had quite a line voltage drop that would blink the lights. Of course the lights would light up, or dim, and cause all kinds of commotion, both in your own home and your neighbors' homes. However, my Dad and mother appeared to be very tolerant. We solved most of those things. We even put in compensation resistors across the transformer primary, so that when the key was up it would throw in a resistor load, so it wouldn't blink the lights. Current was so cheap in those days, no one ever mentioned power bills to me.

"I know our water bill was a dollar a month, no matter what we used. The city found we had a horse and they wanted to raise the water bill, but Dad said `Well, the horse doesn't need any more water than one tree out here.' So they passed that up. However, if you let water run in the street a man came along and you were fined a dollar for letting water run off of your property. Otherwise you could use all you wanted. Our city of Long Beach became very green because water was a fixed charge. A dollar a house."

It was an exciting time to be a radio operator. The aficionados of wireless all shared a common interest: experimenting in this exciting new phenomena that would come to be called radio.

Don's Lost Station

Don entered Long Beach High School in the fall of 1912. The school had a four-year program; entering freshmen were ninth graders. Once other high schools were built it would become known as Long Beach Polytechnic High School. Don's two primary interests were amateur wireless and football. In the course of his studies and school activities that year, he met a number of other "wireless experimenters." They started sitting together from time to time, and by late 1913 had formed the Long Beach High School Amateur Wireless Club.

Since most of them were only 15 to 18 years old, they tended to be, for the most part, long on enthusiasm and short on knowledge. Don was advanced in hands-on experience. He had the best station, too. Most of the others used an untuned Ford spark coil to transmit a maximum distance of 3 or 4 miles. Everyone was self-taught—there were no electronics or radio classes in the school curriculum yet. In short order Don was elected manager of the station that they intended to build at the high school.

Their enthusiasm must have convinced school officials to furnish them with an unused storeroom for their radio station. They pooled resources and dug into their junk boxes. Soon, enough parts had been assembled to build a cat-whisker crystal receiver and a spark transmitter. The station was entirely homemade. No one had funds to purchase any parts but wire. Don wrote the Department of Commerce, requesting that a call sign be issued to the school. He had recently received his own call sign and was familiar with the procedure. He became trustee of the station at Long Beach High School, which was issued 6LN.

Exactly how it was arranged is not clear, but in the fall of 1914 two 60-foot telephone poles appeared on the high school campus. One was erected near the club station and the other one at the far end of the campus. A fan arrangement of wires soon appeared between the poles and 6LN became a formidable signal. Don, already a persuasive individual, undoubtedly convinced the school officials that the poles were a good idea. Seventy-five years later the poles were still up.

America's first wireless organizations were formed in New York in 1909—the Junior Wireless Club, now the Radio Club of America, and the Wireless Institute, forerunner of the IRE/IEEE. Over the next several years "wireless clubs" began to spring up in the larger cities across the nation. Some years would pass before the term "amateur radio" became commonly used.

By 1913 many amateur rigs had improved to the point where stations could work up to 50 miles on a good night. A year earlier Don could barely hear Howard Seefred, 6EA, located in Los Angeles. Now, with the antenna in the palm trees and the galena crystal detector, Howard almost always had a strong signal in Long Beach. Don decided to try working him using voice instead of code.

Howard Seefred was not a typical amateur radio operator. Don remarked in later years that almost all of the amateurs were pretty young. Most older people didn't understand the concept of radio. Wireless was too radically different from what they had been taught about electricity, which was relatively new and difficult enough for them to comprehend. Howard was in his early thirties, very methodical, and he kept a meticulous journal of his wireless experiments. He had studied the modulation problem for some time and knew the problems of impressing a voice on a radio signal and then reconstructing it at the receiving station. In their many contacts using continental code, Howard told Don how to construct a carbon microphone. It would have to be placed in the antenna feed line or the station's ground return, as that was the only known way to modulate a spark transmitter. If it was installed in the feed line, then care must be taken with the insulation as the high voltage present could give a nasty shock if one accidently bumped the microphone while talking. In the more powerful stations it could be fatal.

In a letter to Don, Howard noted: "The diagram is self- explanatory with the exception of the gap which is made of two extremely hard carbons such as are found in dry cells. A close adjustment is necessary for proper operation. The arc must be of purple-blue color to get the best results. If the arc turns to a white glare it shows the carbons are burning." The rotary spark gap was disconnected and Don's straight spark gap transmitter was wired up. The arc would have to be constant while voice modulation was in process. Don followed Howard's instructions and fashioned a carbon microphone which he spliced into the antenna output from the transmitter. He found a piece of asbestos sheet to use as an insulated mouthpiece, forming a funnel shape at the microphone.

Earlier Howard had similarly modified another transmitter of his, placing the microphone in the ground lead. One evening in the fall all the preparations had been made. First he chatted with Don on code and found that signals were very good that night. They switched to voice, and each transmitted briefly. Don had to readjust his receiver, as the transmit frequency changed slightly with the microphone in line. They exchanged signal reports, but Don's modulation was very poor and his signal very weak. Don also had smoke pouring from both his microphone and his power transformer. It seems that amateur radio was not yet ready for voice communications. Howard described his experiments two years later in *QST* (September 1916) and noted: "The writers have experimented three years with this 60-cyclephone and have talked 22 miles to an amateur in Long Beach. They were also heard by another in Pomona, 30 miles distant." Howard Seefred had the only successful phone transmitter in the Los Angeles area prior to World War I. In the period of 1913 to 1916 he played music over his wireless apparatus several evenings a week, which was often heard by other experimenters up to 20 miles away.

Don stuck to the code after that brief experience with phone. Spark transmitters were not a very efficient method to transmit voice signals. One morning early the next year, he worked Charles Stuart, 6CW, in Ventura. By early 1914 the better wireless stations could

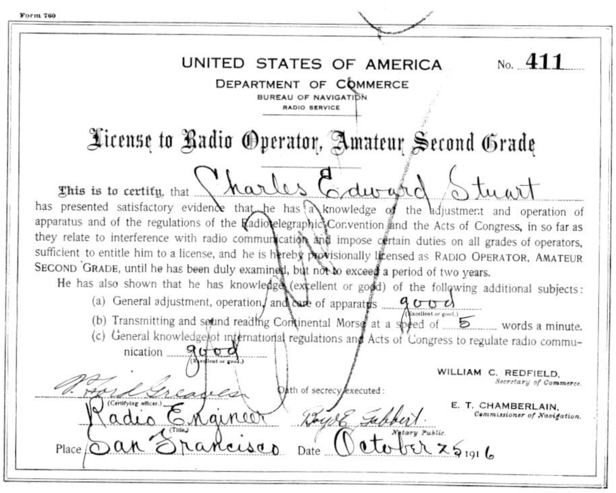

Form 760

UNITED STATES OF AMERICA — No. 411

DEPARTMENT OF COMMERCE

BUREAU OF NAVIGATION

RADIO SERVICE

License to Radio Operator, Amateur Second Grade

This is to certify, that Charles Edward Stuart has presented satisfactory evidence that he has a knowledge of the adjustment and operation of apparatus and of the regulations of the Radiotelegraphic Convention and the Acts of Congress, in so far as they relate to interference with radio communication and impose certain duties on all grades of operators, sufficient to entitle him to a license, and he is hereby provisionally licensed as RADIO OPERATOR, AMATEUR SECOND GRADE, until he has been duly examined, but not to exceed a period of two years.

He has also shown that he has knowledge (excellent or good) of the following additional subjects:

(a) General adjustment, operation, and care of apparatus good (Excellent or good.)

(b) Transmitting and sound reading Continental Morse at a speed of 5 words a minute.

(c) General knowledge of international regulations and Acts of Congress to regulate radio communication good (Excellent or good.)

WILLIAM C. REDFIELD, *Secretary of Commerce.*

V. Ford Greaves (Certifying officer.)

Radio Engineer (Title.)

Oath of secrecy executed: Boyd E. Tabbert, *Notary Public.*

E. T. CHAMBERLAIN, *Commissioner of Navigation.*

Place San Francisco Date October 25, 1916

The pre-war license of Charles Stuart.

get upwards of 100 miles on a good night. The fact that most of this path between Long Beach and Ventura was across water, due to the curving of the California coast line, didn't hurt the signal strength any.

Charles Stuart was the first licensed radio amateur in Ventura and was written up in his local newspaper. He got started in wireless in 1912. Some thirty years later he described the early days. "My elder brother had gathered sufficient parts to build a complete sending and receiving station. His interest in the crude apparatus soon waned and I fell heir to the spark coil, batteries, key, and coils of wire. By rejuvenating the batteries with vinegar and substituting a silicon crystal for the primitive detector, which consisted of a needle carefully adjusted across the apices of two triangular-shaped carbon blocks, two-way communication was established over a distance of one or two miles." This apparatus had evolved from a coherer-decoherer and later an electrolitic detector. Charles Stuart would continue to improve his apparatus periodically and in the thirties he became internationally known.

During the spring of 1914 Don received some correspondence from a number of amateurs on the East Coast. They had formed an organization called the American Radio Relay League (ARRL), and were contacting all licensed radio amateurs. Don had not yet heard an East Coast station on the air, but after reading their material he was very interested in their organization. To date there was no national society of radio amateurs or a publication devoted solely to amateur wireless. Don sent in the required application and became one of the charter members. The first List of Stations of the ARRL arrived in October 1914, and there were already over 400 members.

In the summer of 1914 a sequence of events took place that would affect amateur radio and Don's future career. Although most people in this country were unaware of it, the Great War had begun in Europe. Canada, being allied with Great Britain, joined the war effort and suspended operating privileges of its amateurs. As wireless receiving distances were still quite limited, only those United States amateurs living near the Canadian border noticed their absence.

During the fall of 1914 Don immersed himself in the high school football team. Because of his size, six feet four inches and 200 pounds, he made the first string even though he was only a junior. At the end of the football season he began spending more time working on his radio set. Although it performed well, it was rather modest compared to some of the neighboring installations that had recently been built. Those operators who had regular jobs could afford the higher-powered components and the new receiving tubes (de Forest audions). There were a few more electrical magazines available now, and he occasionally saw an ad for an audion. By doing odd jobs Don saved a little spending money. For the price of $4.25, a sizable sum of money in those days, he could own one. He sent the money to the Electro Importing Company in New York and in due time received his audion.

The audion was shaped like a small light bulb, with a screw base on one end and wires protruding out the other. They were that shape because de Forest had them made by the McCandless Company which made car lamps for Henry Ford, and each one was handmade. The audion looked quite fragile and Don knew that he would have to treat it carefully. He built a shock mount arrangement for the tube with some springs he had in the junk box. Don took care to mount a large knife switch next to it to short the stray RF power around the audion when he was transmitting. A set of batteries on the floor provided the operating voltage for the audion — the A battery lit the filament and the B battery supplied plate voltage. The batteries were expensive and the A battery had to be recharged from time to time. The initial results with the audion were disappointing. Don had connected the audion between the loose-coupled tuner and the headphones of his receiving set. The performance of the audion wasn't much better than that of the galena detector. He chatted with a number of other amateurs, but very few had any experience with the new audions. After a lot of experimenting, Don had an idea. As well as resonating the input circuit, he placed a coil and capacitor in the filament circuit and tuned it as well. After some trial-and-error adjustments, the tube was a definite improvement over his crystal detector. It seemed that there was so much capacity between the filament and the other elements inside the tube that a large portion of the received signal had been shunted to ground. By tuning the filament circuit the energy that was developed across the input of the tube was heard in the headphones. The improvement was only slightly better than his cat-whisker detector had given. At least he didn't have to periodically adjust the audion as he did with the cat-whisker and galena. Previously, each time the operating table was bumped or the door closed sharply, he had to reset the cat-whisker.

The early "spherical" audions were quite touchy to adjust properly. Each one was unique, due to the primitive manufacturing methods used in those days. The element voltages were critical and required a bit of experimenting with each tube until they were properly set. The filament was fragile and, due to the low vacuum in the tube, burned out in a few months. For this reason the tubes were designed with a center-tapped filament. The audion was wired into the circuit with just one-half of the filament illuminated. Once that half of the filament burned out, the wiring was changed to the other half of the filament and the tube could be used for a few more months.

During this period of time Don's father was considered well-to-do. William Wallace had acquired extensive property holdings around the Long Beach area, as well as Signal Hill. He also entered into a speculative land venture at Lake Elsinore, some 60 miles to the east of Long Beach. The next big land boom in Southern California was expected to be in that area. William wanted his son to learn the value of a dollar, so all aspects of

The early audions were spherical. (Tnx Bruce Kelley and the AWA).

Don's radio station were either homemade or purchased from money that Don earned at odd jobs.

One night in early 1915, Don was chatting with Howard Seefred and found out about a new audion receiving circuit. It had been reported by Edwin Armstrong in the *Institute of Radio Engineers Proceedings* (IRE) in January. Armstrong had developed the concept of the oscillating detector while still a graduate student three years earlier, and apparently it was a very successful circuit for improving signal reception. Don found out the details of the Armstrong circuit from Howard and began rewiring his receiver. By feeding back part of the amplified energy out of the audion back to its input, the tube could be induced to oscillate. This "regenerative" circuit greatly increased the sensitivity of the receiver. After that modification, Don was able to hear stations all the way into the Midwest. In short order, Armstrong's circuit became so widely used by amateurs and experimenters that it became difficult to obtain a de Forest audion. The demand far outstripped the limited production of the tubes and it took months to procure a replacement. At this time numerous "independent" tube manufacturers sprang up. They neglected to secure patent rights from de Forest, and some made a pretty good imitation of his audion.

In March 1915, the League sent out the second edition of the ARRL List of Stations. The organization was growing fast Don noted that there were now almost 600 members. In due course the ARRL began printing a monthly publication and in December 1915 the first issue of *QST* (*The December Radio Relay Bulletin*) arrived. Dues were one dollar per year for those who wished to receive the monthly publication, and a majority of the 635 ARRL members elected to do so. After the initial announcements in the first issue, a series of construction articles appeared in the coming months as well as a number of helpful hints. An advancement in tube technology, due primarily to the competition from unlicensed tube manufacturers, called the tubular de Forest Audion, was announced that spring. For those who were particularly motivated to improve their station's receiving ability, a loose coupled-tuner could now be purchased for as little as fifteen dollars.

Howard Seefred was the author of several articles in the first year of *QST*'s publication. The September 1916 issue had an article on phone operation; "Sixty Cycle Wireless Telephone." Howard and his brother were also featured in the "Who's Who in Amateur Wireless" column in the November issue. In the December 1916 issue of *QST*, 6EA published an article titled "Long Distance Receiving With Galena." Don found *QST* to be an exceptionally valuable magazine.

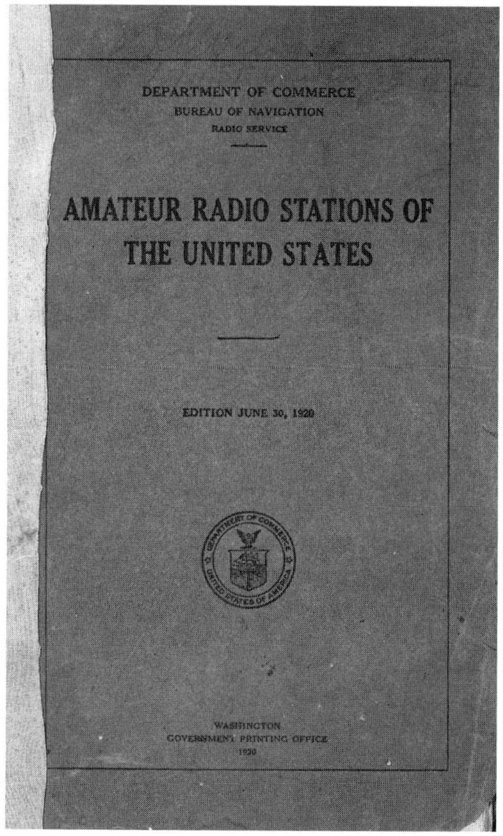

The Department of Commerce Callbook.

The first issues of *QST* were published by Hiram Percy Maxim and Clarence D. Tuska. To stimulate interest in the magazine, they sent many of the early issues free to all licensed amateurs. The first issue was sent to League members and all radio amateurs listed in the government call book. In October 1916 Tuska formed the *QST* Publishing Company. He was a college student in his late teens and *QST*'s office was an attic in his parent's home. His office hours were whatever time he could spare from his studies at school.

Tuska's management policy was to add up the month's receipts, determine how many copies of *QST* he was obligated to print, and ask the printer how many pages could be printed in that quantity with the available money. That method determined the size of each month's magazine. Since *QST* was devoted to amateur wireless it grew rapidly and soon became invaluable to amateurs and experimenters.

By the summer of 1914 Don was one of the more knowledgeable amateur radio operators around Long Beach. He was adept at constructing both the receiving and transmitting apparatus from scratch and he was very proficient with both the Morse (telegraph) and the

International Morse (Continental) code. Don wanted to get a part-time job as a wireless operator, but those jobs were difficult to come by. The minimum requirement was a commercial license and the test was given only at the radio inspector's office in San Francisco. It may as well have been on the moon.

Due to the cost of commercial transportation, Don's only hope of getting to San Francisco was to secure passage on one of the coastal ships that sailed between Southern California and points north. Each weekend Don would go down to Los Angeles Harbor, next to the wireless station (now KPJ), and try to get a job aboard one of the ships. He had easy access to the docks — two years earlier, in June 1912, when his father was vice-president of the Long Beach Chamber of Commerce, William had presided over the opening of the first municipal dock in the harbor that was built with the first harbor bond issue. One day Don found a ship needing someone to clean cabins and he convinced his prospective employers that, because of his height, he would not need to stand on a stool to clean the high portholes. That would be much safer when the ship was under way, they reasoned, and Don got passage to San Francisco as a deck boy on the *S. S. Harvard*. While they were in port he visited the radio inspector, and passed the test for a radio operator, commercial second grade.

Now that Don had a commercial license, on weekends he spent quite a bit of time down at the harbor. He would have to show experience with this license over the next 12 months in order to meet the eligibility requirements to take the first class license test. It wasn't particularly easy for a 16-year old to get the opportunity for wireless experience, but Don was persistent. That winter he was able to fill in on several brief one-day excursions on the *Cabrillo* and the *Hermosa*, which travelled between San Pedro and Catalina Island. That experience was to provide the minimum amount needed to be eligible for the first class commercial examination, and all that he had to do was be able to pass the test. One portion of the exam was on radio theory, and of course, the other was the code. Don secured passage on another coastal ship in the fall of 1915, and again visited the radio inspector in San Francisco. On November 12, 1915, he passed the government test for a radio operator, commercial first-grade license. That license required a code speed of 25 words per minute in Continental and 30 words per minute in Morse. The best thing that could happen to Don now would be to get a summer job as a commercial wireless operator. Imagine having all the fun of operating and repairing a wireless station and getting paid for it! The problem was, you couldn't get the job without the experience, and you couldn't get the experience without the job. Being only 17 didn't help any either. Then an opportunity came along that helped Don's wireless career.

Don still visited the commercial station, KPJ, in San Pedro on a regular basis. (When the 1912 radio law took effect, commercial stations west of the Mississippi were assigned a K prefix to their existing call sign and those east were assigned W.) One day he was at the station, which was receiving normal radio traffic from ships entering the Los Angeles Harbor at San Pedro. There was only one operator on duty, a fellow Don had known for several years. The operator wanted to go out to lunch. Naturally, Don volunteered to watch the store. After some thought, he decided to take Don up on his offer. Don had demonstrated that he was a good operator and it was a slow day.

The operator left for lunch and Don sat down at the operating table. Things were quite slow for a while and then the radio really started singing. A large fire had started up in Avalon, where station KPI was located, and was out of control. Messages were being sent to Don at a rapid pace. Don continued to copy the traffic and retransmit it on the wire circuit whenever there was a short break. A couple of hours later, the primary operator returned. Don was backed up about ten pages of copy. The operator immediately started copying traffic, while Don started relaying it on the wire circuit. Within the hour, they had caught up with all the traffic. No one else ever knew that the regular operator was not at the station during the devastating fire that swept Avalon on November 29, 1915. Don did get a nice recommendation from the operator and introductions to the chief operators of the ships that put into the Harbor at San Pedro.

Nearly 70 years later Don recounted the story:

> *"The KPJ station in San Pedro was on what we now call Terminal Island. There was nothing else there but a few docks and a beer parlor. I liked to hang around the station because I aspired to go to sea any chance I got. The chief engineer was named Baxter. I was sure I wasn't going to tell this story while Baxter was alive. I was there about noon. Baxter said that he wanted to go over and have lunch. I heard later that he liked to go over and play pool and have lunch at this dockside beer parlor. He went over there so all I had to do was take the messages from the ships as they came in. I was sitting there and Avalon called me. KPI was their call.*
>
> *"Avalon had an aerial over there about 150 feet high up the hill, a little bit off the shore. The shore end of the aerial was a pole about 50 feet high. If you*

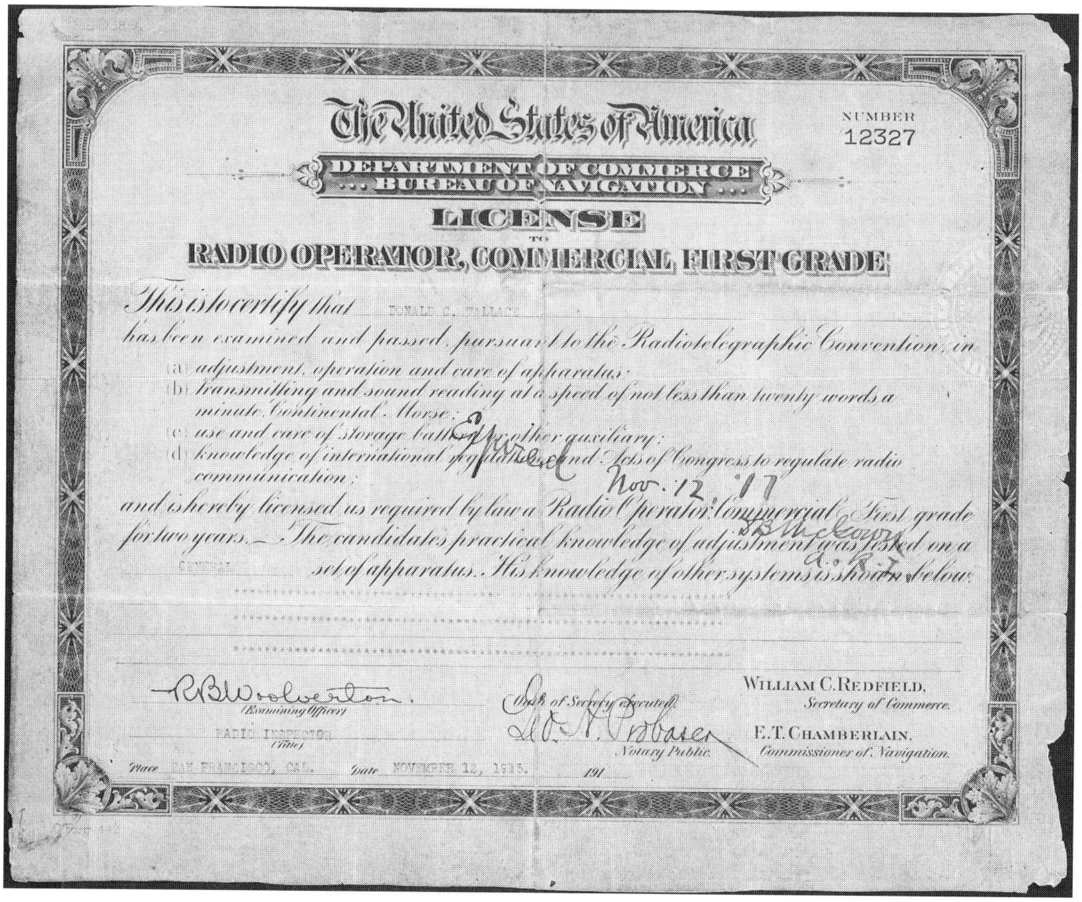

The United States of America

NUMBER 12327

DEPARTMENT OF COMMERCE
... BUREAU OF NAVIGATION ...

LICENSE
TO
RADIO OPERATOR, COMMERCIAL FIRST GRADE

This is to certify that DONALD C. WALLACE has been examined and passed, pursuant to the Radiotelegraphic Convention, in
(a) adjustment, operation and care of apparatus;
(b) transmitting and sound reading at a speed of not less than twenty words a minute, Continental Morse;
(c) use and care of storage battery or other auxiliary;
(d) knowledge of international regulations and Acts of Congress to regulate radio communication;
and is hereby licensed as required by law a Radio Operator, Commercial First grade for two years. The candidate's practical knowledge of adjustment was tested on a GENERAL set of apparatus. His knowledge of other systems is shown below:

Expired Nov. 12, '17

R. B. Woolverton (Examining Officer)
RADIO INSPECTOR (Title)
Oath of Secrecy executed.
Notary Public.
WILLIAM C. REDFIELD, Secretary of Commerce.
E. T. CHAMBERLAIN, Commissioner of Navigation.
Place SAN FRANCISCO, CAL. Date NOVEMBER 12, 1915.

Don's earliest surviving license, Radio, Operator, Commercial First Grade.

go over to Avalon today you will see that pole. It's still there with the yardarm and flags on it. That was the radio pole and the little building under it was the radio shack. Well, KPI called me. The fellow over there, I had seen him because I had been on that run as operator on the Cabrillo *and* Hermosa, *Banning ships. The fellow was hunchbacked; I think that he even used a wheelchair to get to work. It was a nice job for him. He called me, I answered. Avalon had caught on fire. It was burning up.*

"Well, I started writing, in my boyish large handwriting. I had page after page of this continuous story. All I could do was to tell him to go ahead, and I kept writing. This was the story of the fire. After about an hour and a half Baxter came back from his pool game at the beer parlor and saw this pile of paper in front of me. He broke, sent BK, straightened the papers out, sat down, and started typing the rest. I always thought in later years that's probably the reason I could always get a ship job, first choice. I wasn't the official employee of the station, I was just a visitor. I never told a soul about that as long as he was alive. It was a little interesting that the big fire that burned up Avalon, essentially the whole town was wiped out, they did save the radio station because it was right on the edge of the water."

It was the big break that Don needed. The next summer when he graduated from high school, he lined up a full-time job as a wireless operator. The chief operator of the *SS Yosemite* wanted Don as an operator, based on the recommendation from the operator at KPJ. The skipper wasn't so sure about having a 17-year-old youngster aboard ship. This is where Don's persuasiveness came into the picture. Don convinced the skipper that because of his height he would also be able to work on overhead wiring in the radio room without the need of a ladder or stool. Don got the job and proved to be an unusually efficient wireless operator.

The ship worked the West Coast trade. It would steam north to Seattle, drop off its load and pick up a cargo headed south. Every couple of weeks Don got to spend a day or two at home. It was during this period that his code copying skills were really honed. The operators did a four-hour period on duty, and four hours off, around the clock. After three months he could copy incoming messages as fast as any of the other operators could send. Once, while sending traffic he found that he could copy the incoming messages that another operator was receiving. At the end of the summer Don bade farewell to his shipmates and returned home to start school. He had been accepted at Hamline University in St. Paul, Minnesota, his father's alma mater (Class of 1895).

While he was to be away at college, Don decided to store his receiver and transmitter in a safe place. His younger brothers were at an age where they might get into the equipment if he wasn't around to keep an eye on things. Don figured out of sight out of mind. There was a couch in the parlor that had a storage area under the seat. He carefully disconnected the equipment and dismantled it down to major pieces. He placed the oscillation transformer, high-voltage transformer, rotary spark gap, and other transmitter components in the hollow seat. Among these he fitted in the crystal reeiver, loose coupler, the audion circuit, and the long wave coils. Being very meticulous, he placed his logs, QSLs, schematics, 6OC license, call books, *QST,* the pictures of the apparatus, and all other paperwork relating to radio in the same couch. When the lid was put down, nothing showed where his treasure was stored. Next summer the station could be rewired and on the air in a matter of hours.

The Long Beach that Don was leaving was no longer the sleepy little town that it had been in 1904 when the Wallaces first arrived. By 1916 there were nearly 35,000 people living in the city and the major streets in town had been recently paved. Don took the Northern Pacific train to Minneapolis, a three-day trip at that time, and settled into a fraternity house at college. It was the Beta Omicron Sigma Kappa Fraternity located at 823 Snelling. One of his first priorities after enrolling in the fall classes was to visit the radio inspector in Chicago. Don received permission to operate a portable amateur radio station until his new license arrived. He had requested a room on the third floor of the fraternity house to make it easier to bring in feed lines from the antenna that he planned to install. In due course Don's new call sign, 9BU, arrived from the Department of Commerce.

Don immediately got involved in the football team and played the position of center. He was one of the larger fellows to go out for the team and again made the first string. That fall season the Hamline University football squad were the champions in their league. Don, popular with his classmates, was elected president of his freshman class. He also had the leading part in a play in his drama class, which was so successful they toured the state for two weeks. One of the members of the play was

A Sears wireless catalog. (Tnx Windy Windom, W8GZ)

a girl named Bertha Lindquist. Don was taken with her and they began to see each other when time permitted.

A number of Don's classmates also had their amateur radio licenses and they formed a radio club. Enough parts were soon fabricated and procured to build a receiver. The fraternity house had the apparatus set up at the end of the third floor hallway and it became a nightly feature to gather in the hallway and listen to Don's radio. The first broadcast station would not be established for another four years, but entertainment was there to be found if one knew where to look.

The fraternity house was only three stories high, but there was a church with a tall steeple nearby. They ran a wire from the apparatus at the end of the third floor hall out the window, up to the top of the roof of the fraternity house, and over to the top of the church steeple. The wire continued down the street to a tall building some three blocks away. The idea of a fan of wires was given considerable thought, but rejected due to the icing problem in the winter months.

A curious phenomenon occurred at that time for which Don never developed a satisfactory explanation. He recalled: *"Once the antenna was strung we occa-*

sionally could hear conversations on my crystal receiver from the telephone of the sorority house across the street, which had just a normal wire circuit. As the telephone line and my aerial were very close for about three blocks, there must have been a lot of pickup between the wires. My fraternity brothers had great fun with that. Once they heard a fellow call one of the girls across the way and make a date for 7:30 Friday evening. Each of the fellows at my fraternity house proceeded to call the same girl and ask her out for 7:30 Friday evening. No other time would do." All in all, Don and the fellows had a lot of fun with the radio set in the hallway.

Once the receiver was working properly, the fellows turned their attention to building a transmitter. A rotary spark gap was too expensive, but in a few weeks a fixed spark gap rig was on the air from the fraternity house.

That fall some bad news had arrived from home. The financial panic that occurred earlier in the year had sent land prices in Southern California plummeting. Don's father had taken large losses and could no longer afford to pay his college expenses. As a matter of fact, any money that Don could send home would be appreciated by the family. Don visited the local telegraph office and asked for a job. The man in charge was impressed by Don's licenses but doubted that he could handle the work hours along with his college classes. Don convinced the manager that he could work night hours and sleep on a cot by the sounder; he would wake up when the station's call sign was being sent, indicating that a message was coming in. Don got the job at full wages and began sending money home the next week. However, he and Bertha would have very little time to spend together in the coming months.

In mid-December, Hamline had a two-week break from classes due to the Christmas holidays. Don arranged a short leave of absence from work and started packing for the trip back home. It was a three-day trip on the train back to Long Beach from Minneapolis, and he had not seen his family for over three months. The train was to be the primary mode of intercontinental transporation for some years to come. There were less than four million automobiles on the road in the United States, and no paved roads yet existed between cities.

Once Don arrived at home the first thing he noticed was that the couch in the parlor was gone. When he inquired about it Don found that his mother had purchased a new couch shortly after he left for college and the old one had been hauled to the dump. Don was devastated. All of his amateur radio equipment and materials were in the old couch. He spent several days rummaging around the town dump but never located the couch or any of the missing material. Nearly 70 years later Don would still talk about his early wireless station and the disappointment he felt when it was thrown out. He was to develop certain characteristics as a result of losing his first station. Every year or two Don would have a professional photographer take a picture of him at his current station. Several copies were made so if one was inadvertently lost, others would remain. These photographs left a continuous history of his station in the ensuing decades. Don would also carefully store all of his receivers and transmitters in a safe place as they became obsolete. During the next 68 years he would accumulate a considerable amount of equipment.

Don returned to college shortly after the new year of 1917 started. Lacking a station of his own at home, he put all of his energy and knowledge into making the one at the fraternity house first-rate. With some difficulty a used de Forest audion was procured. Due to the popularity of the Armstrong regenerative circuit with the wireless experimenters, they were still hard to come by. The funds of the college students were limited, but they were adept at building most of the required components. Soon an adequate rotary spark transmitter was operating. As Minneapolis-St. Paul was situated in the middle of the country, it was not too difficult to work the fellows on the East Coast. With Armstrong's improvements in receiver design, even the West Coast was occasionally heard. One evening Don worked Hiram P. Maxim, 1ZM (formerly 1WH), the founder of the ARRL. The organization was growing in size and now had nearly 5,000 members, representing about 80 percent of the licensed amateurs in the United States.

While Don was busy attending classes in the spring of 1917, he was aware of the growing conflict in Europe. The British liner *Lusitania* had been torpedoed by a German U-boat off Ireland two years earlier; 128 Americans had been lost, and the saber-rattling by the indignant United States government had been growing ever since. Due to that incident and a variety of political factors, the United States and Germany had severed diplomatic relations in early February. Feeling that the United States would inevitably be drawn into the conflict, Don began making the initial preparations for serving in the armed forces. Don's uncles, Lou and Charles Wallace, had fought in the battle of Manila during the Spanish-American War under General Arthur MacArthur 19 years earlier. Having spent the previous summer at sea as a radio operator, he decided that the Navy would be his preferred service. In April, before the end of the semester and final exams, the United States declared war on Germany.

Within days of the announcement, Don and many of his fellow fraternity members went down to join the armed forces. Most of them joined the Army, feeling solid ground under you was better than being on a boat. Don went to the Navy recruiter and volunteered. Don recalled, *"I was attending college in the spring of 1917 and joined the Navy. The local newspaper in St. Paul said that if you joined west of the Mississippi they sent you to the West Coast for training, and if you joined east of the Mississippi they sent you to the East Coast for training. I cut the clipping out. Everybody in my fraternity house but two joined the Army. "I said `I'm going to join the Navy. I grew up in Long Beach where the ships anchor.' Oh, they didn't like that. They didn't like being on the water, you might sink. They would rather be on land. They were all landlubbers from the Dakotas and Minnesota; they all joined the Army. I wanted a place where you could always get a shower and a clean bunk, so I joined the Navy. I knew the Navy. I went over to the recruiting officer, showed him the newspaper clipping, and I told him `I'm a wireless operator and I have my Radio Telegraph First Class License;' which was the ticket of the day. It was a special: 25 words-per-minute Morse and 30 words-per-minute Continental code. They issued those licenses for a while. "I said, `Should I join now or finish my year in college and then come back?' `Oh, by all means finish your first year,' he said. I thought that was pretty good of that recruiting officer to send me back and let me finish my classes that semester. I was the only one of the group that did complete the first year, so I had the experience of actually finishing my courses, taking the final examinations, which of course is what you go to college for. Then I went down to enlist the day after."*

The last pre-war issues of *QST* carried a call to arms.

Don spent the next two months studying for final exams and got exceptionally good grades. Amateur radio had been banned when war was declared and he had a lot more spare time now. A letter dated April 15, 1917 arrived from J.H. Ashley, District Communication Superntendent, United States Naval Communication Service, Navy Department. It read:

> You are advised that pursuant to the President's War Proclamation, charging the Secretary of the Navy with the closing of all radio stations not operated by the government in time of war, your radio station is hereby ordered closed and dismantled immediately.
>
> The antenna or aerial for your station, whether for transmitting or receiving, or both, must be taken down and all apparatus disassembled.
>
> You are requested to bring this order to the attention of all amateur operators not in receipt of same.
>
> As soon as you have complied with these instructions you will send immediately to this office a letter what action you have taken.

The local radio inspector from Chicago came around to make sure that Don's radio set was inoperable. Both the receiver and transmitter had been dismantled. The antenna was also taken down, rolled up and stored. The pieces of the station were placed in burlap sacks and an official seal placed on each bag.

Don recalled, *"The athletic teams were a bit sketchy that spring. I did a lot: baseball, track, and so on. So many of the fellows had gone; the ones who were left, we did everything. I entered about four events in a track meet. I broke my foot in broad jumping just before the end of the semester. I hit the board crooked and I could hardly walk."*

After the final exams, Don went back down to the Navy recruiter and signed up, broken foot and all. *"When I went down there, I walked straight, didn't limp. The doctor looked at that foot, looked at me, and passed me. I guess he figured that it would get better. It did."*

After the United States entered the war, just a few more issues of *QST*, devoted primarily to recruiing radio operators for the armed forces, were published. Once the September 1917 issue was out, Tuska suspended publication and joined the Army Signal Corps. On its merits as a wireless magazine, the prewar issues of *QST* had a remarkably good performance, especially considering the financial difficulties the ARRL had encountered.

During World War I the term "trench warfare" became widely known. As it turned out, being in the Navy on the water, even with the U-boat threat, was safer than in the Army. Ground engagements in France were to take a terrible toll.

The Great War

Don reported to the recruiting office on June 7, 1917 and was assigned the rank with which he would enter the Navy. *"The recruiter said, `The best I can do is give you second-class petty officer. That's what my sheet shows.' I didn't know the difference, so I said okay. When I got out there to a duty station they couldn't believe that a First-Class Telegraph Operator had been assigned as only a Second-Class Petty Officer. They were giving Chief Petty Officer (CPO) to the first-class operators. There were only a few operators with first-class licenses in the world that were available for the armed forces at the time, relatively few. There were only 5,000 licensed amateurs prior to World War I, for instance."* Less than 100 of those had qualified for a first-class commercial license before the war.

At the request of the Navy, the American Radio Relay League had given a call for radio amateurs to join the armed forces in the spring of 1917, even as the few remaining amateur radio stations were being dismantled. Those with better stations were asked to bring along their radio apparatus due to the shortage of modern equipment at the government stations. This so depleted the ARRL membership and the subscribers to its magazine that *QST*'s publication ceased in September 1917.

When America entered the war her armed forces totaled slightly over 200,000. Eighteen months later there were over 10 million men in uniform.

Don said, *"One thousand amateurs joined the service immediately when America entered the war, and every one of them had an awful lot to do because they were distributed pretty thin over the armed forces. I got out and reported. I wanted to come out here to the West Coast because my parents lived here, and I wired my dad and told him that I would be in on the train in a few days. He came up to San Francisco to see me and we went over to the Navy headquarters where I was to report to their communications center NPG.*

"When we got there I said that I was reporting for duty, and that this was my Dad. The duty officer said, 'Well, you two go ashore, come back in the morning. Have a good evening together.' So my Dad and I went ashore and went out to dinner for the evening, then I went back the next morning. When I arrived the duty officer said, `Go in there on this wire circuit and see what you can do.' "So I went in and sat down with this old-time Navy operator. The circuit that he put me on was NPL. They had a wire relay between NPL in San Diego and NPG in San Francisco. NPG was the head quarters for the whole Pacific, all the fleet, and we handled everything to do with the Navy's traffic and their messages. The operator showed me what he did — I could run the typewriter. Then he showed me the Navy forms they used, and I could do those. Then I started sending a message; I could do that. He went right out and said to the officer,`That fellow can operate just as well as I can. What did you send him in there for?' "`Oh. Well, in that case, pack your sea bag. We got a ship that needs an operator.' So out the operator goes. I relieved a man for sea duty right away. He was an old experienced Navy man, knew ships, that was the thing to do. I shouldn't go on a ship right away, I wasn't a Navy man yet, so I operated that station. Pretty soon Larry Lynde was assigned to NPG also. Larry was from the Los Angeles area, his call was 6SE (formerly JK and 6UG). We had worked several times on 200 meters when I was in high school in Long Beach. Larry and I became Chiefs of Watches. We had eight men under us because we were hams. None of the fellows that came into the station knew communications like a ham did. We were used to establishing communications. No way you could stop us from working somebody [laughter], which was the whole idea. The wire operators that came in didn't know how to work a radio system. Their experience had been to open a key at their end of the circuit and then somebody answered in New York or someplace. The wire operators didn't know radio communications at first. They couldn't tune the transmitter and establish communications. We old ship operators and hams could do it. So Larry Lynde and I were promptly in charge of eight operators to run the whole station for a watch. Larry was a straight ham, but had been a Marconi operator in the summer prior to joining the Navy. We roomed together, they gave us a room right on the base. We had a great time; it was a lot of fun."

For a couple of ham radio operators this was great fun. The transmitter was a 40-kilowatt arc rig. Rather than keying the primary of the power transformer and creating a temporary spark, the transmitter drew a continuous arc. A 500-volt DC motor generator provided the energy to run the transmitter. The two arc electrodes were brought closer to each other until they barely touched and the arc started or was "struck." After striking the arc the electrodes were slowly pulled further apart, still maintaining the arc between them. A rather large distance between the electrodes was possible after the arc was struck. It operated on the principles of plasma physics. The electrodes were drawn apart until the operating current of the motor generator reached 80 amperes, when the full operating power of 40 kilowatts had been reached.

With the arc operating at full power it was considered to be putting out power on a single frequency, although the bandwidth of the transmitter signal wasn't particularly sharp. The frequency was changed by shorting out part of the tuning inductance with relays controlled by the key. The tuning inductor and the antenna formed the transmitter's tuned circuit. The key activated a big contactor with numerous pairs of contacts, each shorting out a small piece of the main tuning inductor, which was very large. This was effectively a frequency-shift keying method which occupied two places on the radio spectrum—the desired frequency when the transmitter was keyed and the undesired frequency, or backwave, where the inverse of the code was present when it was not keyed. The inductor was large in part because of the continuous operation of the arc putting energy into it and also because of the power level with which arc transmitters were operated. Due to the large size of the inductor, arc transmitters usually operated in the region of 2,000 meters (150 KHz), or greater, well below the frequency range of the modern AM broadcast band. NPG's 40-kW arc station operated on 4,800 meters (62.5 KHz) and 8,600 meters (34.9 KHz). Because of this characteristic arc transmitters were never known to have been used on authorized amateur radio frequencies. They were physically too large to operate anywhere near 200 meters (1500 KHz) or the shorter wavelengths.

After Don had been stationed at San Francisco for two months his commanding officer told him that the transmitter aboard the S.F. lightship out in the harbor was not functioning. The new operator who had been stationed out there had been unable to repair it. Asked if he could lend a hand Don replied *"There isn't a transmitter made that cannot be repaired."* Don had the supply boat drop him off during the next scheduled visit. The station had a modest half-kilowatt spark transmitter, and used the call sign JB1. He got the transmitter back on the air within a day. Since the next trip the supply boat would make was two weeks away, Don made a complete set of drawings and operating instructions for the receiver and transmitter, as well as of all the station wiring and antennas. The original manual was left at the lighthouse station and a 75-page carbon copy was brought back for the shore station and kept as a backup. After a two-week stint aboard the lightship, the supply boat picked Don up and he returned to the base at San Francisco.

Don continued,

> *Well, by and by, they posted on the bulletin board 'Volunteers for Manila sign up here.' The Navy was going to put in a 300-kilowatt arc transmitter over in the Philippines, and they wanted volunteers. Most people don't realize that the Navy, at least in my day, was strictly voluntary. They didn't send a man anywhere; you volunteered for it. Consequently, the gripes were relatively few — you had already volunteered to do whatever you were doing. In the first place, there wasn't anybody in the Navy that hadn't volunteered to join. Second, there wasn't anybody in the Navy on any base that hadn't volunteered for that base. Well, Larry Lynde and I wanted to get out of the darned place at NPG. We were tired of handling the communications for the Pacific coast. We wanted to get out and see the world, so we volunteered to put in the 300-kW station in Manila. You see, we decided that we were going worldwide and the Philippines was a possession of the United States at the time. So Larry and I put our names up there on the list. Pretty soon they said, 'All right, you fellows can have two weeks' leave, and when you come back we will send you over to Manila.' They didn't want to send us overseas without our seeing our families for a while first. So Larry and I went home to Long Beach and we had a nice two week furlough. While I was home I went down to see my cousins in Newport. They were girl cousins and we danced down in Balboa one evening and had a grand time. Larry and I had a regular furlough. When we went back up to San Francisco we found out that they had scratched our names off the list. Under our names there was a list of about 20 others that had volunteered for Manila also, and they wanted us at the station in San Francisco.*
>
> *That made us good and mad.*

James Bennett, Don's commanding officer, knew when he had a good thing going. Here were a couple of self-taught radio operators who loved to operate the radio equipment and could fix virtually anything. The 40-kilowatt arc transmitter at his installation would have lots of tender loving care as long as he could hold onto these two young radio amateurs. It would take some fancy footwork on their part to get assignment outside of San Francisco during their term of service in the Navy.

During this period Don decided that he would like to attend the Naval Academy at Annapolis. Aside from his wanting another duty station, Don had the desire to achieve the best performance possible. The academy required very good grades on the part of the applicant, as well as a certain amount of political influence. As Don had both he thought that his chances of being selected were quite good. In a *Western Union* telegram to his father on October 24, Don wired: *"Have Congressman advise you immediately if his appointments or alternates have been appointed to Naval Academy. Can get excellent recommendations from my superior officers who feel sure can pass requirements."*

Two days later William Wallace wired his congressman, the Honorable Charles H. Randall, and received a response the same day. Mr. Randall noted, "Appointments not yet made Annapolis examinations and will write you soon as reach Washington. Glad to appoint son as principal or alternate."

In those days each congressman could appoint one candidate and one alternate to the Naval Academy. Mr. Randall selected his appointments by examination. A preliminary examination was to be held shortly and Don was welcome to enter the competition. If there was one thing Don loved, it was competition. Whenever he had some spare time away from his normal duties, he was studying for the examination. The test was broad in nature, covering mathematics, engineering and navigation, along with a host of other topics. Don passed the preliminary exam with flying colors and was one of the twelve candidates eligible to take the remaining qualifying exam.

At this time he was sent to the Naval radio station at Marshall, California for a few weeks. They were having trouble with the transmitter site at Bolinas and the regular operator was too sick to work on it. Don's commanding officer in San Francisco agreed to loan him out for a while. The 300-kilowatt transmitter installation at Bolinas, KET, was owned by American Marconi and was operated by the Navy during the war. It was an installation to be proud of. The transmitter was one of the largest in the world and often would be heard by ships near the Asia mainland, some 5,000 miles distant. The motor generator filled an entire room and, when transmitting, delivered 600 amperes to the arc. The system power cables were larger than a man's arm and got warm when the arc was running. It took Don and a crew of a dozen men from the manufacturer three weeks to restore the transmitter at Bolinas to operation at the full rated power of 300 kilowatts. The motor generator

CANDIDATES FOR ANNAPOLIS ARE APPOINTED

Two Long Beach Men Naval College and One for Army at West Point.

Donald C. Wallace, son of Mr. and Mrs. William H. Wallace, has been named by Congressman Charles H. Randall as one of a number of Southern California boys to take examination to fill three vacancies at the Annapolis Naval Academy, according to dispatches received here from Washington today.

Arthur N. Ziegler, son of Mr. and Mrs. Frank A. Ziegler, 2201 Pasadena avenue, was also named by Congressman Randall as one of three men to take an examination for entrance to West Point. The father of the young man is assistant cashier of the Farmers' & Merchants' bank.

Harold E. Hedger, another Long Beach boy, was appointed along with Wallace, to take the examinations for entrance to the Annapolis Naval Academy.

DONALD CLARE WALLACE,
Candidate for Annapolis.

A clipping of Don from the *Long Beach Press* in 1918.

was overhauled and its wiring to the arc replaced with heavier wire. The contacts on the main transmitting inductor were carefully serviced and even the feed line connections to the transmitting antenna network were refurbished. Don knew from long experience with his own station that just one bad connection could cause many headaches. Once the repairs were finished, the transmitter was back on line and operating without a hitch.

With the work at Bolinas complete, Don was assigned to Goat Island, California. Along with his regular duties he continued studying for the Academy examination. While he was there Don paid a two-week visit to the installation at Mare Island. After putting the station in proper working order he proceeded back to Goat Island. Upon his return in January 1918, he was promoted to Chief Petty Officer. This was the highest noncommissioned rank in the Navy and was considered to be quite an honor for a young man just 19 years old.

The examination for Congressman Randall's selection to Annapolis was just a week away and Don got a short leave to study for it.

He returned to Long Beach to visit his family and boned up for a few days. The examination was held in the Federal Building in Los Angeles. Don felt good after taking the examination and was confident that he had qualified for the academy. The results were announced shortly after he returned to Goat Island. He had achieved the highest score on the exam and received a telegram from his congressman confirming his selection.

In a letter of commendation, the Communication Chief of the Pacific Division noted:

> Wallace has always been attentive to duty and I have found him at all times to be a very efficient and willing worker. His knowledge of radio apparatus and Naval communications is far above that of the average worker of similar rating. . . .

In the actual process of attending the Naval Academy the men appointed to it were transferred to the school by the Bureau of Navigation of the Navy Department. The transfers were implemented after the candidates were selected from the list of eligibles recommended by the commandants from the different Naval districts. Due to political considerations, the commandants always followed the recommendations of their congressmen. While the Bureau was reviewing Don's service file, a hitch developed in his appointment. It seems that Naval regulations required that candidates must be between 21 and 31 years of age. Those candidates who had not attained their twenty-first birthday by June 1, 1918 would have to wait another year for admittance. On that date Don would still be only 19 years old.

The commandant requested that an exception be considered in this case, which was denied by the Bureau of Navigation. At that point Congressman Randall intervened on Don's behalf. He wrote the Secretary of the Navy and asked that an exception be made with CPO Wallace. On May 4, 1918 Don's father received a telegram from the acting Secretary of the Navy, Franklin D. Roosevelt. Roosevelt's wire was brief and to the point:

> Candidates reserve class Annapolis must be twenty-one. No exceptions.

Don was disappointed at not being appointed to the Academy, especially after qualifying for it and coming so close to attending, but he was not going to let this setback dampen his enthusiasm. He might be too young to attend the Academy, but he was still going to get out and see the world while he was in the Navy and, while he was at it, give his best performance as a radioman. Besides, the Great War was still in progress. He might get assigned to a ship in the European theater of operations and get more closely involved in the war effort.

While Don continued to try for a duty station overseas, he came to know another radio operator quite well. Tony Gerhardt had been a wire operator before the war and was quite proficient with the Morse code. When times were slow at Goat Island, Don or Tony would send as fast as they could with a speed key while the other sat at a typewriter and copied.

This was called burnout and the object was to see who could go faster than the other. During that spring

Don, second row right, and the Goat Island operators in 1918.

they both got to the point where they could copy faster than the other could send, perhaps 45 or 46 words per minute on a speed key, or bug, in both the Morse code and the International Morse code. Don and Tony became close friends during this time. Meanwhile, Don still signed up for an overseas duty station whenever the opportunity arose. In June 1918 he was finally successful.

Don recalled, *"So Larry and I volunteered for everything and the same thing happened each time. As soon as we would see an assignment for volunteers and we would sign up, plenty of people would sign up after us. There were a lot of operators around there to fill these assignments and the commanding officers wanted us to stay around. We were hams, we would always fix the equipment right, and we would make everything work. We were always working something further away than anybody else did — a ship over in Hong Kong or some darned thing. We just loved it. Well, one day a list went up for submarine duty. They needed operators for six submarines that were going down to Panama. Larry and I signed up, and for the first time nobody else signed up after we did. So we were assigned to submarine duty. Well, that was very interesting and we were in the same flotilla, same fleet of submarines, Larry and I. It took quite a bit of getting used to at first. It seemed like every time I went through a bulkhead door I hit my head. They were built for smaller people, you know. After a while I got used to ducking when I went through hatches.*

"One day they issued orders to us from Washington: 'radio silence: you can't use your transmitters.' So Larry and I wet our fingers and sent with the receivers. The receivers were regenerative and used the Armstrong method of an oscillating detector, so they could be heard for some distance on another receiver, like a low-powered transmitter. Touching the receiver's feedback coil changed the frequency of the oscillation, so you could effectively frequency-shift-key the receiver's oscillator and communicate for a short distance. We could talk about 20 or 30 miles on just the oscillator in the receivers. Larry and I would ragchew up to 20 miles or so, farther than we could see each other, of course."

Don spent a total of six months aboard the submarine *O-16* as its chief radio officer and developed a number of innovative ideas during this time. Characteristically, submarines were unable to receive messages while submerged. Don puzzled over this problem. Transmitting was one thing, but one ought to be able to receive messages, even under water. He found that by operating the receiver on an independent battery disconnected from anything else and floating the receiver ground away from the ship's system ground, most of the noise being generated internally within the sub was eliminated. After some experimentation he discovered that by connecting the receiver's antenna input to an insulated wire strung inside the submarine, messages could be received for a short distance, even while the craft was submerged. The Navy brass were quite im-

While aboard the submarine Don used an IP-501 receiver.

pressed with Don's discovery. As Don was already the highest enlisted rank and considered too young to be an officer, he received a commendation from the Navy.

Don's CW skills became even more honed, if that was possible, during those 18 months. All of the shifts of four hours on and four hours off, along with playing burnout in his spare time, made the code second nature. He could send both Continental and Morse in excess of 45 words per minute, and his receiving speed approached 55 WPM.

On the reverse of the operating license was a column for the service record of the individual operator. It was to be noted as "Satisfactory" or "Unsatisfactory." Almost without exception a commanding officer would note either one or the other for the performance of the wireless operator. At the end of Don's submarine service on January 25, 1919 the ship's Lieutenant commander made the notation 'Most Satisfactory' on his license.

Don wanted to get on with seeing the world and after six months aboard a submarine he asked for a transfer. The armistice had been signed on November 11, 1918 and the Great War had come to a close. Although Don's submarine had visited Panama and the East Coast of the United States he had not been able to get into front-line action off the coast of France.

It is estimated that about 80 percent of the amateur radio operators in the United States served in the armed forces during the Great War, most of them as radio operators in France. The contribution of these four thousand young men was incalculable in the war effort. One of the most fortunate circumstances in the history of radio was the fact that, at the time these thousands of trained radiomen were so badly needed, there were some 6,000 amateurs and experimenters in this country. These individuals had been training themselves for periods of typically five years in just the kind of radio activity for which their skills were required. The superiority of Allied and American communications was, in many instances, the deciding factor of the close struggle during the fighting in France. Commendattore Marconi, who had demonstrated the first wireless transmission 20 years earlier, made the following observation of the United States' radio operators:

> American wireless men are exceptionally well qualified . . . A great deal of our communication is carried on with low power and wavelengths in the neighborhood of 200 meters — the exact type of communication to which they are most accustomed.

Some commanders were even more emphatic in their praise. General Russell, Chief Signal Officer of the United States Army Signal Corps wrote:

> On a certain occasion some remarkably clever interceptions of German code messages were made that enabled the Americans associated with the British and French code men to work out a solution for the new code. . . . These code messages are the hardest kind of messages to copy, and in order to be sure of getting the valuable messages, it has been necessary to copy several hundred useless ones per day over long periods of time under difficult conditions and through interference which confuses all but the best operators. The American operators were the only ones who copied all the messages with sufficient accuracy to be useful.

Edwin Armstrong, the inventor of the Armstrong regenerative circuit which was used by all of the bel-

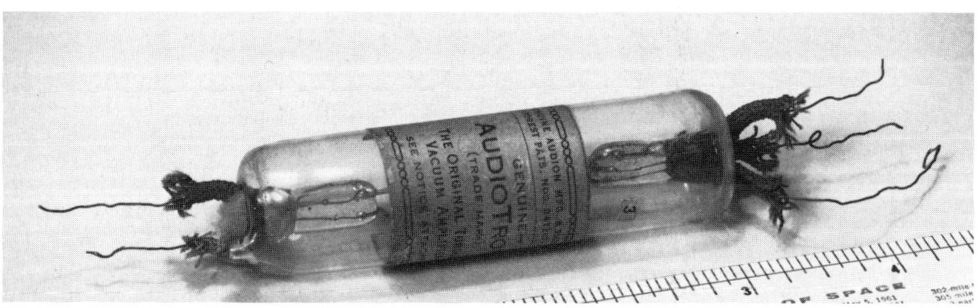

A tubular audion.

ligerents in the war, was president of the Radio Club of America when the United States entered the war and a former radio amateur. He accepted a commission as a captain and was placed in charge of the Army Signal Corps Laboratory in Paris. There he invented the superheterodyne receiver, which one day would come to be almost universally used in virtually all communications.

During the war the tube technology had been rapidly moving forward. De Forest had presented a tubular version of his audion in 1916. The first miniature tube, the "peanut tube" or VT-5, was developed in 1918. That same year the Western Electric Company, using tube technology, had developed a portable transmitter-receiver for the United States armed forces.

"Larry and I would see each other when we got in port," Don continued, *"and take our shore leave together. After a while we got tired of that and we put in for a transfer. I was to be reassigned to the receiving ship in Bayridge, out of Brooklyn, New York."* Don was attached to the *USS Rainbow,* and was in charge of the station WNY. *"There were about 400 operators at the station in New York. By that time the Navy radio schools had really developed operators. There were a few hams among them, but the Navy had sent the new operators to school. They were good telegraphers and they were capable people. About the same capability as someone coming out of the radio school here locally, before he's been on the air. They hadn't been on the air yet and didn't know what to do at the station. Somebody in the group, the hams or old Navy men, was in charge of showing them the ropes. I had been assigned two absolutely green school boys on my submarine a few months earlier. As long as there's one fellow who knows what to do, pretty soon they all know what to do. They're just as good as anybody else. They were smart boys and they had passed the code, passed the technical, and here they were. Once they were shown how to operate the station and worked it a few times they picked it right up. Well, by and by, the new fellows learned all of the equipment and operating procedures fine and they were just as good an operator as you would want to see. There were 400 of those new operators at Brooklyn. I had been upgraded several times and was as high as I could go, being just 20 years old."*

Don was preparing to muster out of the Navy in New York when he met his old commanding officer from San Francisco. They chatted for a while and Don was offered a once-in-a-lifetime opportunity.

A Vibroplex speed key.

President Wilson's Ship

While Don was stationed aboard the *USS Rainow*, he had renewed the acquaintance with his former commanding officer from NPG in San Francisco. Don recalled *"The officer said 'I am making up a bunch of operators for the* USS George Washington, *and I can't find anybody in the whole 400 stationed here who knows anything about arc transmitters. We are putting on a 40-kilowatt arc on the* George Washington *for Presidential use. You're from the West Coast, do you know anything about arcs?' I said, `Oh, there's nothing to them.' `Just the man I want,' he said. So he put me in charge of the* George Washington."

All of the newspapers back in Southern California carried articles about Don's assignment. The story in the Los Angeles Examiner was headlined "Long Beach Boy Is In Charge of Wilson Messages." *The Long Beach Press* carried a story labeled "Young Wallace Wilson's Radio Operator." Other local newspapers carried articles bannered with "Beach Boy On Wilson's Ship" and "Long Beach Youth is Wilson Ship Radio Man."

The *USS George Washington* was the largest ship in the United States fleet. She was a German vessel that had been captured when the United States entered the war in Europe two years earlier. President Wilson was leaving for the Versailles Peace Conference in a few days and the *George Washington* was to be his transportation to France. Don would have less than one week to select operators, see that the new transmitter was installed aboard ship, and have all the details worked out.

Don recalled,

I picked out 35 operators for the trip. I sat down at a table in the coffee room, and there was a bug (speed key) and a buzzer on it. Nobody was using it, so I sat down and started sending as fast as I could. We had tried to send as fast as we could back at Goat Island. Tony Gerhardt and I would try to burn each other up. One of us would sit at the typewriter and the other one would sit at the bug. We would see who would make the other guy holler uncle. Well, we got so that we could type faster than we could send—about 45 words a minute.

So I sat down and started sending as fast as I could. A big tall fellow came along and said `Hey, Chief, do you mind if I copy that?' I said, `No, go ahead.' He sat down at the typewriter and started typing. By gosh, he could copy me — as fast as I could send. Perfect. I said `Do you want to be an operator on the President's ship?' `Oh, sure,' he said. I got his name and address and turned it in. There was one. I kept doing the same thing and pretty soon I had my 35 operators for the trip.

Don managed to select the best of the telegraph operators at the base. Word got around that to be picked as an operator on the Presidential ship was a real feather in one's cap. Included in the 35 operators Don selected was a young man named Fred Schnell. Fred would join the ARRL staff after the war and had a close association with Don in the decades to come.

"They all were fellows that wanted to do it. Now that's a different attitude, you see. They all loved it, every one of them. They were just so thrilled in finding somebody that could send as fast as they could tackle. Every one that I picked out could copy my fastest sending. I can't do that any more, but I could at the time. I had such a skilled bunch of operators on the ship that the reputation of their skill went on for years after the voyage."

Among the things Don had to do was prepare a list of spare parts and equipment for the trip. If something failed during the voyage, the only replacements available would be what was on hand in the ship's storeroom. Along with the typical complement of spare insulators were 250 cathode carbons. A pair of these held the arc between them when the set was transmitting and they would have to be changed daily.

"The Navy gave us a special tugboat that carried all our baggage and the 35 fellows, and took us over to the George Washington. *We reported aboard and the place was just alive with people installing a 40-kilowatt arc transmitter. The transmitter, with its generators and wiring and antennas and everything, was to be installed in just three days; the ship was in port for only three days."* The typical time required for an installation of this magnitude was six weeks. However, the President's trip to the peace conference couldn't be delayed, and the

CPO Don Wallace, Chief Radio Operator for President Wilson.

work schedule had to be compressed to the time available.

The wireless set aboard the *USS George Washington* was being modernized for the voyage. The existing one had been installed some ten years earlier and wasn't up to meeting the demands of all the forthcoming Presidential traffic. On March 4, the *New York Times* carried the following article;

> "New Wireless on President's Ship. Will Enable Him to Talk to Washington From Mid-Atlantic . . . New wireless telegraph and telephone apparatus have been installed on the steamship *George Washington,* in dry dock at the Hoboken Pier of the Morse Dry Dock and Repair Company . . .
>
> It is expected that the President will be in continual communication with Washington until at least halfway across the Atlantic. The wireless telephone, in its present state of development, cannot be used at so great a distance as the wireless telegraph, which has a radius of from 3,000 to 4,000 miles.

The President and his party came aboard on March 4. Along with his wife, and daughter Margaret, the Presidential party included Secretary of War Baker, Chief of the Secret Service Moran, Rear Admiral Grayson and a complement of thirty to forty assorted aides.

"There were hundreds of men," Don *recounted, "everywhere you looked. You could hardly do your own work, there were so many. I watched them and the* George Washington *was pulling out to sea when they were still trying to tune up the 40-kilowatt arc transmitter. They finally got it on the air a little bit, at low power. Then another boat pulled up alongside when we were way out in the ocean, quite a ways outside of the harbor, and took them back to Hoboken."*

The *George Washington* and her complement of escort ships left New York on Wednesday, March 5, 1919. A daily publication aboard ship was the *Hatchet,* which had been started the year before. On March 11 it carried the headline "New High Power Wireless Plant Erected On Ship." As well as describing the new station the article went on to report that a radio compass had been installed. "The radio compass, one of the latest and most ingenious products of science, makes it possible to locate the position of the ship without celestial observations by communicating with other stations. The chief service of the radio compass is in thick weather when such observations are impossible. By its use the *George Washington* would be able to make her way through a fog when close to shore." The President's ship certainly had all of the latest and most advanced equipment available aboard.

"It was our baby now," Don reminisced, *"so we put the transmitter on the air. Our call sign was NEC. We decided that we would send on the even hours and receive on the odd hours. You couldn't hear anything on the whole ship when we sent with the 40-kilowatt arc. Everything else went out of commission." When the arc was started, all of the receivers aboard had to have their antenna inputs shorted. There was so much RF floating around that they badly overloaded and could easily burn out.*

"So we sent the first hour at low power, just barely enough to make the arc go, maybe five kilowatts. At the end of the hour we slowly turned the transmitter power up. Bang! goes a whole lot of stuff. So we all pitched in and fixed the part that burned up during the hour that we were receiving messages. We went on the system of always sending for an hour and receiving for an hour. We used my old idea, that nobody was to ask for repeats. You send it good and the other guy has got to receive it. We haven't time for repeats on this ship. [*Don laughed.*] *You couldn't listen for repeats anyway.*

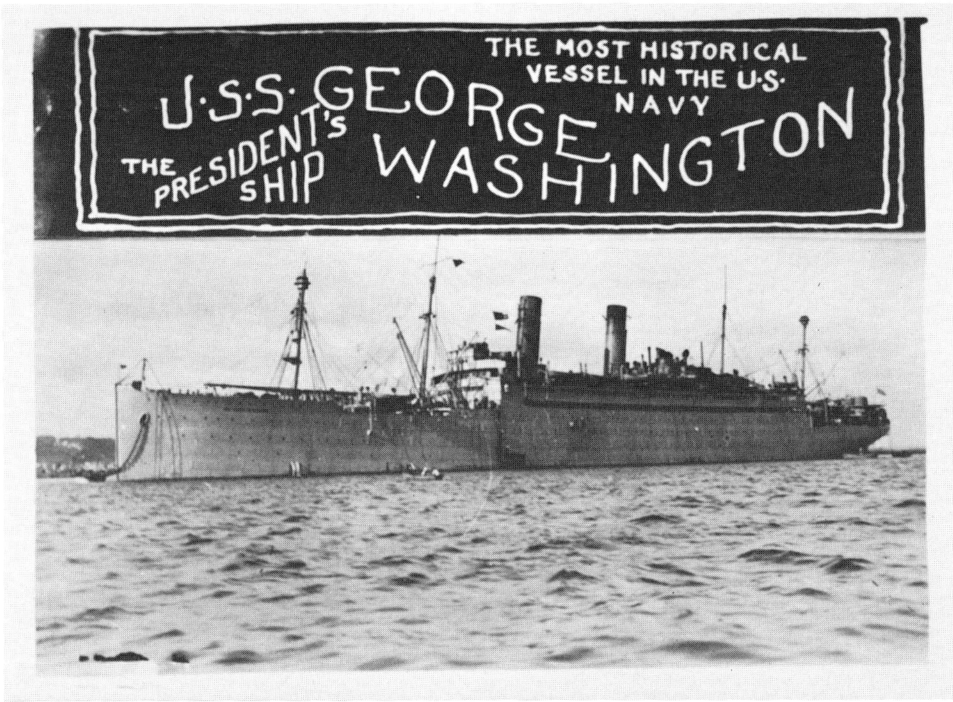

The *USS George Washington* in 1919.

Once the arc is started, you can't break it, you see. It takes a while to start the arc up again. So, we repaired the transmitter while the fellows were receiving. They did the same thing on the off-hour while the transmitter was operating. The maintenance on the receivers was performed during the hour of transmission. We received the shore station at Annapolis." The call sign of Annapolis was NSS and they had a 500-kilowatt arc transmitter which operated on 16,900 meters (17.75 KHz). "*They sent the messages from the White House to us. When we got out to sea we found out that the Alexanderson alternator at New Brunswick, NFF, sounded better. We could copy it easier, so we got them to change to that.*

"*They objected to the change at first. There were several rounds of messages between us. Finally, I said it's just better, we want you to use it. I demanded for presidential use the 200-kilowatt Alexanderson alternator station, rather than the big station that had a 500-kilowatt transmitter at Annapolis, because the code was cleaner.*" Each of the Navy shore stations had a characteristic sound, similar to the unique sound that most amateur transmitters had prior to the war. One Navy transmitter might have a note that sounded rough or mushy, where another transmitter from a different manufacturer could have a sharp, clear note.

Only 20 Alexanderson alternators were ever built. E.F.W. Alexanderson of General Electric developed the technique of generating power directly at the desired frequency rather than needing an arc to supply the energy. It operated on a direct drive principle geared up from a motor. The RF power that was generated in the alternator was directly coupled to the antenna. This principle was much more efficient than that of an arc transmitter, which generates energy across a large portion of the radio spectrum and couples the energy on the desired transmit wavelength to the antenna. The transmitter that Don requested to send messages to the *George Washington* operated on the 13,600-meter wavelength (22 kilocycles, or KHz). The Alexanderson alternator was never used by amateurs because its maximum frequency of operation was perhaps 200 KHz (1,400 meters), still well below the 200-meter wavelength. The one remaining alternator still used commercially as an emergency backup transmitter some 70 years later is in a Swedish military installation and operates on 17.1 KHz (the 17,500-meter wavelength). At these low frequencies, the shore transmitting stations required lots of real estate, due to the tremendous size of their antennas.

"*Well, day and night,*" Don continued, "*we sent on the even hours and received on the odd hours. At the end of the transmitting hour we would slowly turn up the transmit power level until something blew up. Then we would all pitch in and fix the transmitter during the receiving hour. Each time that we repaired it we could operate at a little higher power level. By the time we got halfway over to Europe we were running 40 kilo-*

watts with the arc transmitter. I figured that we might need the full power capability when we got over to France.

"Once we got the power level up part way we blew the bulkhead insulators right out of the wall. That happened the first day out. We had some spare insulators and put them in and we blew them out also. So we opened the doors to the radio room, wide open, and strung Electrose insulators through the doorway. We put up a whole string of them and put the antenna wire in the middle of the insulators. That was our antenna lead-in, out in the weather. The set was exposed to the weather the rest of the trip. We never had a bulkhead insulator that whole trip that would hold this 40 kilowatts of power."

The General Electric Company had contracted with the Navy to send an expert along, in case there was any problem with the radio telephone system that they had just installed. Their representative, Harold H. Beverage, would become widely known in the years to come by the antenna that still carries his name. As Don had quite an interest in antennas, he and Beverage chatted at length about various antenna designs.

The *Hatchet* carried another front page article about the wireless on March 23. "G.W. Radio Phone Breaks World Record On This Trip. Wireless telephone messages were sent daily to the *USS George Washington* on the trip across." The shore station could be heard most of the way across the Atlantic on voice due to the size of their transmitter and their huge transmitting antennas, which covered hundreds of acres. After the *George Washington* was halfway across the shore station could copy only the telegraph messages from the ship.

The telephone mode of transmission was used only when the President or a member of his staff wanted to speak personally to another official back in the United States. The telephone transmitter was a 3-kilowatt arc rig. The big 40-kW transmitter was just too powerful to try to modulate with the limited capabilities aboard ship. The 3-kW transmitter was rather similar to the phone transmitter that Don had built in 1913, except it was somewhat more huskier, and the microphone was water-cooled. Of course, the antenna was full sized, and situated over the ocean it had a perfect ground system.

One day the President wanted to talk personally with Josephus Daniels, the Secretary of the Navy, so Don made all of the arrangements. Secretary Daniels was patched by land telephone lines into the shore transmitter at New Brunswick, New Jersey.

Don set up the spark telephone transmitter, with the microphone suspended at the proper level when the President was seated. As the microphone was in the antenna feed line, and "hot," one had to be quite careful when speaking.

"I got everything all ready to go and the President came in and sat down. We could hear Secretary Daniels quite clearly, with only a slight amount of static. It came time for President Wilson to reply and there was total silence; he was speechless. The President had the worst case of mike fright that I had ever seen. Well, after a few minutes I asked everyone to leave the room. Even the President's personal bodyguard had to leave. The Secret Service Chief wasn't too pleased with that, but it was necessary. Then I chatted with Daniels and assistant Secretary of the Navy Roosevelt for a few minutes. After ten or fifteen minutes Presdent Wilson regained his composure and everything worked out all right. He had his conversation with them." Politicians have overcome this phobia in the decades since the voyage to Versailles.

"We handled all the presidential messages. President Wilson had been a college president at Princeton. He liked the company of young fellows and he would come down and spend most every afternoon in the ham shack, the operating station of the USS George Washington, *sitting in with the boys. I told him it was all*

Secretary of War Baker conversing with Assistant Secretary of Navy Roosevelt with a 3-kW arc phone rig.

right as long as he didn't bother them. So we all became pretty well-acquainted with the President. On the trip we were more interesting to him than talking with his secretary or his staff up in his office.

"We would get an hour's worth of presidential messages from the White House, and we would have about an hour and a half worth of responses to go back. We couldn't send all of the traffic in the allotted time. So when his secretary, the chief of staff, brought down all the messages, I told him, "We are not able to send all of those messages in an hour, we are behind a little bit, and I wonder if you wouldn't tell the President to shorten his messages a little bit?' 'What!' the Chief of Staff sputtered, 'tell the President to shorten his messages?' "I said, 'Well, its just a question. We only have an hour to send them at a time, and we are getting a little behind, and we can't send any faster. The keys won't take it.'

These keys were pretty lumbering big old keys. They had to be big to be able to key the arc transmitter and you could send only so fast with them. 'Just in the interest of keeping up, if you could shorten the messages a little bit, we could keep up and then they would get over to the White House faster,' 'Well, all right." *So after that they were shorter.* [Don chuckled.] *I will never forget the reaction of the chief of staff. This teenager telling the President to shorten his messages."* [Don was twenty years old at the time.]

"I will describe some of the things that we had to do. These keys actuated a large relay which shorted out a section of inductance to change the frequency of the transmitter. With an arc transmitter you don't break the arc, you change the frequency, and you copy the higher frequency. This was the earliest frequency-shift keying. The relays we put in there shorted the inductor in about 15 or 20 different places. These series contacts were great big lumbering things, about as big as a stool. The keys controlling the relays just wouldn't go over about 20 words per minute. No way you could send any faster than that. Our fellows were all good senders, but the relays just wouldn't have kept up. The bulk of the relays was the limitation. They were just too big to mechanically move any faster.

"At first the relays always burned up when we increased the power. They couldn't handle it. So that's one thing we did. We finally put each pair of relay contacts across about a quarter or half turn of the transmitter's inductance. So there were about twelve contacts, each shorting out a half turn of the inductance instead of the whole bunch shorting out five turns to change the frequency of the transmitter. We had pieces of the inductor in every set of relay contacts. That way we could clean the contacts on our down hour when we were receiving messages and have good keying when

President Wilson and his daughter.

we transmitted on the other hour. The modified relay system would hold the whole 40 kilowatts. But the original system wasn't possible to operate at all, even at moderate power, the way they initially put it in. We couldn't have ever got the transmitter on the air with full power."

Realizing that a historical event was unfolding, Don took several hundred pictures during the voyage. The cameras that were available at that time were rather large boxy things and the resulting negative was a six-inch by eight-inch piece of glass, almost 1/8 inch thick. They gave an excellent resolution in the resulting photograph, but the glass negatives were heavy and bulky to store. These photographs give later generations a clear view of Navy life aboard ship at that period of time. Sixty-five years later the author discovered a box of nearly three hundred glass negatives taken during the voyage to Versailles.

The *Hatchet* carried a column called "Wavelengths." Information from the radio room that was of general interest to the crew and passengers was reported there. One column noted Assistant Secretary of Navy Roosevelt's report on Marine casualties in France. At locations like Chateau-Thierry, Belleau, Soissons, Wood and St.

Mihiel, brigade casualties approached one hundred percent. The members of the armed forces who served in the units in France had suffered tremendous casualties in the new "trench warfare." At that time the most striking geographic feature in France was a double chain of snakelike trenches which began on the English Channel and ended at the Swiss border, more than 400 miles away. The great armies had faced each other since 1914 and the titanic struggles which followed were fought on a fantastic scale with millions of casualties.

Besides the Great War, one other disaster struck the world that decade — the influenza epidemic of 1918. Worldwide the virus would claim 22 million victims, over twice the casualties of the past four years of war.

"So we got over to Brest Harbor," Don reflected, *"and we thought it would be a while before the around-the-clock work schedules started again, while the peace conference was in process. The Big Four Allied Powers of the day had this peace conference and they were going to split up the world and make peace. President Wilson wanted to form an early version of the United Nations* (the League of Nations). *He tried to do that; it was his idea. Anyway, the President and his staff went up to Versailles, which is a suburb of Paris, and that's where the peace conference was held. Well, we thought, now we won't have anything to keep us busy for a while. So we organized a baseball team and I was the manager. We were going to play baseball and do a few activities like that while we were in port. The next day, into the radio room comes a fellow with a derby hat, a chain around his hand, and a briefcase padlocked to his wrist. 'The President wants you to handle his messages from the peace conference,' he announced. I said, 'That's a six-hour train ride up there to Paris and then you have to go out to Versailles from there. Why doesn't the President send the messages over to the French shore station at Lyons?' " The French station at Lyons, YN, could run 250 kilowatts of power on the 15,500-meter wavelength (19.3 KHz) and was in regular contact with New Brunswick.*

" 'Well,' the fellow in the derby hat replied, 'the President said they made a mistake in his messages and you fellows don't make mistakes. He wants you to handle his messages.' There were two trains a day from Brest to Paris. So, after that, every day, twice a day, in came a courier with a briefcase full of messages which we sent back to the United States. Consequently, we were on regular duty the whole time that we were in Brest. We had to run our regular shifts around the clock for the next three months, which was unheard of in the Navy when you were in port. There was always at least one operator receiving an hour, and another operator sending an hour, plus fellows for mainteance. With the transmitter modifications and maintenance routines that we had developed on the trip over the equipment didn't fail any more; so we still organized the ball team. I played first base, which I had played in college anyway. We found some awful good ball players. Some had played professional baseball and were pros and there were also some semi-pro players. Some of our fellows were just as good as players on the Angels or the Dodgers. We played the teams that the other ships had formed.

We would get on the equivalent of our 2-meter sets of today. We had little phone transmitters all around. There was one aft, one in the skipper's bridge, and so forth. The transmitters were 20 watts input, and they operated on about 20 frequencies, in the vicinity of the 400-meter wavelength (750 KHz). They had been primarily designed for intra-convoy communications between the bridges of the different ships, so the captains could talk to each other. So we had these little 20-watt sets to use while we were docked. Anyway, we talked to fellows on the other boats with these sets. We beat every ship, of course, but the team that the base had we could never beat."

The 20-watt sets Don used were the first commercially made vacuum tube (carrier wave) transmitters that ran moderate power and would transmit amplitude-modulated voice. The recently developed tubes were a single CW-933 (VT-1) in the receiver and a CW-931 (VT-2), rated at 5 watts dissipation, in the transmitter. These tubes would be the basis for most amateur radio CW transmitters immediately after the war, until transmitting and receiving tubes became commercially available in adequate quantities two years later.

"We played the base regularly, every 3 or 4 days. They liked to play us because we had the best team of any ship. We found out later that when any fellow came through that base they always found out if he played baseball. The clearance fellows liked baseball and they had a baseball team entirely of big league players. [He laughed]. So we played them every few days and we would play the other ships. We could beat the other ships because they were like we were — they would have one or two good players and a lot of college players. It was a lot of fun. We practiced on the ship because we couldn't have much shore leave due to all of the traffic that the President was sending us. That meant operators on duty at all times and a lot of maintenance on the transmitter, so our time ashore was pretty limited."

The *George Washington* had come to Brest with a number of escort ships. The Leviathan was one of their escort ships on the voyage to France. As her presence wasn't required during the peace conference, she loaded up with as many American soldiers as she could carry and returned to the United States on April 17. Among

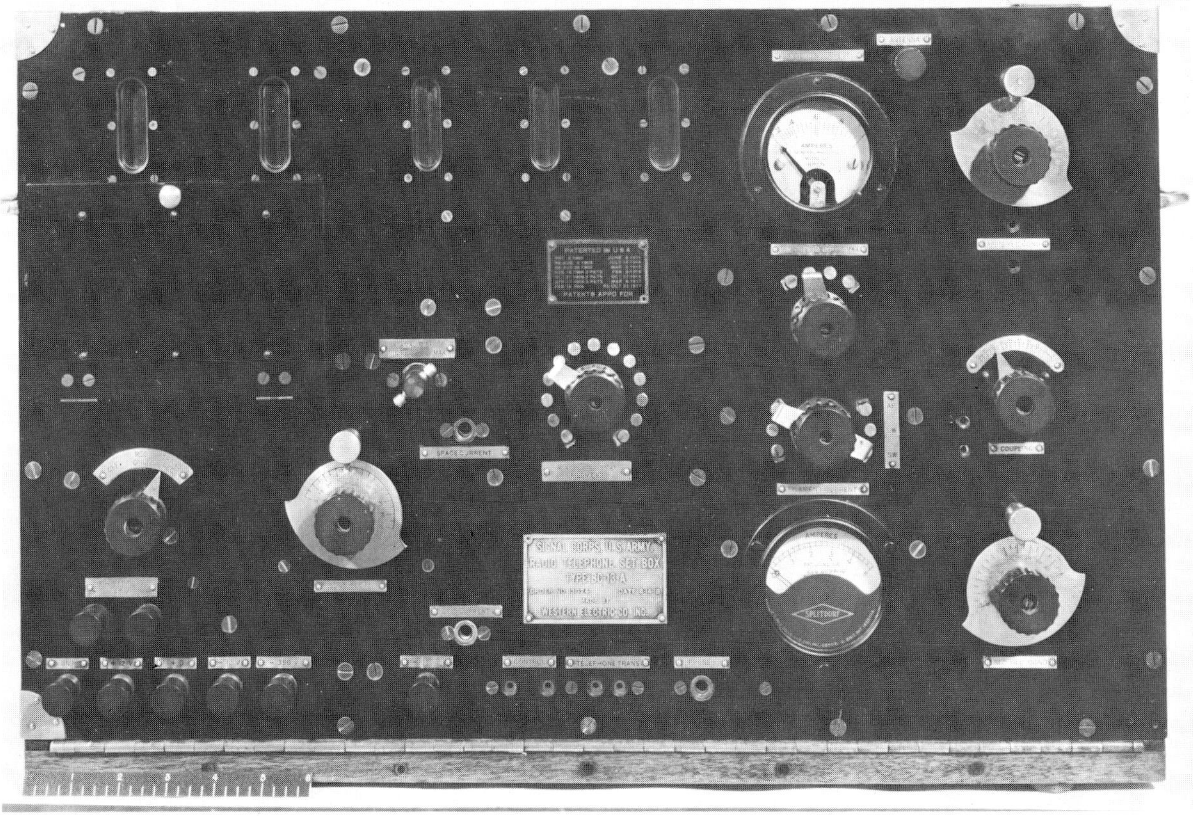

The first carrier wave phone rig, developed by Westinghouse in 1918 for the United States armed forces. (Tnx Bruce Kelley and the AWA).

its passengers was a rising star in the American Army, Brigadier General Douglas MacArthur.

"There was this great big dining room on the George Washington. *This enormous ballroom, which would hold 4,000 people, would also be used when the President addressed the troops on board during bad weather, or for entertainment. We had a dance there the week before the ship returned to New York. The French were very friendly toward Americans, what with us helping them win the war. The ballroom was filled with soldiers on the trip home."*

Once the Treaty of Versailles was signed, President Wilson returned to Brest for the trip to the United States. The treaty reduced Germany's borders, gave away its colonies in Africa, and extracted reparations for all of the loss and damage during the Great War. Even Winston Churchill urged revision of the treaty, seeing in it the roots of another conflict in the years to come. In another two decades the European continent would again be plunged into war.

One Monday, June 30, the President boarded in preparation for the return trip back to New York. The ship was also carrying 6,600 returning American troops. About 50 American soldiers had married French women while they were stationed there and their brides were accompanying them on the trip back. The next morning, at first light, the *George Washington* sailed. Now that the 40-kilowatt arc transmitter had been operating for three months without any serious breakdown, the radio room routine was pretty standard.

During the return trip a number of congressional bills required the President's signature. A pouch containing the bills had been dispatched on the eastbound transport *Great Northern* from New York on June 24. The President's ship, proceeding at 16 knots, was 357 miles off the coast of France when the *Great Northern* met her and the pouch was transferred. This, the first time in history that such operations such as receiving and signing bills in mid-ocean had occurred, was commented on board as marking the record breaking advance in modern methods of communication. The event was reported by the ship's wireless room and relayed from Hoboken to the *Associated Press,* where it was syndicated across America on the wire service.

While they were in the port of Brest, Don had experimented with the new vacuum tubes. The commercially built receiver-transmitter sets could not be modified, as they were government property. However, there was an adequate supply of spare tubes in the ship's stock to accommodate the building of an experimental breadboard CW transmitter. Don applied 750 volts from the ship's generators to the plate circuit of a Moorhead 5-watt tube and it developed perhaps 25 watts of power. Once the circuit was functioning he connected it to one of the ship's big antennas. On the return trip they contacted the Navy station at Otter Cliffs, Maine, while the *George Washington* was still 1,200 miles offshore. That was a record distance for CW transmission at the time.

The trip back to the United States took just under six days. The ship planned to be quite close to the American shore on July 4, Independence Day. The President wanted to address the troops aboard on that afternoon and notified Don that he wanted the speech to be broadcast to the shore stations and nearby ships at the same time. Don immediately began making preparations for the broadcast.

Beverage and the other G.E. engineers had been busy on the trip to Versailles. They had along a prototype CW transmitter that was not yet complete and utilized the new experimental type "P" "Pliotron" tubes, rated at 250 watts dissipation. Later this tube would be known in the Navy as the CG-916 and a VT-10 in the Signal Corps. Two "P" tubes were used in the master oscillator. The output was amplified by 12 "P" tubes connected in parallel. Beverage got the transmitter working adequately after a few weeks in Brest and modulated it with a similar number of "P" tubes. In June a voice contact was established by Don's operators with the shore station at New Brunswick. The fellows at New Brunswick used a bank of 30 "P" tubes to modulate the 200-kW Alexanderson alternator. The communication was of poor quality, largely due to the weak signals and the noisy summertime conditions, but this was the first two-way voice contact that had ever spanned the Atlantic. However, the ship was going to be a lot closer to America for the President's Fourth of July broadcast and Don decided to use the G.E. transmitter for that occasion.

The President would be on deck for the speech and the microphone would have to be camouflaged from him. President Wilson, an excellent public speaker, was still somewhat at a loss for words when a microphone was in sight. Don and his operators were busy around the clock preparing for the President's broadcast. They built up several stages of audio amplifiers with the spare 5-watt tubes in the ship's stock, just in case. Once the lines were run and the CW transmitter was fired up, the engineers calculated that they had 3 1/2 kilowatts of power to the antenna from the G.E. transmitter. Don's radiomen contacted the shore station at New Brunswick for the final adjustments. Their tests were completed and they were ready to go just one hour before the President, flanked by Admiral Greyson, began addressing the troops. Wilson took one look at where the microphone was concealed and edged a few feet away. Fortunately, the extra stages of audio amplification were able to make up for the reduced volume coming from the microphone.

During his speech, Wilson talked at length about the war that had just ended and America's contribution toward winning it. He mentioned the close ties between France and the United States and how he expected this relationship to continue in the years to come. The Statue of Liberty had been a gift from France 33 years earlier and France's participation during the American revolution 110 years prior to that had been instrumental in America winning independence.

The long-wave station at New Brunswick copied the

President Wilson addressing the troops aboard the *George Washington on* July 4, 1919.

The Statue of Liberty , taken from the *George Washington* in 1919.

Presidential speech and fed the received signal into their transmitters. From there it was re-radiated on the wavelengths of 8,000 meters and 13,600 meters. The President's broadcast was the first use of a high-powered CW station using tubes outside of the G.E., Western Electric, or Westinghouse laboratories.

Several of the East Coast shore stations copied the broadcast directly from the *George Washington,* and ships up to 500 miles distant also reported good reception. This was the first broadcast by a President on the high seas to his countrymen. The complete text of the speech was put on the wire services and printed in all the major newspapers across the country. The ship arrived in Hoboken two days later and was greeted by Governor Smith of New York and Mayor Hylan of New York City. There was also a parade for the returning veterans of the war.

When the celebration was over Don had his pictures developed. Of his almost three hundred photographs from the voyage, 30 were sold at a quarter each to the Western Newspaper Union Photographic Service. Several photographs were run in the *New York Times* and other newspapers across the nation. At this time Don also requested two weeks leave from the ship. He would be leaving active duty soon, and wanted to get enrolled in college back at Hamline. On August 21, 1919 Don arrived in Minnesota to continue his education. He was officially released from active duty on September 9.

The New York skyline in 1919.

Back to College

When Don returned to the Minneapolis-St. Paul area, he started fall classes at Hamline. He considered attending the University of Minnesota at Minneapolis because their football team played a tougher schedule than Hamline's, but he didn't have time to transfer his previous college credits and enroll before classes began. Don returned to his old fraternity house at 823 Snelling and settled in for the fall semester. Then he began looking for his radio apparatus which had been dismantled in the spring of 1917, some two years earlier.

"I moved back into the fraternity house," Don said. *"I was sad to hear that some of my fraternity brothers would not be returning from the war. Most of them had served in the Army in France and had a pretty tough time of it. I went looking for the burlap sacks that the radio inspector had sealed two-and-a-half years before. They had all of the parts of my transmitter and receiver of my 9BU station in them: the loose coupler, the oscillation transformer, high voltage transformer, the rotary spark, the audion, and so forth. Well, the sacks had been opened and all the pieces scattered to the winds. Some parts were in various projects that the fellows had put together. Fraternity brothers can do that you know — what belongs to one belongs to all. I gathered up the few parts that were left for another set."*

Although the war had ended some ten months earlier, the ban on transmitting by American amateurs was still in effect. Receivers however, had been allowed in the United States since the previous April. There wasn't much radio activity on the air. The first broadcast station was still more than a year away; the only signals heard originated from stations sending commercial and government traffic and virtually all of them used the Continental code. The transmission of voice signals outside of the experimental laboratories was still very rare.

Hiram Percy Maxim and the American Radio Relay League led the effort to lift the postwar ban on transmitting. *QST* resumed publication in June 1919. At the urging of *QST*, ARRL members wrote hundreds of letters to their congressional representatives asking for the reinstatement of amateur communications. Secretary of Navy Daniels had written President Wilson two months earlier asking that oceanic and overseas communication be controlled by the Navy Department. This surely would have been the end of amateur wireless. Whether President Wilson was influenced by the afternoons he spent in the radio room of the George Washington chatting with Don and his radio operators, most of whom were licensed before the war, by the political pressure brought to bear by the League and its members, or a combination of the two is not known. The end result was his decision to reinstate amateur radio. On September 26, 1919 the director of the Naval Communications Service announced the removal of all restrictions on radio amateurs and the resumption of authority by the Bureau of Navigation of the Department of Commerce.

QST was the voice of the ARRL and had first been published in December 1915. In September 1917 its publication was suspended after just 22 issues had been sent out. Seven months after the Great War, in June 1919, publication resumed. Its editor, Tuska, had to go into debt to publish the last prewar issues. After the war he wanted to go into the manufacturing business. The League borrowed enough money from its members to acquire the magazine and enable Tuska to pay his debts. Without *QST* to keep the ARRL membership informed and provide revenue the League might have faded away. Without the ARRL to formulate and implement policies for the amateur's common good during the critical period ahead amateur radio could have been severely curtailed or have even vanished in this country.

Those amateurs who had been licensed prior to the war found that their licenses had expired. The term for a license at that time was still just one year. During the hectic months at the close of 1919, the Department of Commerce made special arrangements for the relicensing of amateurs. In the rush, no provision was made for amateurs to retain their prewar call signs. Allocations were made on a first-come, first-serve basis. The radio inspectors in each district were empowered to authorize temporary operation while the licenses were being processed. Each passing day saw dozens of stations coming back on the air.

For those who are too young to have listened to a receiver during this era, the experience is perhaps best described by Arthur Budlong in "The Story of the American Radio Relay League":

> "King Spark! Grown now to full maturity, developed and perfected by years of pre-war experience, it

reached its highest peak in the succeeding eighteen months. Glorious old sparks! Night after night the mighty chorus swelled, by ones, by twos, by dozens, until the crescendo thunder of their Stentor bellowings shook and jarred the very universe! A thousand voices clamored for attention. Five hundred cycles' high metallic ring. The resonant organ basso of the sixty cycle `sync'. The harsh resounding snarl of the straight rotary. Character: Nervous impatient sparks, hurrying petulantly. Clean cut businesslike sparks batting steadily along at a thirty word clip. Good natured sparks that drawled lazily and ended in a throaty chuckle as the gap coasted downhill for the signoff. Survival of the fittest. Higher and higher powers were the order of the day. The race was on, and devil take the hindmost. Interference. Lord, what interference! Bedlam! Well, it could not be utopia."

"*When I went back to Hamline after the war,*" Don recounted, "*I set up an amateur station in the hall of the fraternity house. I got a room on the third floor again. They had only one room up there, but it had four beds in it. I didn't want to clutter up the room with the radio set. The hallway was big so I set up a minimum one-wire aerial and this receiver I had built up after the war. It was very similar to my station before the war. We strung a single wire between our roof and the church steeple down the street a couple of blocks.*

"*My station was constructed of all homemade stuff, of course. Commercially made receivers and transmitters were virtually nonexistent. The ones that could be bought weren't anywhere near as good as what you could build if you had some radio background. It was essentially the ones you made that had any performance to speak of. My homemade equipment was calibrated with pencil marks on the dial where you wanted to tune and all that.*"

With the lifting of restrictions on amateur radio Don began work on his new transmitter. The easiest and least expensive way to get back on the air was with a rotary spark gap. Don was still sending home to his family what money he could spare, so extra spending money for a radio set was in short supply. Many of the parts he used were homemade. He started working on the station after the football season was over and finished building the transmitter just as his new call sign, 9DR, arrived on December 11, 1919. He had built a spark gap rig for code transmission and also a 30-watt voice transmitter utilizing a VT-2 five-watt tube (from the spares on the *USS George Washington*?). This was one of the first phone rigs specifically built for amateur radio use. At that time there were perhaps a half-dozen amateurs in the country that had the capability to transmit phone signals.

Most amateurs used rotary or straight spark gap rigs in their initial operation after the war. Elmer Bucher, an engineer working at Marconi (which became RCA after 1919), had published *Practical Wireless Telegraphy* in 1917. It became the standard text for radio classes during the war and was widely reprinted. Upward of 80,000 copies were in print and virtually every wireless experimenter or amateur had read a copy or had one on hand for reference.

As soon as Don had returned to Minneapolis he renewed his relationship with Bertha Lindquist. She was in her last year of college. They had exchanged letters during the war but hadn't seen each other for over two years.

Don made first string that fall on the Hamline football squad. They were undefeated and the Tri-State champions that season.

Don also got a job at night as a wire operator for Great Northern Relay. What little spare time he could find he spent reading technical journals and the current radio magazines. *QST*, as well as *Pacific Radio News* and *Amateur Radio News* were available. *The Proceedings of the Institute of Radio Engineers* (IRE) was also a good source of information. Don began to realize the tremendous advances that had been made in the field of radio during the war.

General Electric and Westinghouse had done a lot of work the past few years with the methods of evacuating vacuum tubes. The new tubes were no longer the fragile devices in which the filament burned out in a few months. However, tubes were still very difficult to procure. Five, fifty, and even 250-watt transmitting tubes were now in existence for those who had some influence and could afford them.

Marconi controlled the Fleming patent on the vacuum valve and in 1914 the Marconi Wireless Telegraph Company had brought an injunction against the de Forest Radio Telephone and Telegraph Company, restricting the sale of the de Forest audions. The suit claimed that de Forest's tube was an infringement of the Fleming patent. However, Marconi couldn't produce audions as they infringed on de Forest's audion patent. A stalemate. The result of the suit and countersuit was to make audions very scarce. Marconi could make Fleming valves but nobody wanted them. In response to the demand, a number of "independent" manufacturers sprang up and tubes such as Cunningham's Audiotron could be procured through this black market. Due to the war effort, de Forest supplied audions to the armed forces in 1917 and 1918. After the war the injunction

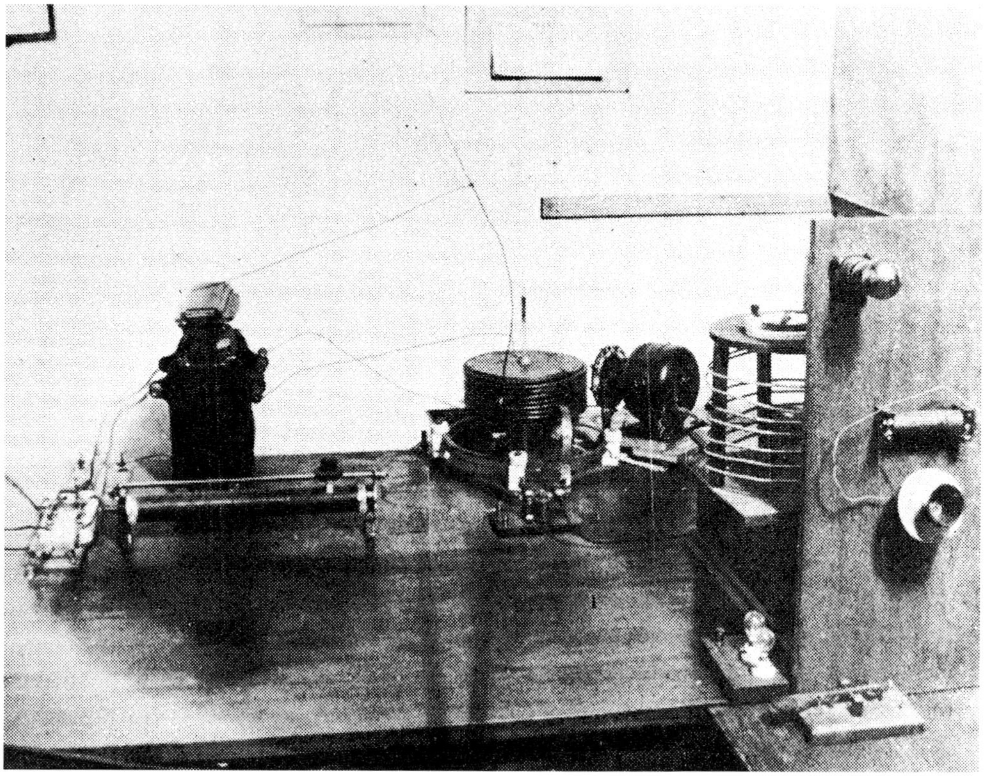

Don's radio statio, 9DR, at the fraternity house in 1919. From left: slide rheostat, high voltage transformer, capacitor, rotary spark, gap, oscillation transformer. At the top of the panel is the VT-2, and the real rarity, the microphone.

against him again became operative. Until the suit was settled an interim compromise was reached; the Marconi tubes would be manufactured by a third party, Moorhead, and licensed by de Forest. The names of Marconi, de Forest and Moorhead were printed on each tube and Moorhead paid royalties to both parties. Gerald Tyne in *The Saga of the Vacuum Tube* provides a detailed description of the entangling suits and litigation. However, it will suffice to say that the manufacture of radio tubes would be quite limited for the next two years.

In the fall of 1920 the ads for the de Forest audion in *QST* noted, "Now free from all restrictions." In the midst of this litigation Western Electric had come out with the 205B vacuum tube, which was a VT-2 stamped as a 205B and intended as an audio amplifier for telephone wire repeater circuits. The VT-1 and VT-2 tubes were manufactured by Western Electric for the armed forces during the war. Western Electric had bought the audion patent rights for repeater applications from de Forest prior to the war. After the war they stopped selling VT-1s and VT-2s and their ads for the 205B tube contained a note which stated; "sold only for uses other than the transmission and reception of radio messages." Virtually any amateur who procured a coveted 205B disregarded the announcement. By the middle of 1920 there were perhaps 200 amateur stations in the country that had a tube transmitter operating with CW transmissions. Only a small fraction of these operated on phone.

That summer a radio telephone link was installed on Catalina Island between Avalon and Long Beach, some 26 miles away. Although there were over one thousand telephones in Long Beach, owned mostly by the affluent, there were just a few in Avalon. The half-mile deep channel between Long Beach and the island had discouraged attempts to lay a telephone cable in between. Tube technology had progressed to the point where a voice could easily and reliability be imprinted upon a radio signal at moderate power levels. For the radio/-telephone link the Department of Commerce had authorized two wavelengths; 470 meters (638 KHz) from the mainland to Catalina Island and 400 meters (750 KHz) in the opposite direction, which allowed full-duplex operation. At Long Beach the radio link connected to the regular telephone lines via an operator. Telephone service began on July 16, 1920. Most people who used the telephone on the island were unaware that the call was going out over a radio set and that anyone who had a receiving apparatus could listen in. With the improvement in receiving technology, most days the radio link was readable in the Los Angeles Basin and throughout Southern California. It was about the most interesting station to be heard by a wireless enthusiast as broadcast stations were not yet in existence.

"Often during the evening," Don reflected, *"after sunset on the West Coast, the Catalina Island radio link came in clearly and the fellows were fascinated listening to these conversations. It was really quite exciting once in a while because the people didn't know they were on the air. There would be a wife calling for her husband over there saying 'You better get home soon' [laughter] and various things you can imagine. Well, the fellows all liked to go up to my receiver and tune that radio telephone station in. I let them use the set all they wanted to."*

Don had to balance his available time between school classes and homework, football practice and games, amateur radio activity, the work schedule, and his fiance. Bertha Lindquist graduated from school that spring and in August of 1920 found a teaching job in Buffalo, Minnesota. Along with all of his time commitments that fall, Don found himself doing quite a bit of commuting to Buffalo on weekends.

Harding had been elected President of the United States, and took office that spring. Among his appointments was Herbert Hoover as Secretary of Commerce. The Department of Commerce was the patron saint for amateur radio. Mr. Hoover made it clear that he was a wholehearted supporter of amateur radio and encouraged it all that he could.

At this time the big battle of spark versus CW was forming. Spark signals were rather broad and, due to this spreading of energy the typical distance of reliable communications was somewhat limited. Many of our amateurs had used the new CW transmitters, which utilized a 5-watt tube, in the fields of France during the war. These were the same little portable sets that Don had used on the *George Washington* for communications between the ships in the convoy. The main variation between the Army, Navy and Army Air Force versions of the radio was the type of power supply. With their meager power output they performed as well as, or better than, many of the big spark rigs had before the war. Tubes were still quite difficult to come by due to the continuing litigation between Marconi and de Forest. For many amateurs, a tube procured through the military or their surplus system was the only one they would possess for the next two years.

Due to their power being concentrated at a single frequency, CW signals were quite difficult to tune in properly compared to spark signals. Articles on CW

Don on the gridiron at Hamline in 1919.

receivers began to appear in *QST* in 1920. In June 1921 *QST* published an article by John Reinartz, 1QP, who had developed a superior CW tuner. The next March an article on an improved CW tuner, also by Reinartz, also appeared in *QST*. Of the 5,700 amateurs currently licensed, those that got their hands on one of the precious tubes would start experimenting with CW receivers and transmitters.

In the summer of 1920 Don transferred from Hamline to the University of Minnesota at Minneapolis and joined the DKE Fraternity at 1711 University. *"The University of Minnesota was across the street from the DKE house. In fact, the armory was right across the street and I liked that because after football or basketball practice all I had to do was take a quick shower and come right across the street. It was one of the reasons I picked them when I was being rushed. The university wasn't far either. It was two blocks to the radio station that Professor Cyril M. Jansky, Jr. had put together, 9XI, on the third floor of the electrical engineering building. That was one of the first amateur experimental licenses issued after the war because he was experimenting with radio circuits. He also had all of us who wanted to add to the station bring our equipment to school. I had brought some of my equipment over there too. So I operated from the university, besides having a workable station in the DKE fraternity house. Well, we put together a pretty good station at the university."*

Professor Jansky was one of the noted radio instructors in the country. Only a few colleges offered radio courses of any nature and the University of Minnesota was quite fortunate to have him on their staff. The year before, Professor Jansky had developed the undamped wave transmitter. Jansky was quite pleased to have a former Navy operator with a first class commercial telegraph license in his classes and soon had Don teaching his radio class on Fridays.

"Jansky was delighted to have me in his radio classes," Don reminisced, *"although I was taking a major in money and banking, business law, things like that. I had always heard that engineers were poorly paid and I was in the age when engineers were not considered the same status as they are today. Today engineers are well-paid and looked up to; in those days they were almost looked down upon.*

"I don't know why, but the generation before mine was all ministers. That was the ambition of many people then, to be a minister. If you were a minister you were the top man in your community. They had the prestige and the good life and everything that went with it. In my day the engineer was not important; the businessman was suddenly the important man. So I was taking business courses but my heart was over there in the radio classes.

"Jansky taught one of the few radio college courses in the country. I don't think there were over three or four colleges in the whole country that offered radio classes at the time. So I was very fortunate. I didn't have the calculus and mechanical background as the other students had. There were twelve students in his classes and they all became famous. One of them developed the first pictures by wire system for Bell Labs, another one became the QST *technical editor and later started his own company, another one became president of the biggest radio receiver outfit of the day. I think virtually all of them became millionaires because of their knowledge of radio when the industry was just beginning.*

"Jansky's plan was that I had experience as a ship operator. As far as we knew at the time, I was the only one that had a radio-telegraph first class commercial license in the state of Minnesota. Each Wednesday night I would have a different one of these twelve students assigned to me and we would keep the station on the air all night. After we finished the broadcast that evening we'd operate the ham station. We would take turns sleeping in the cot provided there and our job was to keep 9XI on the air all night.

"Initially, we had a spark transmitter and a tube receiver at 9XI. The receiver's bandwidth was very broad by today's standards, but in its day it was awfully sharp. We had a standing joke that it was so sharp maybe we could tune the sopranos out and leave the rest of the singing, because we didn't like to listen to sopranos."

That summer Don had found a job with more flexible hours. He worked part time as a wire operator at the North American Telegraph Company. It was also closer to the university, so he spent less time traveling between school and work. From every paycheck he still sent whatever he could afford back to his parents in Long Beach.

QST began a column labeled "Calls Heard" in 1920. Shortwave listeners from across the country and around the world would send in lists of the stations that they had heard. It immediately became very popular and most amateurs who subscribed to *QST* thoroughly scanned the reports, looking for their call signs. The newsstand copies of *QST* did not contain this column; one had to subscribe to the magazine for this information to be in their issue. Due to their big stations with high antennas and the large amount of activity, 9XI and 9DR began to appear quite frequently in the "Calls Heard" columns.

That fall, more tube ads began to appear in *QST*. Montgomery Ward and Company put out a catalogue of high grade radio supplies, similar to the one Sears, Roebuck and Company had resumed printing after the war. The number of licensed amateurs exceeded seven thousand and was increasing daily.

The ingenuity of amateurs during this era should not be underestimated. In February 1921 *QST* carried an article about an amateur who had an advanced station that operated spark, CW, and radio telephone, yet the operator was living in the most reduced circumstances. The lad was but seventeen and had taken a full-time job to help support his family. He had constructed every last detail of his station himself. Using discarded test tubes and the filaments from burned-out light bulbs he assembled vacuum tubes. He had built a mercury vacuum pump to exhaust the tubes and they had performed as well as any manufactured ones that were not yet available to the public. The total financial outlay for his complete station was twenty-five cents—the price of a pair of combination cutting pliers. An effort of this nature was the true spirit of amateur radio.

Don recalled, *"In the fall of 1920 Jansky assigned me to be there at the station one night a week. So I showed up every week and operated the ham station all night. 9XI was a 1-kilowatt spark station. We had put together a rotary spark gap transmitter that summer. Earlier in the evening I put on a broadcast which was the first regular broadcast in the Northwest. We used a 5-watt tube transmitter (with a VT-2) for the voice*

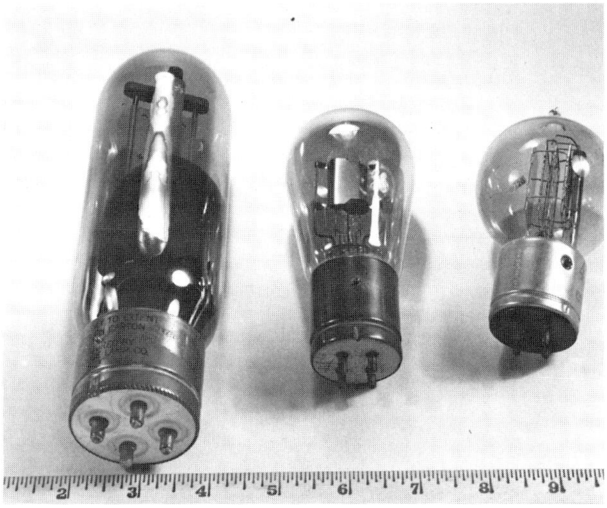

A VT-2 and UV-202 (5 watters), and a UV-203 (50 watts dissipation).

broadcasts, which was put together that fall as a class project. Back east they referred to Minnesota as the Northwest. Out here in Long Beach we would call the Seattle area, Washington and Oregon the Northwest. Minnesota is northwest from New York, in other words. The western states were so thinly populated in those days that no one considered them worth mentioning.

"Pretty soon the radio inspector came around and told us that we would have to get an assigned broadcast call sign from the Department of Commerce. We couldn't continue using 9XI; that was an amateur experimental call sign. Our broadcasts had become regular and were no longer experimental. So after Februray 1, 1921 we began to use the call sign WLB for the university broadcasts and we continued using 9XI for amateur radio work. The only other broadcast station that we knew of was KDKA Pittsburgh and that was put up by a ham named Conrad, 8XK. He had built an amateur station with a pair of 50-watt tubes. He just ran it as a broadcast station and called it a Westinghouse station. So they had the honor of being the first in the country with regular programs thanks to Conrad. The University of Minnesota, 9XI, was second." Early in 1921 another CW transmitter at 9XI was built. Professor Jansky had obtained several VT-2s from the Signal Corps and the students used them in the experimental amateur transmitter.

"Jansky's wife would bring out a few records to the university station every week. She would go down to the music store and get some records, much as the stations do now. I'd play those during the evening for a while. In the meantime I had called up a few people and got somebody to come by and give a talk. Among the speakers were the president of the university, the mayor, and some representatives from the state legislature. They would come down and give us a ten-minute talk for the radio audience. So we developed a suitable radio audience. The people that listened to us had mostly homemade sets because there wasn't anything else available. We developed a nice clientele and a bunch of fan mail."

The broadcast boom was just beginning. Dozens of small broadcast stations were coming on the air and the public was becoming aware of this new entertainment available. Many of the stations used but a single 5-watt tube to transmit and on a good night they could be heard up to 100 miles distant. Commercially built receivers were in great demand, but only a few had been built. The first radios which had been specifically built for broadcast reception had became available that year.

Receivers with names like Paragon, Grebe, Crosley, Zenith and Aeriola began to appear in magazine ads. Some of the early models didn't necessarily work as well as the better homemade sets and they were quite expensive. But for those people who were not handy at building electrical projects the commercially made radios were quite suitable. Among these manufacturers was R.H.G. Mathews, 9ZN, who had earlier formed the Chicago Radio Laboratory. Using the suffix of his call sign he originally referred to his radio line in the ads as the Z-Nith. A short time later the now-familiar name of Zenith began to appear.

In late spring, Professor Jansky received a call from the General Electric Company. They were going to open a store in Minneapolis and were looking for someone to run the radio department. G.E. inquired as to whether the professor had a particularly gifted student he could recommend for the job and Jansky recommended Don Wallace. In June 1921, Don became manager of the radio department of the Peerless Electrical Supply Company at 227 West 5th St. which was owned by G.E. His salary was negotiated for $125 per month plus a commission of 2 percent of his gross sales.

The radio department of the new store soon became very busy. Don was an excellent manager and also a good source of information for customers. Along with the broadcast boom there were large numbers of radiomen returning from the armed forces who were getting their ham licenses. By June there would be over 10,000 licenses issued by the Department of Commerce. The enthusiasm of the amateurs was at a peak. The ARRL convention that summer had an attendance upward of 1,500 hams, representing nearly 15 percent of the licensed amateurs in the United States.

In December 1920 RCA introduced two Radiotron

Don (center) and Professor Jansky with a student in 1922.

receiving tubes; the UV-200 and UV-201. They were about $5 each and would be used in many of the early tube receivers.

The next spring RCA had announced the development of a new transmitting tube. The UV-202, or 5-watter, was available for $8. That August the UV-203, a 50-watt tube, was introduced. The 50-watter was for those of more substantial means; one could grace your station for the sum of $30. That amount of money represented a week's salary for many people.

After Don left active duty he had remained in the Naval Reserve. During a two-week cruise the previous summer he had been commissioned as an Ensign. Now that Don was over 21 the Navy no longer had any qualms about promoting him to an officer. He went on a second two-week cruise in August 1921 and then wound down his Naval Reserve activity.

That fall found Don with a full load of classes, playing on the U of M Tri-State championship football team under Coach Doc Williams, operating two radio stations, and working full time. Don and Bertha also got married on November 24, 1921. Soon he became an expert at budgeting his time. "*At the end of the WLB broadcast program my last year in school*," Don recounted, "*I would call my wife in Buffalo and talk for five minutes. I would sign off the broadcast and then talk to her with the broadcast transmitter. I had put together a receiving set for her down there in Buffalo where she was staying. Bertha and I had been going together about five years and it was high time to get married. We were getting older. I had been to war, got back, and it was time. She was teaching school in Buffalo and she had a second-story room in a nice home. The set was hooked up there and she would tune in the university broadcast. Pretty soon, every Wednesday night, all the neighbors came to her house to listen to the broadcast program. It was the only station they could hear and they particularly enjoyed the five-minute talk I'd give her at the end of the program, my personalized talk to her. That talk became so popular that it developed an even bigger audience than the program itself had ever developed. All the mail and comments we got, that was the most popular part of the program. I had already signed off the broadcast part. Talking to Bertha afterward, that was what fascinated the people more than anything else on the program. So you might say that I was in the early days of broadcasting. I broadcast most of my two years at the University of Minnesota.*"

In November 1921 the ARRL sent Paul Godley, 2ZE, to Scotland in an attempt to hear American radio signals from Europe. Paul had developed the Paragon receiver and he was considered to be one of the foremost receiver experts in the country. He set up an antenna at Ardossan that was several hundred feet long but just ten feet high. This antenna had been developed by Harold H. Beverage, 2BML, at Otter Cliffs, Maine. It was referred to as a wave antenna, typically being one wavelength long, and reduced unwanted atmospheric noise dramatically. During ten rainy and dreary December days, on a soggy Scottish moor in a rented tent, Paul heard 27 stations from the United States

In the first week of November the ARRL had selected those stations that would participate in the trans-Atlantic tests. Stations heard over 1,000 miles distant from their transmitting location were invited to participate. Earlier in the year the station at 9XI had been supplemented with a new transmitter that employed four of the Signal Corps VT-2 tubes. Don, operating 9XI and running 150 watts input, qualified for the competition.

A group of engineers in Connecticut, headed by Edwin Armstrong, wanted to transmit the first signal heard across the Atlantic. In the last two weeks of November they rebuilt the station of Minton Cronkhite, 1BCG. A 108-foot high antenna mast was added to the existing 75-foot one at his location so a better antenna could be erected. Armstrong provided the three 250-watt tubes that were connected in parallel for their transmitter. With 2,000 volts from a DC generator on the plates, the power input was probably in the vicinity of 2-3 kilowatts. Godley heard 1BCG and telegraphed the long-wave station at Caernarvon, Wales, MUU. They passed the information to the long-wave station at New Brunswick, WII, and Armstrong was notified that a message should be sent by Cronkhite. That year 1BCG, operating on 230 meters (1300 KHz), was the first and only station to pass a message to Godley in Scotland, who was set up as a receiving station only. A British operator, W.R. Burne, 2KW, also copied the complete message from Cronkhite to Godley.

Don and several of his classmates took turns manning 9XI during the ten days that the trans-Atlantic test was in progress, but their 150 watts on the 300-meter wavelength (1000 KHz) wasn't heard by Godley. After Godley's successful receiving experiment, the wave antenna that he used for receiving came to be known as a Beverage antenna.

The trans-Atlantic tests had demonstrated that propagation, the refraction of radio signals off the ionosphere and back to earth, was a reality on the short waves. Nicola Tesla had first observed the phenomenon two decades earlier at his laboratory in Colorado. The long-wave stations had accomplished trans-Atlantic communication by means of ground waves, but operation at those low frequencies required huge antennas and enormous amounts of power.

Two months later, in February 1922, Minneapolis was hit with a tremendous ice storm. All telephones lines leaving the city, and most of those within, were useless. Don, operating at 9XI, handled outgoing messages and a mountain of incoming traffic for City Hall as well as for his neighbors for a 3-day period. Boyd Phelps, operating 9ZT at the Minneapolis courthouse, provided a similar community service. After three days, weather conditions finally improved enough for the utility companies to repair the downed telephone and power lines. In the years to come amateur radio was to provide similar public service assistance.

Throughout the 'twenties the great majority of amateur radio operation was traffic handling. Only a few stations would have adequate equipment and antennas for long distance (DX) contacts. Standard telephone calls between cities were quite expensive and simply not possible in bad weather. Mail went by train and a penny

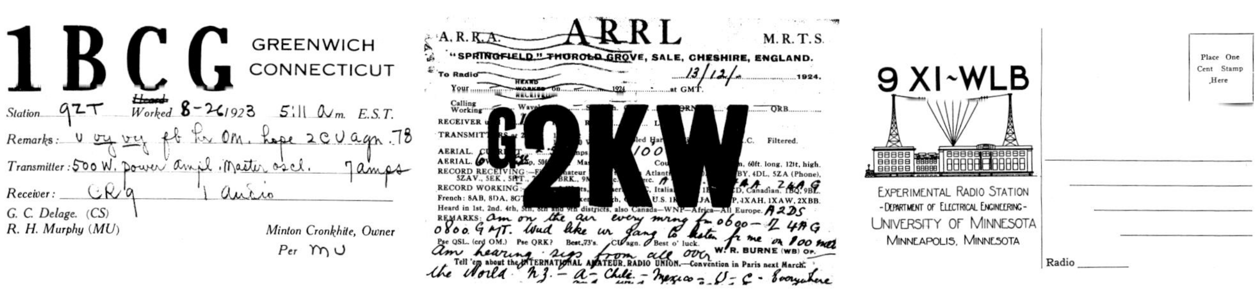

The QSLs of 1BCG, (G) 2KW and 9XI.

postcard could take a week or more to reach its destination across the country. In traffic handling, the amateur accepted a message from a local citizen and passed it over the air to an amateur radio operator closer to the message's destination. The message might be relayed from station to station several times through the night until it reached the desired city. The amateur at that city filled out a radiogram, which was mailed like a penny postcard, and dropped it in the mailbox. The local mail delivery was usually within a day. In this manner the first contact between Hawaii and mainland United States occurred on January 18, 1922. A.H. Babcock, operating 6ZAF in Berkeley, worked Cliff Dow, 6ZAC on Maui. Without any fanfare, the two stations passed ARRL message traffic between their cities.

That spring the Radio Corporation of America (RCA) put out a catalog entitled "Radio Enters the Home. How to enjoy popular Radio Broadcasting; Complete Instructions and description of apparatus." The catalog also contained the latest tube that RCA had developed the summer before, the Radiotron Model UV-204. The tube was rated at 250 watts dissipation and could be purchased for the modest sum of $110.

Due to the proliferation of broadcast stations and broadcast listeners (BCLs), voluntary quiet hours were being observed in the evening by many amateurs. Between 8 PM and 10:30 PM, when most broadcasts took place, amateurs were asked by the ARRL to refrain from transmitting. The operating wavelengths of broadcast stations were located very close to amateur wavelengths, which resulted in excessive interference when an amateur transmitter was used near a broadcast radio. After 10:30 at night, when the general public went to bed, the amateurs would come on the air. Meanwhile, Don was getting ready to move out of the DKE fraternity house. Bertha's contract with the school district in Buffalo was up for renewal at the end of the spring classes, and she returned to Minneapolis. She and Don rented an apartment at 1830 Stevens and moved into it in June. Don's first son, William Wallace II, was born on August 24, 1922, exactly nine months after they were married.

The summer of 1922 found Don spending a lot of time working on his amateur station. During the longer days the radio propagation was not as good and the noise level was higher. The longer and warmer days also made it easier to do antenna projects. Don had selected a location that made it easy to put up a good antenna outside. In a note to *QST* in the fall of 1922, Don described his station:

The present station of 9DR was erected about two months ago. The aerial consists of two wires, flat top, between two apartment houses. Height above ground about 70 feet and height above apartment house about 20 feet. A 9-wire fan counterpoise 10 feet high covers a great portion of the area between the two apartments.

A radial counterpoise served to form a capacity ground, which was very effective in rocky or sandy soils with poor conductivity. It had to be of sufficient size to have considerable capacity to ground, therefore, it should cover as much ground area as possible yet be free from self-resonance effects. Its height was not critical but should allow clearance to walk underneath it and yet not be high enough to reduce the effective height of the antenna.

The transmitting set can be noted at the right hand side of the picture. Two 50-watt tubes are used in parallel as oscillators, current for the plates being supplied from the AC power mains rectified by two kenotrons. A kenotron was G.E.'s and RCA's name for a rectifier. Each UV-217 kenotron would put out 150 watts of DC power for a UV-203 amplifier tube. For the low-power CW stations, the UV-216 kenotron would put out 20 watts of DC power to run the UV-202 amplifier tube.

Radiation is at present 5 amperes, although the

RCA published *Radio Enters the Home* in 1922.

Don's radio station, 9DR, in early 1922.

radiation was as high as 6 1/2 amperes before the aerial was raised above the apartment house. The plate current is 200 milliamps and AC is used on the filaments. Reversed feedback connection as used by a great many amateurs is used with a small variation. Both counterpoise and ground are used and the entire set is designed to operate as a relay transmitter, the changeover from receiving to sending being remarkably quick. One switch does the entire operation with a very short throw. The bulk of the receiving is done on a single circuit receiver in combination with the GE AA1400 Detector amplifier, using Western Electric phones. If at any time the signals become unreadable, the Radio frequency set to the left of the picture is used. This set is still in the process of development, but is working very satisfactorily at present. Its main disadvantage is due to the sharpness of tuning, so it is only used when the single circuit set will not bring signals in properly. (This receiver would become widely known in the years to come!) *Either set can be tuned with shutting off the other, so that the change from one set to the other can be effected in a very few seconds.*

9DR is a one man station, being situated in the corner of the parlor, more or less to the detriment of the aesthetic proportions of the parlor, as one might judge from looking at the picture.

The operating hours are from 9:30 to 11 PM on about five nights per week. Considerable relay traffic is handled. A 2-way communication has been maintained with well over 60 distant stations during the last four or five weeks. Hundreds of cards have been received during the same time from as far as the corner states of the United States. During one 10-day period, reports were received from all districts.

Stations worked during the last few weeks include 2FP (New York), 5EK, 5DI, 5ZG (Tennessee/Texas/Oklahoma), 7ZU (Montana), 8AB, 8AM, 8AXB, 8BEF, 8CAZ, 8VY, 8ZA, 8ZZ (Ohio/Michigan).

9DR has had four separate locations for the past year, using a 5-watt portable set , but now has settled down at the present location. It is believed that 9DR will continue handling its share of Northwestern traffic.

Shortly after sending the station description off to *QST,* Don received a new call sign assignment, 9ZT. Apparently the ARRL was suitably impressed with Don's station. The February 1923 issue of *QST* carried a picture of 9DR/9ZT along with the station description.

Don had a good income coming in from his job at the store and, with his ability to buy items at wholesale prices, his station continued to improve. This would be a source of almost constant discussion between Don and his wife in the years to come.

The Hoover Cup Station

Don graduated in June 1922 with a degree in business administration. His father, William, now the vice president of the First Exchange Bank in Long Beach, wanted Don to follow in his footsteps and become a banker, but fate was to take him on a different career path.

His interest was in radio and he elected to remain as manager of the radio department of the Peerless Electric Company in Minneapolis for General Electric. Recalling those days Don said, *"Professor Jansky and I remained friends all of our lives after I graduated from college. Jansky apparently liked me so well he suggested we have lunch together once a week. He would come down to the store where I worked to meet me. The G.E. Company had hired me back while I was still in college, the year before, on his recommendation. Jansky and I would have lunch every Tuesday and talk over radio related events. Well, in the years to come he became the ARRL director of the Dakota Division.*

After all he was the trustee of 9XI, and he made a fabulous director. In fact, he was so highly regarded that he was also appointed to the Hoover Commission and the Federal Radio Commission."

Amateur CW stations were becoming more common and they could work other stations farther away than spark stations. The main limiting factor was hearing other CW stations through the racket created by local spark rigs. One evening Don worked Donald Snow, 6RD, in Mountain View, California. Snow was running 50 watts CW with a Leyden jar (chemical) rectifier. Snow wrote on his QSL card, "Good work OM, glad to get ur sigs. Please listen for 6's, and QSL on getting my sigs."

By passing the higher amateur theory and code requirement in the fall of 1922, Don had received a special call assignment. *"9ZT was my call sign. The stations that had a Z prefix had special privileges. They were officially allowed up to 220 meters* (1360 KHz), *where the general class license amateurs were only allowed up to 200 meters* (Their band was 150 to 200 meters, or 2,000 to 1,500 KHz). *In other words, we had what today is considered extra class licenses. The incentive plan was already there. If you could pass the code, I don't think it was very fast, about 20 words per minute, you could apply for a Z call sign, and have the extra 20 meters of operating wavelength* (150 to 220 meters). *Everybody used it anyway, but 9ZT also had the use of 375 meters* (800 KHz). *At that time everybody felt that the longer wavelength was better."*

The number of licensed amateurs after the war grew at a tremendous rate, and by mid-1922 there were over 14,000 licensed stations. The radio laws of 1912 had remained intact, but in 1923 their interpretation by the Department of Commerce was changed so that only those wavelengths specifically mentioned were authorized for amateurs. Undoubtedly the emerging commercial interests had brought pressure to bear for this reevaluation of the radio regulations. Those amateurs with standard licenses were authorized to use the 150-200 meter wavelengths. For those few like Don who passed the amateur first grade test, the 200 to 220-meter region was added. Any extra authorized wavelengths, such as 375 meters, were noted on the individual's license. The experimental stations with the X-suffixes did not have any wavelength restrictions.

In an effort to help keep the amateurs within their authorized wavelengths, the ARRL set up Official Wavelength Stations (OWLS) among its members. The great majority of amateurs had home-built equipment and no accurate way of calibrating it. OWLS stations such as Don's had the new General Radio wavelength meters. They would usually operate on just one or two wavelengths during the evening or night and give their wavelengths often during transmissions. In this manner those stations monitoring could calibrate the dial of their receiver and adjust the frequency of their transmitter accordingly.

In a typical correspondence with Don, ARRL secretary Warner noted:

> Station 9BAB, operated by Issac Johnson, Waseca, Minnesota, is in trouble with some BCL's (broadcast listeners). We are interested in ascertaining whether or not he has been operating on legal wavelengths. Will you be good enough to drop me a line and advise what you know about this station. In particular, can you definitely say that you have heard it exceeding 200 meters, etc.

In a brief letter Don noted the following wavelengths used by 9BAB: 211 meters and 205 meters. He then helped Johnson get back below 200 meters with the rest of the general class amateurs. The ARRL performed a real service for both the amateurs and the BCL's in trying to keep the amateurs' transmitted signals within their authorized wave lengths.

In the fall of 1921 the hams first collided with the broadcast listeners in a big way. At that time other radio services commenced efforts to appropriate large slices of the amateur radio spectrum. The ARRL was kept busy by helping the amateurs fend off these intrusions. In 1922 de Forest estimated the American listening public at over one million. He shrewdly predicted that there would be five million BCL's in two years, and 20 million by 1927.

During 1923 Don's amateur station was in a state of continual change. The receiver was redesigned and rebuilt several times. The early spherical audions were initially replaced by de Forest type T tubular audions, and then later with some of the new lower noise and higher gain receiving tubes that had become available, RCA's UV-200 and UV-201A. Since the patent litigation between Marconi (now RCA) and de Forest was finally settled, tubes were being manufactured at an incredible rate. So many people were building radios that distributors couldn't keep tubes on their shelves.

The two 50-watt tubes in Don's transmitter were starting to get a bit soft and their output power was diminishing, so he replaced them with the latest transmitting tube available from RCA, the UV-204. It was rated at 250 watts dissipation, and would run a kilowatt input when pushed to the limit. This was a commercial version of the "P" tube in the G.E. transmitter that he had used on the *George Washington*, with end fittings added instead of just wires protruding from the glass envelope. The UV-217 Kenotron rectifiers had been designed specifically to compliment the UV-203 50-watt tubes and couldn't supply the higher power that the UV-204 transmitting tube required. Don built a Leyden jar rectifier to supply the DC power for his transmitter. The Leyden jar chemical rectifiers had been described in the February 1921 issue of *QST*, "An Electrolytic Rectifier for CW", and their implementation had been refined in subsequent articles. Don's transmitter was redesigned to a Hartley circuit with the antenna tapped right into it and the high voltage isolated with a blocking capacitor. He had the local power company install a separate five kilowatt pole line transformer for his house so his lights wouldn't blink when he was transmitting.

Don had built from scratch his transmitter, receiver, and even the insulators supporting his aerial. He also had developed an optical ferrometer to check the efficiency of his transmitting tube, which was accomplished by visually comparing the ferrometer to the redness of the plate anode of the UV-204. Don tuned up his transmitter with an RF ammeter in the antenna feed line by adjusting the tuning condenser for maximum RF output. Counterpoise wires for the antenna were added all around the lot until his automobile would barely clear them when he pulled in the driveway. The counterpoise wire system served to act as a ground connection.

Although a few high-powered broadcast stations running several kilowatts had been installed in the New York area by mid-1923, most broadcast stations across the country still ran just moderate power, in the vicinity of 50 watts or less. They were usually assigned a wavelength of 250 to 400 meters (1200 to 750 KHz). Don's special call sign let him operate a transmitter in the frequency range where these broadcast stations were located. That year he worked several of them.

"Late at night 375 meters (800 KHz) was better for certain distances," Don reflected, *"like 700-800 miles to 1500 miles. What I would do is get on the air on 375 meters, and I would have a broadcast station from Atlanta tuned in, for instance. From Minneapolis it was about 1500 miles away. I would get on 375 meters and call the broadcast station periodically. It was usually after midnight and I seldom bothered anybody near me. Well, the engineer at the broadcast station was always a ham, and sometimes he would be monitoring the various wavelengths. If the broadcast station had a receiver turned on sometimes I would get their attention and talk to them. They would break their program for a minute, we would have a little chat, then they would go back to their regular program."*

By 1923 Don had worked all states but one, Wyoming. Some 11 years earlier Arizona had achieved statehood, bringing the total in the United States to 48. One night Don was tuning his receiver and heard KFDF, which was a broadcast station located in Casper, Wyoming. Don called them on the 375-meter phone, and they eventually noticed him calling. KFDF stopped their broadcast and exchanged pleasantries for a few minutes. The engineer then relayed to his local audience the details of Don's station and the weather in Minneapolis. Don received a QSL from KFDF shortly after that. The ARRL did not offer the Worked All States (WAS) award for another 15 years, but 9ZT is thought to have been the first to qualify for it.

That year Don was on the air almost daily for a couple of hours, with the exception of when he had the rig apart or the antenna down for repair. According to his logs, 9ZT passed over 2500 messages in the 12-month period. Almost anyone who was on the air that year either heard or worked Don. A newly licensed youngster with the call of 6BPI in Phoenix, Arizona, logged 9ZT and sent the report in to the *QST* column "Calls Heard." In the decades to come that young man's call became K7UGA, and Barry Goldwater would become well known to the amateur radio community as their staunch supporter in the United States Senate.

In May the Long Beach Radio Club and *Radio Journal* held Trans-Pacific tests, which were coordinated with the Wireless institute of Australia. The outstanding stations were announced as 6JD, 6AWT, 3YO, 5AEC, 7BJ, 9AUL and, of course, 9ZT. Don was also starting to appear in the various radio magazines more frequently. In the section of "With The Amateurs" the September *Radio News* carried an article labeled "Station 9ZT, Minneapolis, Minnesota." It described Don's station and antenna installation, and the improvements on it he had recently made. Sixty years later a copy of the article was sent to the author by Vince, W8DB.

Don attended the second annual ARRL convention, which was held in Chicago that year. The consensus of the amateurs at the convention was that Don would have an excellent chance to be selected as the Hoover Cup recipient for 1923. Two years earlier Herbert Hoover, the Secretary of Commerce, announced that he would award a cup annually to the best amateur station in the United States. In 1921 Louis Falconi, 5ZA in New Mexico, won the cup and in 1922 Walter Ostman, 2OM in New Jersey, took the honors.

After the convention the Minneapolis Tribune carried the following article:

> Minneapolis to be in triangle for radio test. D.C. Wallace's station will link with others in Texas and Connecticut.

The article went on to mention the continuing CW vs spark controversy, which was then in full swing.

A few weeks later, Don received a card from Leon Deloy, 8AB in France. They had met at the convention and chatted about the possibility of amateurs establishing trans-Atlantic communication on the short waves. During the Versailles Peace Conference four years earlier Don and his radiomen aboard the *USS George Washington* had made the first two-way Trans-Atlantic voice communications by a government station (or anyone else!), which took place on the long wavelengths. Deloy expressed regrets at not being able to visit Don and his station in Minneapolis. Don was located too far northwest for Deloy's limited itinerary while he was visiting the United States.

Earlier that year Professor Jansky had been appointed a member of the Hoover Commission. The commission was studying the issue of mandatory quiet hours, or hours when amateurs would not transmit. The interference with broadcast listeners was at an all-time high due to the many amateurs that had disregarded the ARRL suggestion of voluntary quiet hours. Late that year the Hoover Commission voted to implement the mandatory quiet hours of 8 PM to 10:30 PM local time.

A lot of notable radio activity occurred in 1923. That summer a ship went up to the North Pole and spent the winter there, frozen in the ice pack. A team of scientists wanted to study the Arctic first hand, and the ARRL had been invited to send a radio operator along. Don Wallace was their first choice for the trip, and he was offered the position. However, his son Billy was not yet a year old, and Bertha put her foot down. Don had to pass on the trip. In late June, Don Mix, 1TS, left with the ship *Bowdoin* as their radio operator.

By saving a little of his income each week, Don and Bertha were able to accumulate enough money for a down payment on a house. Don selected a lot that seemed to be in a good radio location and commissioned MacDonald Builders Incorporated to build their house. It was completed in late November, and the Wallace family moved in on Thanksgiving weekend.

Don had been working for Peerless Electrical for more than two years. One day, the owner wanted to have a talk with him. It seems that Don was a tremendously successful salesman. So much so, the owner claimed, that Don was making more money than he was. Don's contract was up for renewal and was going to be rewritten. His salary would be $125 per month plus one percent of his sales, essentially half of what he had been making. Don recalled *"The president said I was making more than him and he would have to cut my salary in half. He was quite convincing and even had a tear in his eye; but I thought it was all just a performance."* Don stayed with the store and continued to make a good salary, but after that he was somewhat disillusioned with the Peerless Electrical Company.

The overall economic outlook in the United States had improved in the past year. Unemployment was low and consumers now had more buying power than ever before. The stock market started an upswing that continued for the next six years. Don no longer had to send money home, as his father had paid off his debts and was now on solid financial footing.

That fall the big push in amateur radio was to establish two-way communications across the Atlantic. U.S. amateurs had been heard in Europe the previous year, but a two-way contact was yet to be made. The output power of Don's UV-204 transmitting tube was falling off, so he replaced it with the latest Radiotron tube available, the UV-206. This tube, considerably more expensive than a UV-204, had a higher power dissipation, 500 watts, and would comfortably run 1-1/2 kilowatts input.

On November 27, 1923, Fred Schnell, 1MO, worked Deloy, 8AB in Nice, France using the 110-meter wavelength (2700 KHz). Schnell had been one of Don's operators on the voyage to Versailles 4-1/2 years earlier. John Rheinartz, 1QP/1XAM, who had designed the transmitters in use at all three locations (See *QST*, June 1922), worked Deloy the same night. Don had moved down to the 110-meter wavelength also but his antenna mast had blown down that day. His antenna was draped across his roof and the tree out front, so he had to content himself with listening to Schnell and Reinartz work Deloy. Don got the mast back up the next weekend and eight days after the historic trans-Atlantic QSO, on December 5, he was heard by Deloy. Conditions were poor that night and Schnell had to relay Deloy's message to Don in a 3-way contact. Don and Deloy made 2-way contact the next month. Three days later, on December 8, K.B. Warner, operating 1MO, worked British 2KF in London, for the first Anglo-American contact. The Atlantic had been bridged by CW, and so rang the death knell for spark.

The commercial and government radio services sat up and took note of the recent developments in amateur radio. Amateurs had achieved trans-Atlantic communication with a small fraction of the power that government stations required. It seems that these "useless" wavelengths to which the amateurs had been consigned were turning out to be the most valuable. The amateurs now had to contend with serious attempts by other services trying to grab their territory.

In December Don received one of the experimental licenses that were rarely issued, 9XAX. Typical amateur radio licenses were constrained by authorized wavelengths to transmit, maximum power input to the transmitter, and in some cases the hours that transmission was allowed. Don's 9XAX license read, "Any power, any wavelength, any time." Only a few of the "X" licenses were ever issued, and the purpose of the license was for experimental development and testing of radio equipment. It was not to be used for standard amateur contacts. Each year Don had to justify his use of the license to the radio inspector in Chicago in order to have it renewed for another twelve months.

On April 28, 1924 Don received the following letter from the Secretary of Commerce:

> The Board of Directors of the ARRL by unanimous vote have decided that you are entitled to the Department of Commerce Cup for 1923 in recognition of the notable efficiency of your radio station and your activity in amateur radio work.

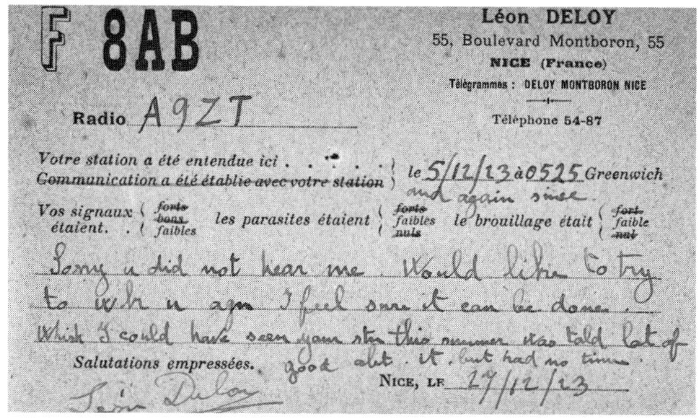

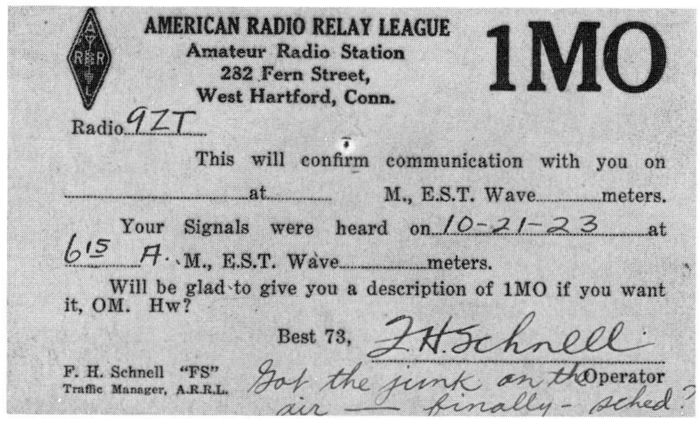

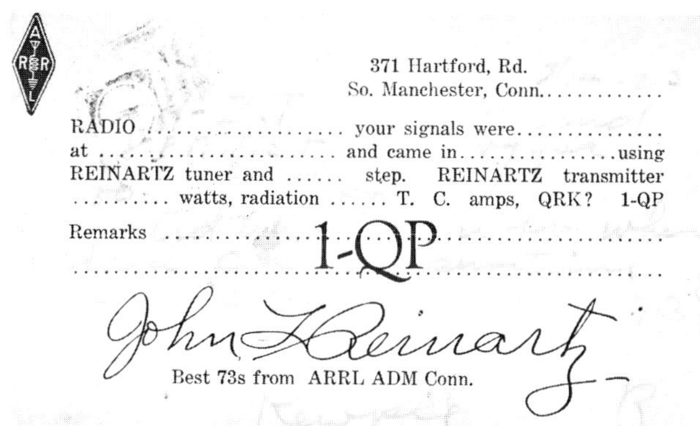

Leon Deloy of France and Schell and Rheinartz in Connecticut made the first trans-Atlantic contact in November 1923.

> "It gives me very great pleasure therefore to present you with the cup which has been sent you by express. I also desire to express my hearty congratulations on the success of your work.
>
> Yours faithfully,
> Herbert Hoover."

The "Hoover Cup" inscription states "The best all around amateur radio station in the United States, the majority of which is built by the amateur himself." The May 1924 issue of *QST* had an article entitled "Wallace

Wins Hoover Cup". The highlights of the article are as follows:

9ZT is the near realization of a lifelong ambition; namely, to have a workable and useful amateur station (Don was just 25 years old). Compromises have to be made in any set, compromises between efficiency, practicability, workability, time of construction, and pocketbook. 9ZT was built and assembled with considerable thought, having in mind the compromises necessary.

It is part of the obligation of an amateur to be always in commission, somehow, some way. His masts may blow down, transformers burn out, and minor mishaps may occur. 9ZT has been in commission always. Scarcely any day has gone by in the entire twelve months that this has not been true.

The entire set, so far as is practical, has been made by the operator himself, even to the erection of poles, adapting of transformers, winding of coils, and cutting of plates.

The question has often been asked where the time for operation comes from. This was planned long ago. First by a location close to work, just one and one half miles from the center of downtown Minneapolis. The electrical efficiency of the station suffered thereby (by local noise), but it was this or no radio at all. By going to bed at nine and getting up sometime after midnight for two hours, an average of eight hours sleep per night helps in fitness for the day's work.

Don Wallace with his Hoover Cup in 1924.

The transmitter utilizes one 250-watt Radiotron tube. The meters and tube are in clear sight of the operator; one glance tells all, and this feature has no doubt saved the tube on more than one occasion.

It has been found poor policy to change waves, and except in rare instances only two waves have been used, 215 meters (1395 KHz) and 115 meters (2600 KHz). French 8AB was worked on 115 meters.

Several months ago 9ZT received special permission to use low waves, but in the last week of December 9XAX was received as a call for this type of work.

The transformer is an old burned-out 3-kW line transformer that was saved from the junk man. The 6000-volt tap is used into a 124 jar rectifier, lead and aluminum. A faint glow is noticed when the whole room is dark. The solution for the rectifiers consists of ten gallons of battery water, two pounds of Twenty Mule Team Borax, and a teaspoon of household ammonia.

This receiver seldom picks up local interference although the night of the WNP test relay 6BPZ (California) fairly wiped (ARRL) President Maxim (1AW) off the map. 1XAM (Reinartz) is worked on 56 meters (5350 KHz) with ease.

9ZT has been reported from Alaska, New Zealand, Australia, Hawaiian Islands, Mexico, Panama, South America, Puerto Rico, Cuba, WNP, England and France. Every province of Canada and every state in the U.S. has been worked.

To put up a 100-foot mast is easy. To have it stay there is different. In November 1923, the gang helped put up a 100-foot mast. It blew down on Thanksgiving Day, as the Wallaces moved into their new home. Using the 50-foot stub, 9ZT shot up to 114 feet. It lasted a few days and then blew down the day of the trans-Atlantic tests (with 8AB). The present mast of 85 feet was put up and it seems a joke next to the others; however, it stays up. The other mast is out front, and due to the wife's suggestion it was only made 60 feet high.

The flat-top antenna (consists) of six wires 60 feet long. Eighteen-inch plate glass insulators are used throughout. Using a broken three-cornered file, a brace and plenty of turpentine, glass is drilled easily. The shape of the counterpoise is more or less radial, like spokes in a wheel, making a covered circular

Don Wallace, 9ZT, in 1923.

The transmitter's Leyden jar power supply in the basement.

area 170 feet in diameter, thanks to the kindness of neighbors.

Don's neighbors were indeed very kind. All of the antenna wire he had put up undoubtedly was not well received by the neighborhood at first glance, but Don has proved to be a very persuasive individual. The power supply for Don's transmitter occupied most of the basement. Rectifiers were in the early stages of development. The kenotrons that provided DC power for his 50-watt tubes were inadequate for the 1-kilowatt rock-crusher he now had.

The 124-jar rectifier that Don mentioned was spread out in a 6-foot by 6-foot area on the basement floor directly below Don's transmitter. The electrolytic "slop jar" rectifiers had been developed three years earlier and any amateur that ran several hundred watts or more of input power with his transmitter had a similar system installed, or had access to a high-voltage DC generator. Each jar was the standard one-quart canning size Ball jar. Thirty-one jars made up one quarter of the full-wave rectifier. The output was 4,000 volts that ran up through the ceiling of the basement and through the floor of radio room where the transmitter was located. The basement had to be kept locked so that little Billy (now a toddler) wouldn't accidently be exposed to the high voltage when Don was transmitting.

The Hoover Cup was big news and many radio magazines of the time carried the announcement, along with the station details. The April 12, 1924 edition of *Radio World* carried the article: "Hoover Cup Winner Gives Tips On Transmitter Work." The article noted,

> "Aside from its value as a relay station for handling message traffic, this station, 9ZT, is regarded as an exceptionally valuable unit for emergency communication and has frequently been of aid while telephone and telegraph wires were out of commission." The photos accompanying the article were of Don at his station, and the Hoover Cup.

Killian Lansingh, 1BAN, wrote Don March 24 and remarked, "Just a few lines to congratulate you on the winning of the Hoover Cup, which I noticed in the radio section of this morning's newspaper." Most of the newspapers across the country that had radio sections, and there were many, probably published a similar announcement.

At this point, most amateurs used a single 5-watt

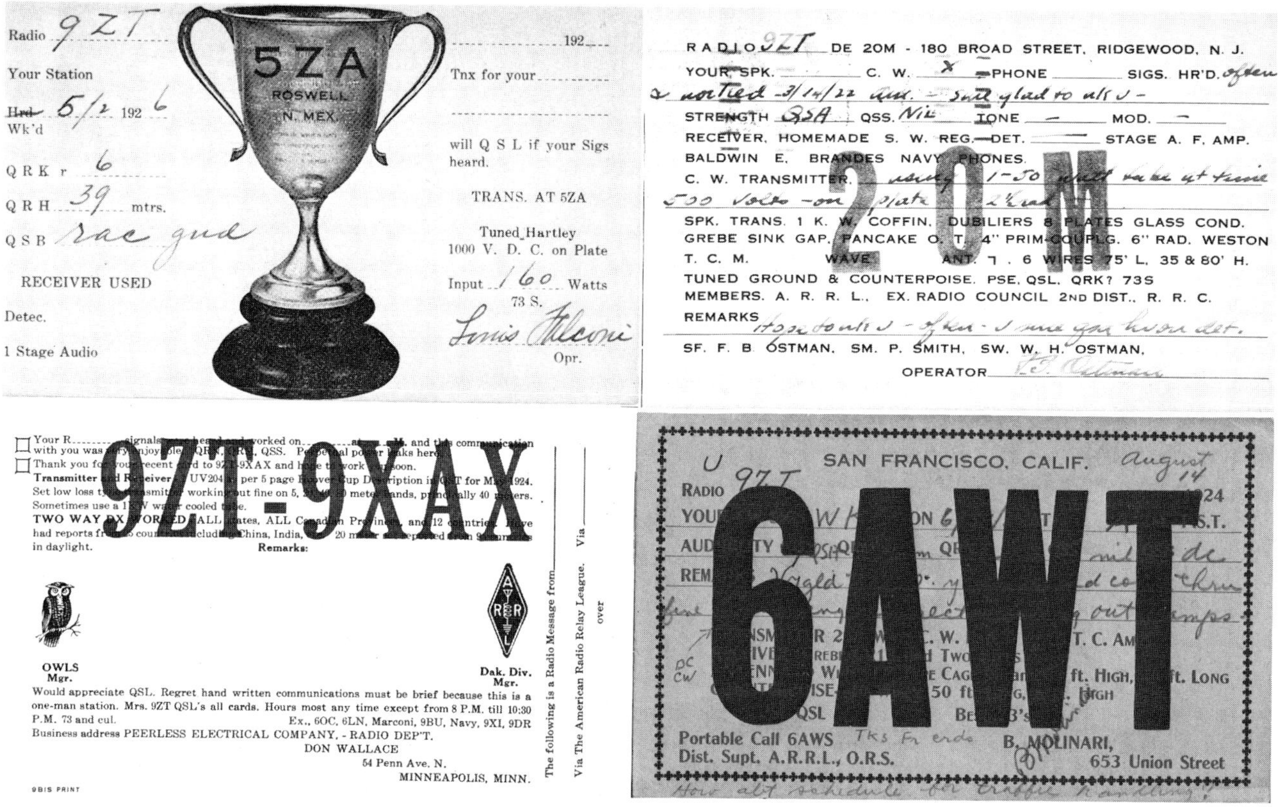

The QSLs of the four Hoover Cup winners.

transmitting tube. For most households, the larger tubes and correspondingly more expensive transmitter components were simply not affordable. ARRL headquarters was issued a club station license that year, 1MK. The station they built used four of the 5-watt tubes in parallel and was conservatively operated at a typical input power of 20 watts.

In order to understand the motivation behind the drive to continually upgrade and improve one's amateur radio station, a report written 15 years later by Fred Schnell is quite helpful. In his Planning Committee Report for 1938 to the ARRL Fred wrote:

> "One of the most important departments of *QST* was "Calls Heard." The information fired the interest of the amateur. That department delineated the indicated performance of an amateur station better than any other method. The one objective which was growing up in the mind of the amateur was a louder signal at a greater distance than that of his neighbor in his own locality. The spirit was competitive and it was educational. The amateur whose station was regarded as being heard louder by more stations over greater distances was, let us say, "Final Authority."
>
> Every community had a Final Authority, and it was to him the less successful amateurs came with their questions and problems. Final Authority reigned only as long as he was able to show better records and better performance of his station.

Being the Final Authority in the Minneapolis area, and the northwest, was a role that Don clearly relished.

During these events Professor Jansky continued to visit Don on Tuesdays. Don recalled, "*We would have lunch together and talked over what we could do for amateur radio. So we decided on having a convention in Minneapolis. I became the ARRL director of the Dakota Division in the summer of 1923. K.B. Warner* (secretary of the League) *appointed me to the position. During our meetings Jansky and I would talk over matters that concerned the division. It was very fine advice. Between the two of us we did a lot of good things, we thought, for the League and kept things going on a very even keel.*

"It seemed to us that the ARRL wasn't a very

democratic organization because K. B. Warner appointed all the division directors, who are equivalent to our division managers of today. In other words, we were his cabinet. We told him what we thought about various matters, and then he did as he pleased. There wasn't any say so by the membership at large except by coercion. There was no election of their specific division director. We decided that the league ought to elect their division managers, and I wrote a letter to the dozen or so division managers and suggested that we all vote ourselves out of office and have a general election. We submitted this vote to K.B. Warner and he went along with it. The basis of the ARRL division manager system was created and is still in existence.

"One of the things I said was that I did not think that the amateur who ran the division should be in business. I thought it was possible that some of the division managers might use the position to further monetary interest, and the league put that stipulation in there. I thought they ought to be elected and changed periodically, which they still do. So they modified the original proposal a little bit, they said you could be in the business in a position where you don't make decisions. For instance, you (N6AW, the author) *would be eligible. Whether you are an amateur or not there is nothing to further your job* (as a Senior Engineer at Hughes Aircraft) *by doing anything with amateur radio directly. It enhances your knowledge, but that is all right. I would not be eligible. I would be now because I am retired. But when I was in business as a manufacturers rep, I did not want that kind of man to be eligible. They voted it that way."*

Don was doing quite well at Peerless, even with his reduced salary. He had also become widely known by this time. Over 60 years later Guy Blencoe (W4HVU) still remembered Don well. "In 1924, when I was a sophomore in high school in central Wisconsin, I drove 150 miles to a radio parts store in St. Paul, Minnesota. I gingerly inquired about parts I could get to make an amateur transmitter without having to spend much. I had already built a receiver which worked well enough to whet my appetite to transmit. The tall handsome and accommodating man at the counter was none other than Don Wallace, of 9ZT fame, before he came of W6AM fame. After listening to my query about ham parts, he went into a back room. He was gone about 15 minutes, but he arrived with an armful of 'stuff.' He had a capacitor made of two Bakelite plates about 3" by 6" , with alternate sheets of foil and wax paper; a coil made of edge-wound copper ribbon; an old key; a pair of headphones (Murdoch as I remember) and an Electro Importing Company catalog. He said if I wanted parts

Fred Schnell, 1MO, made the first trans-Atlantic contact in 1923.

Don Mix, 1TS, operator of WNP aboard the *Bowdoin* in 1923/24.

The American Radio Relay League, Inc.
Hartford, Conn., U. S. A.
TRAFFIC DEPARTMENT APPOINTMENT
DIVISION MANAGER
DAKOTA DIVISION

This Certifies that D. C. WALLACE is appointed to this office. Until further notice he is invested with all authorities of that office and is subject to the rules and regulations of The American Radio Relay League, Inc., in the performance of the duties of the office which he holds.

It is his duty to co-operate with all radio amateurs in his territory in such manner as will best further the work of The American Radio Relay League, Inc.

Dated this ______ day of ______, 192__.

F H Schnell
Traffic Manager.

Don's division manager certificate.

listed in that catalog, to write to him as he had quite a stock of parts but no catalog at that time.

"I was scared to death that the price for all this was going to be beyond my limited resources. When I inquired as to the price he answered, 'When I find a young man as enthusiastic as you are, and so far from a source of supply, just take it along and be my guest; the stuff came from my junk pile in the back of the store that I keep for situations just like this.' I nearly jumped for joy and thanked him to the best of my youthful ability. That introduction to a radio ham made a great impression on me and significantly influenced my whole career. In a month or so these parts and some others that I acquired became integrated into a ham station and I went on the air as 9CAV. That simple beginning permitted me to ultimately become a Colonel in the Signal Corps. Upon retirement from the Army I became assistant manager of the National Communcations System."

The plans of Don and Professor Jansky had been taking shape in the ensuing months. The first Dakota Division Convention was planned for the late fall in 1924. Don recalled, "*While I was division manager I put on a convention. Jansky and I cooked it up and we did things I don't think have ever been done before or since. Every month, for the entire year before, I sent out a postcard with the date of the convention and where it was. On it I had some feature; 'Don Mix just returned from the North Pole where he operated WNP, the* Bowdoin, *in the Arctic expedition. Don Mix says he's coming to our convention to give us a talk, so you want to be sure to hear Don Mix.' Next month I had another, 'Fred H. Schnell, Traffic Manager of the ARRL, says he'll be at our convention. Fred is the first one to talk to Leon Deloy, 8AB, in France. Fred did it on 110 meters which was the first trans Atlantic communication in the high frequencies. Anyway, Fred says he'll be here and everybody come to see him.' The next month another little yarn like that — only one thing on each card.*

"Then I published a program which was pure profit. I called some fellows up and in a matter of an hour I had

sold enough advertising to finance the convention to where we had the whole thing for $3, including the banquet. When we got through I found out we had some more money left over so I gave everybody a free dinner at the end besides. It was a tremendous success.

"We had half the ARRL members of the Dakota Division attending that convention. Now if you had half the members of any division today, you couldn't get them into a city; the percentage was fantastic. The postcard a month—of course they were penny postcards—did it. But the ads were so profitable in the program, that we had all this money left over. We gave them a meal they didn't even bargain for, free. Now that's unheard of too."

The *Bowdoin* had returned two months earlier, and Don Mix was the featured speaker. Aftewards MacMillan predicted that no expedition would attempt to go north again without radio equipment along.

"So the early days of radio were such that I feel I had a little bit to do with the present organization. I think it is better because of it and as a part of that they decided to have an election for Section Control Manager (SCM). So I was promptly elected as SCM for Minnesota, and operated in that position in a minor capacity. Jansky was promptly elected as division manager by the new system, and he went to Hartford, Connecticut for the ARRL meetings and contributed his bit. We continued to have our lunches every Tuesday, so Jansky was well informed of what went on the air through me, and what to do. I think Jansky and I had a lot to do of the progess of the ARRL in the early days.

"Maxim used to come to the conventions; he was a great old man. He was one of the most wonderful personalities that ever existed. K.B. Warner was one of the most skillful men I ever knew. Those two fellows organized the League and they had such an idea and were so capable that I just can't say enough good about them both. A lot of people didn't like K.B. Warner; he was a little bit of a recluse. He wasn't hail-fellow, well-met, by any means, but I liked him a great deal and thought, 'Well, we are certainly in good hands as long as he is here,' and we always were.

"When I was a division manager, I suggested a life membership, that we ought to be able to buy a life membership in the ARRL. I bought one for the University of Minnesota Alumni Association (UMAA) for $50, so I am a life member. I get the publications and solicitations for the scholarship funds that go along with it.

"I thought it would be a good thing for the league, but Warner said the bookkeeping would be too complicated. It wasn't until 1972 that they had it available. I am in the first group of course. I think I paid $125 for it. But that was one of the ideas I tried to get over at the time, that we ought to have a life membership. Of course the funds, if properly administered, will do quite well. The league under Dave Hooten (two decades later) *had a fabulous record of properly investing their funds. They always made more money under Hooten, the treasurer, than they needed. So that is what kept the dues down so low for many years. There seems to be a ceiling at which they can charge advertising and serve their purpose. But Hooten always made a lot of money on the side with the league funds and they always had a lot of money on hand. He was a grand fellow and I knew him well. He finally retired some time back."*

Between his radio activity and working at Peerless Electric Company, Don had become well known in Minneapolis. He had become widely known nationally in 1924 because of winning the Hoover Cup, and all the publicity that came with it. He enjoyed the attention and proved to be an able writer, judging by the articles he published in the coming years.

International Contacts

Winning the Hoover Cup multiplied Don's opportunities. The *Minneapolis Tribune* was interested in having him write articles each week for their Sunday page of "Radio News and Features." Commencing in March 1924, and for the next three years, Don wrote a weekly column consisting of an average of two articles. At this time he also entered negotiations with his wife, Bertha.

It had come to his attention that his wife spent his income, plus 5 percent. She was somewhat put out at the money Don spent on the amateur radio hobby and felt that it should go toward household expenses. He and his wife agreed that the money he made from writing articles could be spent on his hobby. He allocated Monday evenings to writing the weekly columns for the Tribune, and from time to time he would prepare an article for one of the assorted radio publications, usually *Radio* or *QST*.

"In the spring of 1924," Don reflected, *"I started writing articles. I would sit down and make 45 bucks in an evening, which was big money in those days, just writing about something that I was interested in. I never had to look for any particular subject, I just wrote about things that interested everyone. For instance, G.E. started college graduates at $100 a month in those days. That was their standard rate. At the store they were paying me $125 per month plus 1 percent of my sales, never knowing that the one percent in the course of six months would make me more money than the president. Radio hit, you know. G.E. cut my commission when the contract ran out. Cut it in half. It still gave me a salary of right under their president with the commission figured in, so I started writing for extra income. I wrote every Monday night. On that evening no one was supposed to bother me and I wrote for the* Tribune. *After a while it became 12 newspapers as my column became popular. In time I also wrote for* The Christian Science Monitor *and several of the radio magazines. I wrote three columns every Sunday. They were tickled to death to have my articles, so they syndicated them. I got $5 for wherever they appeared. That was at least $15 a week, usually quite a bit more. Even with the extra income I had enough sense to keep my job at the store. People sometimes go a little bit nuts when they are writing. They think they can write all the time, but they can't. The job fed my information center. Whenever anybody asked me a question at the store I scratched it on a piece of paper and threw it in a drawer. On Saturday, we always worked on Saturday in those days, I took all of the paperwork home. On Monday night, I stacked all the papers into piles and wrote articles about the various topics the papers contained. I would write an article about the biggest pile first. People would ask me questions all day long at work. I probably had the only commercial first class telephone and telegraph licenses in the area. They were the only first class licenses that we knew of in the whole state. Consequently, a lot of fellows stopped in to see me and talk about radio.*

"It all started because I wrote an article for a magazine and got a check. My wife said, 'Well, you ought to spend that money on your ham rig.' No matter what income you make your wife always wants to budget it, you know. So my budget was $15 a month and I built a super ham station with that amount. When you sell something from the station the money goes back in your budget. Then, when I wrote articles, she said that I could spend that money on ham stuff. Well, pretty soon I thought, what the heck, I made a few hundred dollars a month writing and that's the idea.

I generated more income from what I wrote than I could spend. No way I could spend that much money unless I deliberately bought everything in sight, which I didn't care to do."

At this time about three-fourths of the household broadcast radio receivers in the United States were homemade. Broadcast transmitting stations had proliferated and most communities across America had at least one small station within receiving range, if not several. Atwater Kent, Magnavox, Tuska (the former publisher of *QST*) and a number of other manufacturers were now building radios at a furious pace, but they couldn't keep up with the demand. The public was fascinated with this new entertainment. In the rush to market radios, many manufacturers did not spend a lot of time perfecting their design. Many of the better home-built radios still outperformed the commercially available ones and were a lot less expensive.

Don's initial articles for the *Tribune* described amateur radio and local radio activities in the Minneapolis area. Considering the large quantity of people who were building their own broadcast radios, Don felt that a number of them would progress into being shortwave listeners (SWLs), and eventually get into the amateur radio hobby.

In the following months Don's columns took a reader with virtually no experience and explained how to read a schematic, solder wires, and gave them a basic understanding on how a radio receiver worked. The articles discussed how to build a simple receiver for use with headphones only, and then the steps to add an amplifier and speaker to it, so the entire family could listen to the radio. They also covered the fundamentals of antennas and how to put up one suitable for broadcast station reception. Don stressed the danger of antenna wires installed near electrical power lines and the precautions that should be observed. Then he went on to describe how to install an antenna that would stay up through the snow and ice conditions of a typical Minnesota winter.

In his articles Don proceeded to explain the pros and cons of various types of batteries and their proper maintenance. The cost of batteries to power a receiver often exceeded that of the radio, so a column on how to build and use a battery charger followed shortly. Internal AC power supplies in broadcast radios would not become available for another five years.

As the readers progressed with their receivers, Don's articles delved into the more technical aspects of radios. They described how to build a receiver that tuned very sharp. This approach was for those people who had two broadcast stations close in wavelength to each other that came in strong at their location, so one station had to be rejected. Subsequent columns followed on how the BCL could reduce interference from transmitting stations in the reader's locality. Don's articles proceeded to describe the typical sources of static and noise in broadcast reception and how to reduce them in the reader's receiver. The common sources of noise and static within one's home and community were also covered, as well as how to minimize them. Now that the readers were well-versed in the operation and adjustment of their radios, Don's columns explained how to troubleshoot one's receiver for various problems that commonly occurred and the best way to effect repairs.

In later articles Don discussed the advantages of the common receiver types: the Tuned Radio Frequency (TRF), the Superheterodyne (developed by Armstrong in 1917), and the regenerative receiver. His columns also described how to build and adjust each type of receiver and how to improve the performance of each particular set. Many of the *Tribune*'s readers modified their commercially made radios and found an improvement in their operation and performance. Large numbers of others built and aligned their own radio sets.

Then Don introduced his readers to the phenomenon of propagation. His columns explained why, at broadcast wavelengths, the best hours for reception are after sunset and usually improve as the night progresses. They also described how to locate the weaker broadcast stations situated some distance away and the most productive time to look for them, which was during the wintertime when the nights were longest and local noise the lowest.

In another series of articles Don explained the basic operation of a tube and how to determine if it is functioning properly. The various tube types were described as they became available and the advantages and disadvantages of each in a particular receiver circuit were also discussed. Another article showed the reader how to build a simple circuit for testing the emission from common types of tubes. Don thoroughly enjoyed doing the column for the *Tribune* and his articles covered a wide range of radio-related subjects. Occasionally he would be sent a newspaper clipping of one of his articles from someone in another state. The *Tribune*'s "Sunday Page of Radio News and Features" became so popular that other newspapers across the country often reprinted it.

Meanwhile, the size of Don's family was getting larger. Don Wallace Jr. was born on June 24, 1924. Nine days later Don penned this note:

> *What am I in this world for? I hope to be loved and respected by my community whether my 'community' be five, five thousand, or five million. I hope to leave this world a better place to live in, some of this improvement being due to my efforts. In my own way and in my own chosen field, things are going to be helped along, to a happier, more pleasant status for all."*
>
> My Creed D.C. Wallace July 3, 1924

After his columns became widely published, from time to time Don would prepare an article for *Radio* magazine. Their August 1924 issue carried one titled: "Filament Filter Systems." The two-page article described how to reduce the transmitted noise in amateur and broadcast transmitters and Don was pictured in the center of the first page holding his Hoover Cup. Three decades later Harold Greenwood, W6MEA, cut the pages out of his magazine and sent them to Don. His enclosed note remarked, "I thought you might want this for your

The Wallace family in 1924; Don, Bertha, Bill and Don Jr.

scrapbook. That's a pretty good looking guy in the photo!"

One of the local amateurs, Boyd Phelps, 9BP, was a close friend of Don's. They had attended Jansky's classes together and had spent many evenings operating 9XI and WLB. Boyd was the previous ARRL Dakota Division Manager and had been trustee of the 9ZT call sign for two years at the Minneapolis courthouse prior to its issuance to Don. Once Don had his station set up at his new house Boyd and a number of the fellows often came over to the Wallace's on Saturday evenings; usually around midnight, well after Bertha and the children were asleep. Don and the fellows would sit up and operate the 9ZT radio set until dawn. Usually a couple of stations in Europe would be raised during the night and several in Australia or New Zealand would be contacted near sunrise. (See appendices 15 and 16) This was the forerunner of what would later become Don's annual open house.

At that time the feedpoint impedance of an antenna wasn't well defined. There were a lot of theories around, forwarded by engineers and mathematicians, but each antenna installation had its own unique characteristics due to the relationship of the antenna's height above ground, its proximity to trees and buildings nearby, and any number of physical objects that created variations in its radiation pattern. Repeatable antenna measurements and related data were hard to come by. The rule of thumb for amateurs was that one adjusted one's transmitter for the maximum antenna current (also referred to as maximum smoke). The RF power that the transmitter generated and did not transfer to the antenna for radiation was reflected back to the transmitter and dissipated as heat. With a powerful transmitter such as Don's this could be a considerable amount of energy. Don had rebuilt his power supply with one of the recently developed mercury vapor rectifiers in recent months and he could now run upward of 2-1/4 kilowatts of power with the UV-206 in his transmitter.

After one particular night of operating with the fellows at Don's, Boyd Phelps noticed the ground wire that went from the equipment and out the window to the ground stake was rather warm. He asked Don to key the transmitter while he went outside with a cigar. One of the other fellows took a picture of Boyd lighting his cigar from 9ZT's smoking-hot ground wire.

Don had acquired a mercury arc rectifier that summer. The 124 rectifier jars of electrolytic solution were very inefficient and with the UV-206 power tube the slop jar system had been overloaded. At $35 the newly developed Mercury rectifier seemed like a bargain. The recti-

Boyd Phelps lighting a cigar from the 9ZT ground wire

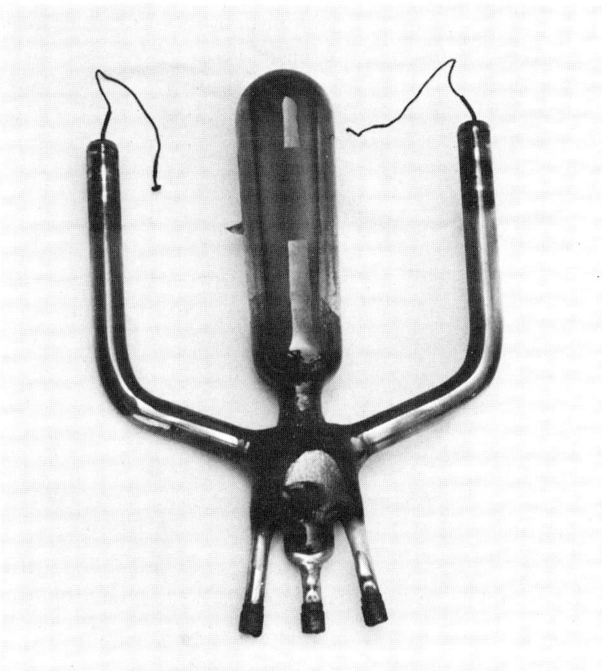

A mercury vapor rectifier

fier was shaped like a "W," about 12 inches across by 18 inches high and two inches thick. The high voltage was brought in at the upper points of the W. The low points of the W were energized with 88 volts AC from a transformer. The tube was tipped slightly and the internal puddle of mercury flowed across the 88 volt potential and it sputtered for a few seconds until the chamber in the middle point of the W filled with mercury vapor. The tube was then tilted back to its original position for normal operation. Due to the power handling capability of the tube and the large amount of heat generated, all but the uppermost portion of the tube with the high voltage leads attached was immersed in oil. This was accomplished by suspending the tube over a five gallon can filled with motor oil. The mercury arc worked so well that the electrolytic rectifier system was retired from radio use and Bertha could again use the jars for canning. Another advantage of the mercury rectifier was that it had been relocated to a corner of the radio room and now the children could play in the basement without fear of getting near the high voltage.

At that time there were less than a dozen amateur radio stations in the United States with the capability to run a kilowatt or more of power and that also had a good high antenna. Consequently Don received hundreds of letters with signal reports from all over the world. A letter dated December 24, 1924 from A.L. Lord in New Zealand noted: "You have been heard here working Z2AC (New Zealand). Signal strength equal to locals on my 1 valve (one tube) receiver. Congratulations on your excellent transmitter. It is sure reaching out as it is heard here often." (See appendix 16)

The next week Don received a letter from G.W. Smithson in Auckland, New Zealand. In it he reported: "I heard you on about 80 meters (3750 KHz). Your signals sure have kick, OM! They simply roared in . . . You're easily the most QSA (strongest) of any American hams I have ever heard." With his high-power station and good antenna Don did put out a tremendous signal. The surprising thing is that the hams on the West Coast were not stronger out in the Pacific. A number of the fellows in California had very good stations also. One of them would win the Hoover Cup for 1924.

By the end of 1924 international contacts were still rare for most hams but they were not unusual for a handful of powerful stations like Don's. Although amateurs in the United States were assigned a call sign with a number followed by two or three letters, this system was inadequate for international contacts (QSOs), since most other countries used a similar method of call sign allocation. Initially United States stations started adding A (America) during the early international QSOs. For example, when Don worked Deloy (8AB) in France, 9ZT signed A9ZT and 8AB signed F8AB. The self-assigned prefixes for Canada (C) and Great Britain (G) fell easily into place with this system. When United States stations started working Australians and South Africans a conflict arose, as those countries also used the A prefix. Soon after the United States stations began using the U (United States) prefix. On foreign QSL cards to Don his call was listed as 9ZT, A9ZT, and U9ZT. This transition took about one year. The official call assigned to Don by the Department of Commerce was still 9ZT.

The magazine articles describing Don's station were widely reprinted and he received a lot of mail requesting additional information. A letter from Roger Burnham simply asked: "Would you please send me an official letter from your station? I wish to keep it as a souvenir. I saw you in *Popular Radio*." In fact, Don became so widely published in the years to come he was never sure how many magazines his articles appeared in. The number of magazines that published articles he wrote or those that described his station is thought to be somewhere around twenty, including those that no longer exist such as R/9, *Short Wave Craft, Radio World, Amateur Radio News, and Western Radio Journal.* Radio subjects and material were so popular in the mid-twenties that even the *Christian Science Monitor* periodically reprinted Don's articles.

Some of the letters that Don received were from surprising places. *QST* was widely read in Europe and in those days many issues apparently worked their way throughout the European colonial empire. Aside from the hundreds of letters from stateside hams, Don would receive queries about receiver design from such unlikely locations such as Palestine, British India, Burma and Borneo. In those days, the sun never set on the British empire. Often the operator would remark that he only had one tube available and inquire as to what would be the optimum receiver design with that particular type.

Don appeared in the November, 1924 issue of *QST* in the section of "Who's Who in Amateur Wireless." Along with his photograph a one page story traced his amateur radio activity for the past 15 years. The last paragraph of the story noted:

> As operator of 9ZT/9XAX, Mr. Wallace is too well known for comment. 'Radio for me until 90 years old,' Don says!

As we will see, he almost made it to 90, and he operated an exceptional amateur radio station the entire time.

In early 1925 mandatory quiet hours of 8 PM to 10:30 PM were imposed for most amateur stations by the Hoover Commission. This was the prime time for broadcast radio audiences and seemed to be the best compromise between their interests and the radio amateurs. Although Don was exempt to the regulation with his experimental license, 9XAX, he still abided by the rules. Besides, the majority of his operation was between midnight and dawn when most broadcast listeners were asleep.

The Hoover Commission was also considering what were the best wavelengths for broadcast station use. Jansky and Don conferred at length on this issue. After considerable thought, and drawing from his experience of operation on most wavelengths from very low up to the short waves, Don decided that the best wavelengths for a local broadcast audience were in the vicinity of 350 to 400 meters (850 to 750 KHz); certainly not below 200 meters (1500 KHz) or above 600 meters (500 KHz). Jansky persuaded the other members of the commission to follow that advice and all of the broadcast stations were assigned to that region as their licenses came up for renewal. Some 65 years later the 200 to 600 meter range (540 to 1620 KHz) was still the current United States AM broadcast band.

In 1925 the Hoover Commission also recommended new wavelengths for amateur use. Since commercial interests had noticed the previously unwanted amateur wavelengths, the commission felt that to try to hold on to all of them would be hopeless. Broad regions around the 80-, 40-, 20-, and 10-meter wavelengths (3.5, 7, 14, and 28 MHz) would be left to the amateurs. Only CW transmissions by tube transmitters would be allowed on these wavelengths. The amateurs would also have regions around 5 and 2-1/2 meters (60 and 120 MHz), as well as the unwanted wavelengths beyond.

During the broadcast boom it was inevitable that a few mischievous amateurs would have a little fun at the expense of the BCLs. Perry Esten (now W6PN) lived in Rochester, New York, and was licensed as 8BOX. He had built a CW transmitter that operated in the 200- to 400-meter range and could modulate it for voice operation. That year at a previously empty spot in the broadcast band, around 400 meters, a station appeared that played music by the Royal Grenada Band. Occasionally speeches by Grenada officials were given between the periods of music. Enthusiastic broadcast listeners reported the reception to local newspapers where it was dutifully published, to the delight of Perry and his associated "broadcasters." After a number of 2-hour long shows Radio Grenada disappeared as mysteriously as it had appeared, never to be heard again.

Due to the widely published accomplishments of the CW stations, many amateurs were in the process of changing over from spark signal to CW reception. The January 1925 issue of *QST* carried the article "The Neutrodyne CW tuner at 9ZT." Since the first international QSOs the year before, most amateurs had seen the way of the future and had rebuilt their stations for CW operation. By this time less than 5 percent of the amateur radio stations continued to use spark transmitters. Consequently the short wave listeners, who outnumbered the amateurs, also had to rebuild their receivers also. Receiver design changed dramatically from the traditional method used for spark reception in order to tune in the amateur's (relatively) narrow CW signals properly.

Perry Esten, 8BOX (now W6PN).

9ZT's 5-meter kilowatt.

Always looking to improve his receiver's performance, Don found that when the brass bases of the UV-200 and UV-201 tubes in his receiver were removed, the circuit losses were reduced. The February issue of *QST* carried a reprint of an article from the *Minneapolis Sunday Tribune*, "Celluloid Supported Coils" by D.C. Wallace. This article demonstrated how to further reduce the losses and improve the performance of one's receiver. The trans-Atlantic contact between Schnell and Deloy the year before led to amateur experimentation on the shorter waves. Don and a number of other stations participated in the 20-meter daylight tests which were sponsored by the ARRL and the results were reported in the February issue of *QST*. Frank Jones, 6AJF in California, copied Reinartz in Connecticut solidly for some time in the early afternoon. In what would be the salvation of normal family life for many radio operators, daylight trans-continental communication was now a reality.

That spring Don had a brief QSO with JXCX in Japan. Another United States station, 7HG, had worked JUPU in Tokio (Tokyo) the year before, but the Japanese operator was never identified. Japan was just coming out of being a feudalistic society and most of the limited amateur radio operation there was undercover. Don's contact with JXCX was one of the first known where a QSL from Japan was actually received.

The success of the 20-meter transcons led to further amateur work on the short waves. The Experimenters section report of the May issue of *QST* featured "The 1-Kilowatt 5-Meter Set at 9ZT." Don had built the first high-powered transmitter to operate in the short wavelengths. A UV-204 that had been taken from his Hoover Cup transmitter was used as the transmitting tube and the set was used to broadcast for two 45-minute periods each day. The article noted: "The layout is in the usual 9ZT haywire style with lots of air circulation and lots of (power) input."

A few months later the ARRL forwarded to Don a letter received from B. A. Ostroumoff at the Nijni-Nowgorod Radiolaboratory. Ostroumoff noted: "The 9ZT (5-meter) transmitter work is really estonishing! We get informations about your work from *QST* only." The May issue of *QST* also contained a story about the Hoover Cup winner for 1924, Bartholomew Molinari, 6AWT in San Francisco. This was the fourth and last Department of Commerce (Hoover) Cup issued.

Scientists had learned the value of having radio communications equipment aboard on Arctic expeditions from Don Mix's operation of WNP two years earlier and they decided to continue the practice. In 1925 Reinartz was the radio operator that accompanied the Navy-MacMillan Arctic expedition. They also added another schooner to the expedition to accompany the *Bowdoin*, the *Peary*. The July 24 headlines in the *Tribune* stated "Wallace Hears MacMillan's Radio" and "Wallace Hears Talks of *Bowdoin-Peary*." The articles carried the current status of the explorers and the messages they sent.

The Third National Convention of the American Radio League was held in Chicago that summer. In was held at the Edgewater Beach Hotel from August 18-21, Tuesday to Friday. Accommodations were available for those wishing to take either the amateur or commercial license examinations at the hotel during the convention. Talks were given on vacuum tubes, receiver design, interference phenomena, high frequency transmission and the transmission of pictures. The featured speakers included Fred Schnell, 1MO, R.S. Kruse, F.E. Handy, 1BDI, Professor C.M. Jansky, 9XI, and D.C. Wallace, who gave a talk on the operation of power tubes on 20 and 40 meters.

Due to his radio operation in the Navy, and as a commercial wire operator, Don had developed a concise operating style. In his amateur radio activity he noticed that there were not yet any standard operating procedures that had become widely used. For instance, when a station called CQ, indicating that he wanted to initiate contact with another station, he might send the letters CQ a couple of dozen times and his call letters a similar amount. Once another operator had tuned this station in, the wait could be substantial until he stood by and listened for a station to call him. Don decided that a three-times-three CQ was sufficient. That is, to send CQ three times, your call sign three times, and repeat the whole sequence twice more, then listen for stations calling you. He mentioned this idea to the stations that he worked, and also dropped a note to the ARRL. The "Calling Practice" column of the October 1925 issue of *QST* noted:

> Don Wallace was responsible for the good suggestion that has now become a part of League operating procedure that we adapt a `three-times-three' CQ.

Over the years a number of Don's other suggestions would become a part of League operating practices.

After the first trans-Atlantic contacts two years earlier, Don realized that a lot of fellows were hearing other stations better than he could. Initially, Deloy in France could hear Don's signal, but Don couldn't hear Deloy very well above his local power line noise. In his initial attempt to work France, Fred Schnell in Connecticut had to relay Deloy's response to Don. Determined to do better, Don began investigating ways to improve reception at his station.

Don carried his receiver, batteries and all, around the neighborhood in an attempt to locate the particular noise source that was causing his problem. He was unable to pinpoint any particular source in the vicinity of his house. The only conclusion that Don came to was that once the receiver was 50 to 100 feet away from power or telephone wires, the noise started diminishing rapidly. Then he placed the receiver in his automobile and drove around various areas of the city. Don came to the realization that the noisiest location in Minneapolis was in the neighborhood where he lived. Two large 110,000-volt power lines met near his home in the shape of a "V" and bracketed his house. The power line voltage was stepped down by transformers on the poles and the resulting 13,000-volt and 220-volt power lines were the noisiest of all. Not one to be easily discouraged, Don placed his receiver at different locations within his house. He found that the front of his house was less noisier than the back. It turned out that his attic proved to be the quietest location of all and he installed the receiver there. As it was not practical to move the transmitter upstairs to the attic, Don left it in the radio room. To operate his station it was necessary to run downstairs to transmit and back upstairs to receive. In this manner Don worked Deloy of France in January 1924, just six weeks after the historic trans-Atlantic QSO between Schnell and Deloy. After several days this practice was abandoned, as it was very hard on both Don and the stairs. No doubt Bertha and the children slept better after that, being spared having to listen to the chief operator pounding up and down the stairs in the middle of the night.

Don put the receiver back in the radio room and puzzled over how to reduce his local noise, the factor which limited his station capabilities. In the coming months he decided to try shielding the receiver itself. Don fashioned a copper box four feet long, 12 inches deep and 10 inches high. The entire receiving set, including the batteries and spare plug-in coils, was built into the box. The lid fit snugly and the only openings in the enclosure was where the antenna lead came in and the headphone wires came out. This shielded box tremendously reduced the extraneous noise in Don's receiver.

In the spring of 1925 Don decided to try a separate receiving aerial in order to further reduce unwanted noise. Remembering his discussions with Beverage aboard the *George Washington* five years earlier, Don put up a wire that was 250 feet long but only 50 feet high. With that antenna the noise level was lower in relation to the desired signal than with the high antenna. Don would have tried installing it a bit lower, as Beverage had suggested, but due to the houses and streets the antenna wire crossed, that proved impractical. From time to time Don lengthened the receiving aerial until it was 750 feet long. It seemed that optimum

Don's station, with the shielded receiver, in 1925.

performance occurred there. Longer lengths were tried, up to 1,200 feet, but the signal-to-noise margin continued to diminish after 750 feet. Except in rare cases Don's other higher antenna usually provided a superior transmitted signal.

This antenna experimentation process required several months and by November 1925 Don's receiving system was working the best it ever had. Local stations often remarked that they could not copy the foreign stations that he worked night after night. That month Don was on the air 16 nights and worked Europeans each of them in a total of seven different countries. Up to this time, 25 different countries had been contacted by Don at 9ZT. Prior to his reducing the noise from the power leak only the loudest foreign stations could be copied. Now stations in those same countries could be heard even if the overseas operator was using a receiving tube as a low-powered transmitter (typically less than 5 watts). Many of them remarked that Don was the first ninth district station worked and often he was the first United States station they had contacted. Now that his station was at its peak performance, Don spent a Monday evening typing up the description of his noise investigation and his receiver modifications, along with his final conclusions, and sent it to *Radio* magazine. A short time later the story appeared in the May 1926 issue, "Improving 40 Meter Foreign Reception."

One of the California stations Don worked from time to time was Loren Bristol, 6DN. Sixty-five years later Loren (now KE6QF) remarked, "I worked Don when he was 9ZT/9XAX. Somehow he managed to get authorization to run as much power as he wanted. I think that was quite a bit, as he was always plenty loud in here. I was so impressed with his signal that I put his QSL card on the wall."

One evening when Don had his transmitter running full-bore, Bertha passed by the stairs to the attic and noticed a sparking sound. She investigated and found a number of rolls of old wire hanging in the attic. There was so much stray RF floating around when Don transmitted at full power that the coils of wire arced between each other. After that, Don saw that they were kept carefully separated to reduce the fire hazard.

An amateur a few hundred miles to the east of Don was also doing a lot of antenna experimenting. Lorence (Windy) Windom, 8GZ, also had a modestly powered amateur station. Windy had installed a high zepp antenna which was fed off-center, instead of the traditional center-fed manner (the zeppelins, or dirigibles, commonly used this antenna configuration, hence the name). Although he had a UV-204 and could run several hundred watts of power when he wished, Windy began to experiment with transmitting at low power. He built an oscillator using a UV-199 receiving tube that ran less than 1-watt input. On the evening of December 30, 1925 the propagation conditions were particularly good and Windy put the low power transmitter on the air. That night 8GZ worked Australian 5BG with a total measured power of .567 watts to the UV-199, including the filament. His "Windom" antenna would be described in *QST* in another two years.

In early 1926 the ARRL and the Wireless Institute of Australia (WIA) held a trans-Pacific test. They were interested in just how well United States stations could copy the chaps from "down-under." The WIA broadcasted a 500-word message from Australia and those amateurs who copied the message were to relay it to ARRL headquarters. 9ZT was one of the most distant stations to copy the WIA transmission and receive the certificate they offered.

That spring Brandon Wentworth, 6OI in California notified the ARRL that he had contacted stations in, and received QSLs from, all 6 continents and asked that an appropriate award be issued. In April the ARRL queried its readers in *QST* if any others had received QSLs from all 6 continents. Don looked through his cards and sent the 6 QSLs to ARRL headquarters in Hartford, Connecticut. A total of six amateurs appeared in the first listing of Worked All Continents (WAC) members in the June 1926 issue and *QST* listed the additional members for the next six months. At the end of the year those members that had submitted proof of contact became WAC Charter members and 6OI was issued award number one. Don had worked the 6 continents in 1925, and received the QSLs late that year. Had he thought to inquire, he may have found out that he had been the first to confirm 6 continents.

Since his graduation from college, Don and Bertha had somehow found the time to take at least one college class each semester. Considering Don's schedule of work and amateur radio operation, it was a fine example of

The low power rig of Windy Windom, 8GZ.

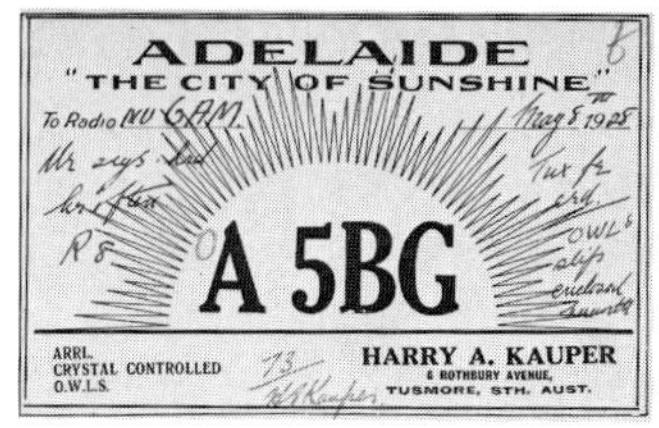

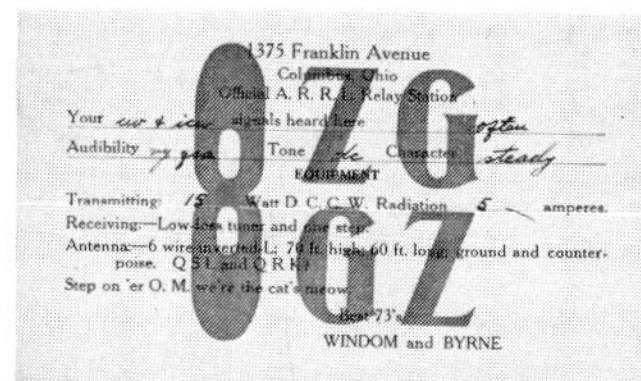

The QSLs of 8GZ & A5BG.

what could be accomplished in with just a little spare time. The "Strays" in the February 1926 issue of *QST* observed "Mr. and Mrs. 9ZT speak Swedish, French, German, Latin, Spanish, and English."

Being a businessman, Don saw quite an opportunity at hand in the Wall Street stock market. He had dabbled rather moderately at first and made a surprising return on his money. As Don invested more money he kept getting larger and larger returns. By mid- 1926 he was in an economic position that could be described as rather comfortable for a young family man who had graduated from college just four years earlier.

That summer the citizens of Minneapolis hosted a dinner for Herbert Hoover, the Secretary of Commerce. Don and Bertha attended the dinner, which was held at the Nicollet Hotel on July 20. Don and Professor Jansky, being the most prominent of the amateurs in the community, were seated with their wives at the head table with Mr. Hoover. The had plenty to chat about, as Professor Jansky had been appointed to the Radio Commission by Mr. Hoover in 1923, the year Don had received the Department of Commerce Cup that Secretary Hoover sponsored.

Since his employment at the Peerless Electrical Company five years earlier Don had been doing a booming business in the radio department at the store. Both amateur radio receivers and broadcast radios used similar components. With his series of articles in the *Tribune* and the occasional articles that appeared in the assorted radio magazines, people would travel some distance to ask his advice about their particular radio and

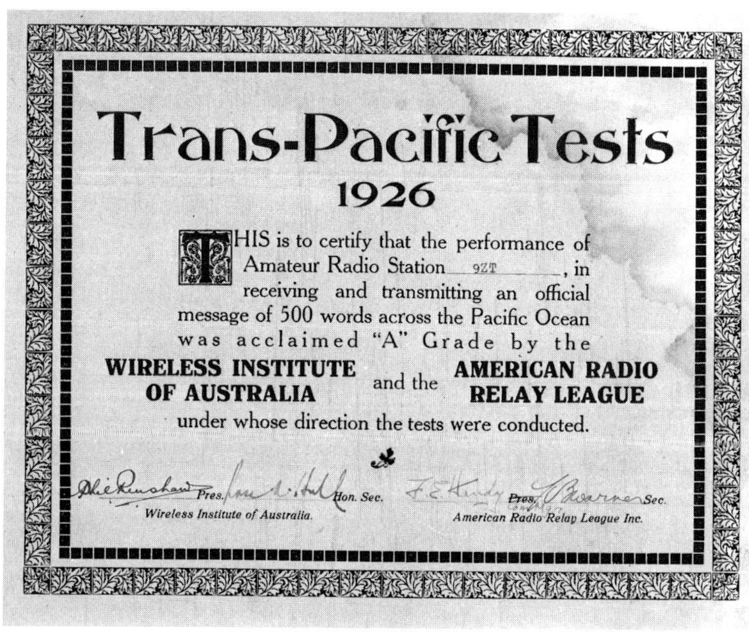

Trans-Pacific Tests
1926

THIS is to certify that the performance of Amateur Radio Station 9ZT, in receiving and transmitting an official message of 500 words across the Pacific Ocean was acclaimed "A" Grade by the WIRELESS INSTITUTE OF AUSTRALIA and the AMERICAN RADIO RELAY LEAGUE under whose direction the tests were conducted.

Pres. Hon. Sec. Wireless Institute of Australia. Pres. Sec. American Radio Relay League Inc.

Don's WIA certificate.

Molinari's 5-kw rock crusher used a pair of UV-204s.

then buy the components for their soon to be constructed or modified receiver. All of this success was not lost on George Q. Hill, the national sales manager of the National Radio Company.

George had lunch with Don one day and offered him a job. National Radio needed a sales manager for the western states and Hill offered Don the job. It was such a good offer, with the combined salary and sales commissions, that Don had to seriously consider it. However, the job would require traveling at least a week each month throughout the western states. In the end Don decided to stay at his job with Peerless Electric. In parting George remarked that Don should contact him if he ever changed his mind.

A few months later the president of Peerless Electric had a meeting with Don to discuss Don's extraordinary salary. The radio sales at the store were so good that Don's salary was again exceeding the president's. Now that Don's contract had expired the president wanted to rewrite it so that Don's income was $125 per month plus one-half of one percent commission of Don's sales. This would reduce Don's salary to half of what it had been the previous two years and one-fourth of his original commission in 1921. Don reflected *"Well, again, the president said I was making more money than he was and wanted to cut my salary in half. He had a tear in his eye and all that, but I finally realized that it was all just a performance. It seemed that I wasn't going to go anywhere in my career aspirations as long as I stayed in that position with Peerless. So I thought it over and then I called George Hill."*

Don talked the matter over with Bertha and then made the arrangements with George. He would be handling the accounts for Sylvania, Turner, Day-Fan, and Peerless in the 11 western states. Don realized that his business travel time would be greatly reduced by moving closer to his sales territory and decided to return to Long Beach, where he had grown up. They put their house in Minneapolis up for sale and it sold in August 1926.

Packing up 9ZT was a monumental undertaking. All of the local fellows turned out to help take down the antenna and masts. Don crated up his station and all of the associated goodies and shipped them out to Long Beach by train. Bertha figured they would be also taking the train, but she was in for a surprise. Don decided to drive the family across the country. It would be a marvelous adventure, he assured her.

One of the local amateurs snapped a picture of the Wallace family as they were leaving and sent the following letter six years later:

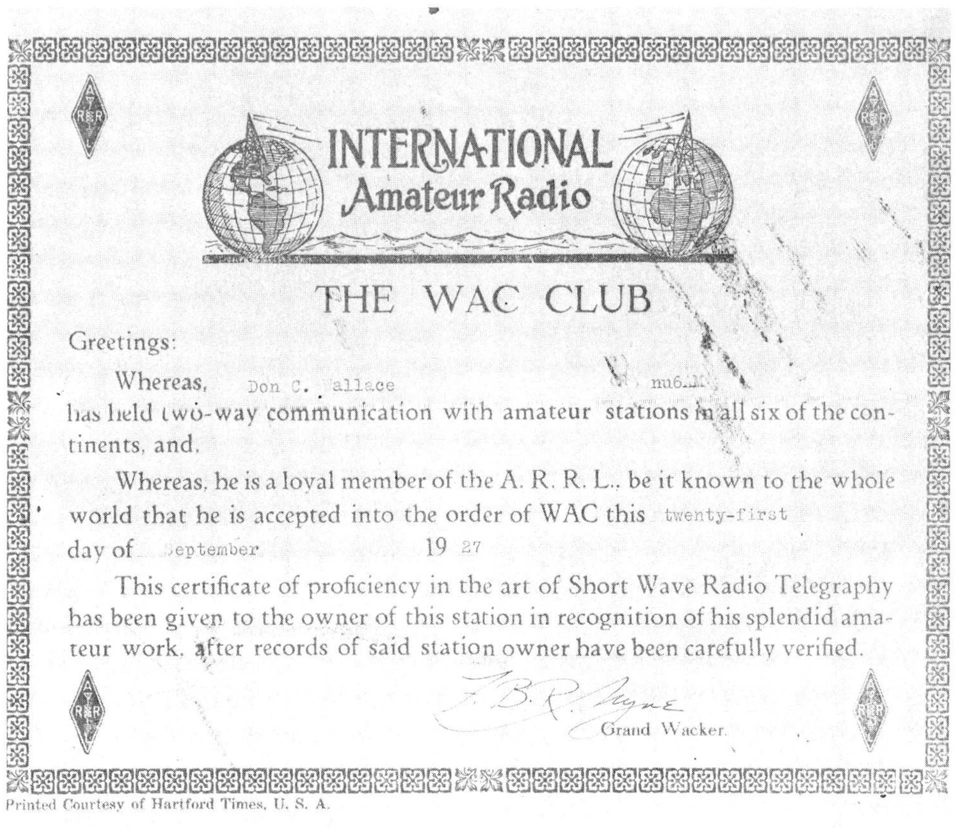

INTERNATIONAL Amateur Radio

THE WAC CLUB

Greetings:

Whereas, Don C. Wallace nu6[illegible] has held two-way communication with amateur stations in all six of the continents, and,

Whereas, he is a loyal member of the A. R. R. L., be it known to the whole world that he is accepted into the order of WAC this twenty-first day of September 19 27

This certificate of proficiency in the art of Short Wave Radio Telegraphy has been given to the owner of this station in recognition of his splendid amateur work, after records of said station owner have been carefully verified.

F. B. R. [illegible]
Grand Wacker.

Printed Courtesy of Hartford Times, U. S. A.

In April 1926 Don was a charter Worked All Continents Club member.

The UV-201, and with its base removed.

"Dear OM Wallace. Went rummaging thru some old drawers and came across the enclosed prints. The occasion was when you moved from Penn Ave. in Minneapolis to your present location in Long Beach. You were busy trying to get the various members of the family loaded in the car when I asked you to pause long enough for a picture. I am sending another copy to Tom Edmonds, 9CRW. The two of you did a perfect stranger a big favor, just when I needed it most. I had come across the St. Anthony's Falls Bridge, up Third Ave. and stopped in to see you at Peerless Electrical. We proceeded over to your house, where I met Mrs. Don, and saw the old 9ZT outfit, with the 900 ft skywire, the big receiver cabinet housing storage batteries and everything within the shield (very radical at the time), and I especially recall the method of starting the rectifier that you devised. You would pull on a rope with your left hand and then applied your right hand with a vigorous pumping motion to the end of a rod which was link-coupled across the room to the arc and rocked it when pumping was done at the operating position. The boiled owl parties at your house on Saturday nights, when a single CQ would line up 15 or 20 Aussies, was to my mind one of the real features of your old station. Mrs. Don sure had to put up with a lot in those days, and I suppose that things are much the same yet.

After visiting your home that first afternoon you had to go back to the office and dropped me off with the chairman of the welcoming committee of the Twin Cities Radio Club. There Tom Edmonds asked if I had eaten yet. As a matter of fact, the malted milk you bought me was the first food or drink that I had in over a day, and the amount of gasoline in my flivver tank when I rolled up to your office did not exceed one quart. (A flivver was an inexpensive automobile, usually a Model T Ford.) Things were pretty black when I hit Mpls. I told Tom I hadn't the price of a meal. We met 9CBA, Art Tabraham from Chicago, and the three of us went to supper. Afterwards we went to where the flivver was parked and found that I had accumulated two flat tires. Those fellows helped me patch up the tires and loaned me 10 dollars apiece. Do you blame me for having such a soft spot in my heart after that kind of a reception from fellows who absolutely were perfect strangers and who only knew that I was a ham, down on my luck. Three days later I got a job and paid back the loans advanced two weeks later. If you ever come east I would be very glad to have you look me up and I will see if I can't show you what results your helping hand has had in ham ranks around here. The Bloomfield Radio Club is ready and willing at all times to help out any ham who might find himself in the same predicament as I.

Walter A. Cobb, W2CO."

Don finally got the family packed up and started out for California and it was indeed an adventure. No highways or paved roads existed yet between Minneapolis and the West Coast and most of the existing dirt roads simply marked the one mile square sections of land. The first automobile trip that crossed the nation had taken place just 14 years earlier and the rural roads had been slightly improved since then. Two hundred miles was the most that could be traversed in a day, and that was unusual. The journey in their 1926 Buick sedan took the Wallaces up through South Dakota and across Wyoming to Yellowstone National Park and then south across Utah to Arizona and the Grand Canyon. Betty Jean, the third and last of the Wallace children, was just nine months old at the time. Her milk had to be kept in a cooler lashed to the running board on the side of the car. The last stretch from Phoenix to El Centro was on the old wooden slat road, consisting of railroad ties laid side by side and cabled together. Two weeks after they started out from Minneapolis the Wallaces, dusty and tired, arrived in Long Beach.

California Kilowatt

The Wallace family arrived in Long Beach in September 1926. They moved into a temporary residence at the Huntington Arms Apartments on Ocean Boulevard. Don made arrangements to rent the apartment on a month-to-month basis. Long Beach had dramatically changed in the 10 years of his absence. The city's population had increased to more than 100,000 people and many downtown area streets were now paved.

While Bertha was unpacking and getting the children settled into their new home Don started his new job as sales manager of National Radio Company for the 11 western states: California, Oregon, Washington, Arizona, Nevada, Idaho, New Mexico, Utah, Colorado, Motana and Wyoming. He would be required to travel from a week to 10 days each month visiting his customers.

A monthly journey over such distances had to be by train, of course, and on arriving at each major city Don would visit whom he considered to be the best distributor of radios or their components in the area. His goal was to convince them to sell the Day Fan radio line, not yet well known. Don could be very persuasive. The distributor he considered to be the best in virtually every city he visited ordered a quantity of Day Fan radios. Day Fan received so many orders for their radios in the coming months that they often filled a boxcar with each order, which also yielded the most favorable shipping rates for the product. The trucking industry was still in its infancy and most shipping across the Unites States was by train. At this time Don acquired the nickname of "Boxcar Wallace."

In Long Beach he set about getting his station back on the air. He secured temporary operating permission from the radio inspector in San Francisco while he was being issued a new call sign by the Department of Commerce for the sixth district. Since he had an amateur first grade license Don was able to choose his call sign from the available 2-letter suffixes. Scanning the list of call signs, 6AM caught his eye. Don liked the sound of 6AM and selected it. It would be both a good CW and phone call. He erected two temporary masts 20 feet high on top of his apartment building, with the manager's permission, and strung a zepp antenna between the masts. He ran a feed line from the antenna in through the window of the spare room of his apartment, where his rig was set up. Within a week Don was on the air as 6AM but at reduced power. Most buildings at that time had just the 2-wire "twisted pair" for electrical wiring, insufficient for his transmitter power requirements. Soon the ARRL appointed 6AM an Official Wavelength Station.

The Wallaces had hardly arrived in California and gotten their bags unpacked when Don had to leave town for a few days. There was a radio show in San Francisco and he had to go up and represent his new products: the National Company, A.H. Lynch, and Day Fan. The ARRL was one of the sponsors of the show. The October 1926 issue of *QST* carried a brief paragraph describing the get-together. It noted, "Among the many prominent visitors were piCD8, and kingpin of them all Don Wallace, 9ZT-6AM." The same issue of *QST* also carried an article entitled "A Radio Picture Demostration." At the last show of the Northwest Radio Trade Association in the Twin Cities (Minneapolis is northwest from New York!), of which Don Wallace was president, the transmission and reception of pictures was demonstrated. With special permission from the radio inspector in Chicago, Don used the call 9ZT for the demonstration. None of the participants suspected how this early version of television would impact the country in another two decades, once the technology was available to adequately support its development and the American consumers were in an economic position to afford television sets. The main components of this early television receiver accompanied the Wallaces to Long Beach and were discovered by the author 60 years later.

In 1926 the Southern California Chapter of the *Institute of Radio Engineers (IRE)* was being formed. Don had become an associate member of the *IRE* in Minneapolis on April 11, and due to his prominence as 9ZT he was invited to become a charter member. Two years later he was elected Chairman of the Southern California Chapter.

In his first few weeks back in Long Beach it became apparent to Don that it was much more difficult to hear Europe from California than it had been Minneapolis. Consequently, it would also be a lot tougher to work them. The first thing Don set out to do was to further

Don's Hoover Cup receiver was rebuilt to meet the more demanding requirements of European reception on the West Coast.

improve his receiver's performance. Before modifying it, he decided to take a commercial receiver and try adding some of the changes that he had in mind. Using a standard 201A tube (a variation of the UV-201), he built a preselector for the Browning-Drake 30-meter (10 MHz) receiver. The addition of a knife switch allowed the circuit to be switched in and out of the antenna input to the Browning-Drake. The modifications were successful and Don found Europeans were easier to hear after that, but still a far cry from the strength that they were in Minneapolis.

Don's arrangement with Bertha was still in effect—any income derived from writing articles could be used for expenses related to his hobby. *QST* did not pay for material submitted to them; they felt that the author's seeing his article in print was sufficient reward for his efforts. Don recalled, *"I had started writing for Radio magazine all of the time. They paid me two-and-a-half cents a word, which was big money in those days."* Consequently, *Radio* magazine received numerous articles from Don.

Due to his long ARRL association, Don periodically sent articles in to *QST*, even though it was not particularly sound business practice. He spent a Monday evening documenting the modifications of his design and sent the story into *Radio* magazine, where it appeared in the February 1927 issue. Simultaneously he prepared another article on how to cosmetically improve the appearance of a home built receiver. "Subpanel Dial Mounting" also appeared in the February issue.

Within two months of moving to Long Beach Don found a suitable house to rent, located at 279 Molino, while he and Bertha were looking for a permanent home. They got their family settled in and Don put up two antenna masts, each 65 feet high. He put one out in front of the house and another at the back of the lot. Between them he strung his trusty zepp antenna. Don brought the antenna feed line in to the covered porch at the back of the house, where he set up his rig. He still had to operate the transmitter at reduced power, however; it would take a few weeks to get the electrical company to bring in the high-current 220-volt power line for his transmitter.

Don was hardly back on the air and learning the ropes of his new job when the *Los Angeles Times* tapped him to write radio articles for the Sunday edition of their newspaper. His reputation in Minneapolis had preceded him to Long Beach and commencing with the November 7 edition of the *Times*, Don's articles began to appear regularly for the next two years. Interestingly enough, the *Minneapolis Tribune* continued to print Don's radio column until the next February, presumably from the backlog of material he had submitted prior to his departure. Not to be outdone, in the March 22, 1927 issue the *Long Beach Press Telegram* ran the headline: "Don Wallace, Radio Expert, Joins Staff of Special Writers." Beginning the next Sunday Don wrote two articles a week for the *Press Telegram* for the next two years. The articles were of a similar vein to those he had produced in Minneapolis: fundamentals of radio, improving its operation, noise elimination, and antenna installation. At this time fully half of the broadcast radios in the country were still home-built.

Don had two projects simultaneously in progress in the closing months of 1926. The first was to build a portable transmitter so he could talk to Bertha each evening while he was traveling. Although none were in existence yet, except for the low-powered type developed for the military in 1917, this did not prove too difficult a task. The second, and more formidable task, was teaching Bertha the code.

"I had a portable transmitter in my suitcase" Don recalled. *"I was issued the portable call sign 6MA. To operate a transmitter away from your home you had to get a portable call sign separate from the one you were issued for fixed station operation. I had a strong suitcase and I built the set into it. There were no portable radios commercially available in those days. The first one I made in 1926 operated on all four bands, 80, 40, 20 and 10 meters, with plug-in coils for each band. Once when I checked the suitcase on the train, it got all smashed up. That was just a short time after I had the transmitter working properly. The porters weren't particularly careful with heavy bags, you know."* After much work with Don, Bertha learned the code to a limited extent. Remembering those days he recounted, *"Well, at first Bertha wasn't too fast with the code. When I would arrive in a city I would get set up with the portable transmitter. I had an antenna installed up on the hotel's roof and I would be on the air at the scheduled time that we had previously arranged. I would have to send quite slowly. Bertha would have out a list of code characters that I had drawn up and write out each character of the code that I sent. Afterwards she looked up each letter of the code that she had copied and wrote out my message. Of course, this took quite a while. She was game, though, and would stick in there with me until she copied everything. I would send the itinerary of where I was staying, where I was traveling to the next day, that sort of stuff."*

Although she was not particularly inclined to proceed, once the portable transmitter was in operation, daily code classes of a half-hour in length were implemented by Don (when he was in town). One evening another couple had arrived and the Wallaces were preparing to join them for a dinner engagement. Don, realizing that Bertha had not yet had her code practice session that day, immediately sat down and proceeded with her lesson. Dinner for all was delayed a half-hour. After a period of resisting the inevitable, Bertha finally threw in the towel and decided to learn the code, which she accomplished in just two weeks.

Bertha passed the amateur license examination and was licensed in May 1928 as 6EN. Don remarked, *"When she got her license she could answer my replies and let me know how the children were, if any emergencies had come up, important correspondence that had arrived, and so forth. That was the only time she got on the air, to work me in our daily schedules when I was traveling. She didn't work anyone else if she could help it. Consequently, when she responded in the code it wasn't particularly fast. Once we made contact, our schedule took most of the rest of the evening to complete."*

After the first portable transmitter had been smashed on the train, Don built another, more rugged version. *"The next transmitter I built was for 40 meters only. I built that one in 1927. I think that the only reason I stopped using it was that the war came along. So I used it up until the end of 1941. Complete with the call book in the lid and with everything else, it weighed 68 pounds, including the aerial. I was a big fellow then, I weighed about 220 pounds, so I didn't mind the weight. I worked home from portable locations for 15 years and only missed one schedule. I had forgotten to charge the relay batteries in my Long Beach station before I left on a trip, so I missed that one schedule with my wife. After that I put a trickle charger on the relays. Then they were run it off the power to the main transmitter.*

"Regarding my 200-watt portable transmitter, there was a big article in Radio ("Home Contacts for the Traveling Man," Oct. 1928), *and later another one in* QST ("The Travelling Man's Portable," Oct 1931) *describing it. Some time after* QST *ran the story the Navy wrote me for details on it. They wanted to develop a portable radio program of their own. It was an awfully new concept then. I did a lot of things that were new then.*

"I had a zepp antenna made out of loop wire, which was an exceedingly flexible wire. The coupling coil of the transmitter's power output tube had two small

capacitors connected to the aerial output jacks, and when the feed line from the antenna was just a little bit too long I had to have series variable capacitors so I could trim the antenna for the proper match to the transmitter. I adjusted them by the brilliance that the light would glow; that was my antenna meter. The indicator light had three positions of sensitivity; no short, partial short, and dead short. If I couldn't tune the antenna on the dead short, I would use one of the other switch positions to adjust the transmitter for maximum output.

"When I arrived at the hotel each evening I would ask for a top floor room. I wouldn't stay at that hotel after that if I couldn't get one. When I got in the room I would open the window and throw the curtain out so I would know which room was mine. The outside fire escapes were always open, so I would climb out the window and walk up to the roof with two balls of string tied together. I would throw the string over the sign on top of the hotel. There was always a sign on top of the roof about 50 feet high. When they were thrown the bottom ball of string would unroll and then the top one would unroll. I would hook the aerial on the string and go to the other side of the sign and pull up one end the aerial. I would have a whale of an aerial up there. Then I would bring the antenna lead-in down the outside of the hotel and to my room, the one with the curtains showing, and in through the window. I would estimate what length of feed line that I needed to reach the operating position I had set up and then go outside to pull up the other end of the aerial to fit the feeders. I had a good aerial on top of sometimes a 25-story hotel.

"After the antenna was in place I would open the lid of the suitcase. There was a key all hooked up, the call book, and the headphones in place. If I was in a hurry, I could get on the air in as little as three minutes. The receiver was battery-powered. I was always receiving a signal from a 3-phase full-wave unfiltered CW transmitter. At home my transmitter had six transformers in it. By its sound I could tell it instantly, a 300-cycle note riding on the signal. It was so strong I didn't have to have much of a receiver to hear that signal. The whole receiver that I built took a space about the size of a small shoe box.

"The transmitter had a 35-pound transformer to power it and it put out 200 watts, more or less, depending on the line voltage at the hotel. The transformer was 800 volts output and it had a 400-volt tap on the same winding. So the voltage for the oscillator and buffer amplifier came off the inside taps and the 800 volts for the power tube came off the outside of the winding. In the morning, once I was finished operating, I always worked a few other fellows you know, I would

Don operating the portable transmitter that he built in 1927.

go up on the roof and roll up the wire and string and be finished in 15 minutes."

Don had worked to keep the size and weight of the portable transmitter to a minimum. Any weight savings that would not cost operating performance was important. As he wanted to carry the portable rig on his business trips, the lighter and smaller it was the better. Don reflected, *"Well, I think I made the transmitter just about as light as I could and still run 200 watts with it. Each night, after two hours of operation the transformer would start to smell, so its size was about right."*

A young ham in Falls Church, Virginia read Don's articles that described the portable transmitter with great interest. Bill Hunton had received the call 3AG in 1927 and worked Don from time to time. Little did Bill realize at that time that when the opportunity arose two decades later he would adapt a similar method of portable radio operation.

While Don was developing his portable transmitter and had his main transmitter modifications coming along, he was also looking for a good location for his house. He had to balance the site selection between the family considerations of community and schools and what was a good antenna location. The area that the Wallaces had lived in while they were in Minneapolis was noisy and Don knew he would be spared a lot of grief if he could find a high location that was situated some distance away from high-voltage power lines. Don had grown up just a mile north of where the family was now living in a rented house, so he was very familiar with the city and surrounding area. As a youngster Don had often ridden his bicycle over to the Dominguez Hills, located just five miles to the northwest. They rose several hundred feet above the surrounding terrain and would be an idea spot for an antenna. Unfortunately, there were no suitable lots for sale over there. One family owned most of the land in that area and were not interested in selling any of their property.

Next, Don looked to the west 10 miles. There, on the Palos Verdes Peninsula, the land rose 1200 feet above the ocean. That location was no doubt a superior site for Don's planned antenna farm. The Vanderlip family had acquired virtually the entire hilltop in 1913, about 16,000 acres, when land prices in the area were depressed. Just a few homes had been built so far. The Palos Verdes Peninsula was 20 miles south-southwest of downtown Los Angeles and considered rather remote for those people that had to commute into the city on a daily basis. As Don was a manufacturer's representative and would be doing a lot of business out of his home, as well as frequent traveling, the distance was not an obstacle. The Vanderlips had several home sites for sale, a number of which would be quite suitable for amateur radio purposes. The problem that developed in Don's plan was that the Vanderlips insisted on a number of deed restrictions for any property that they sold. They were developing one of the first planned communities and they insisted that the outside appearance of the property be approved by an architect. Don knew that no architect would approve what he had in mind, so he returned to Long Beach to look for a home site.

Another good antenna location that caught Don's eye was Signal Hill, in the middle of Long Beach. Don's father had owned quite a bit of property there prior to the financial panic of 1916. Since oil had been discovered there in 1923 the property was no longer affordable. Besides, Signal Hill was rapidly becoming covered with oil derricks. Unfortunately, his father had to sell his Signal Hill property to meet other financial obligations when the price of land was depressed. The current owners of the property had become millionaires.

Rancho Los Cerritos is located in the central part of Long Beach, just two miles north of where Don grew up. The terrain there has a gentle rise, perhaps one hundred feet above the surrounding area. The Virginia Country Club, where Don's father, William, Sr. was a charter member, had moved to its current location there just five years earlier, in 1921. It was here that Don purchased a lot for his home, just two blocks down the street from the country club. His nearest neighbor was a block down the street and one could see several miles unobstructed in all directions. Due to the country club's location nearby, Don felt that this area would become very affluent in the years to come. He decided that the antennas at his home should be as unobtrusive as possible, while still meeting his amateur radio requirements, of course. "*I took it up with my architect* (H. Roy Kelley) *on what would be the least visible structure to put antennas on at my house in Long Beach. The fellow advised me to put up telephone poles. People are used to seeing telephone poles around the city. I had already taken it up with the city attorney before I bought my lot. I could put up anything that I wanted to. So I went down to get a permit from the city to put the poles up and they said 'You don't need a permit.' They had never issued one. I said 'Well, I want one anyway. I will pay for it.' They said 'Come back tomorrow.' In the meantime, my next door neighbor, who lived over a block away, came over that afternoon while I was at the lot. Billy Deb was a contractor and he told me that the permit department downtown had contacted him and wanted to know if they should issue a permit to me. Now he was a contractor, and he would get permits from the city every few days or so. He knew the ins and outs of dealing with the city. I told him that I wanted antennas for a ham station. Billy talked it over with me and then said, 'I'll talk to Fisher.' So the next day I went down to city hall and they issued me one. I paid for it. It's a good feeling, you know. Now they have an ordinance against towers beyond a certain height. You have to have an approval, permits, and so on. I don't know anybody who has been issued a permit since the new ordinance.*"

In November 1926 the first issue of the ARRL *Radio Amateur's Handbook* was published. It was the brainchild of F.E. Handy, 1BDI, the ARRL Communications Manager. As it was not known if the amateur community would accept the handbook, the ARRL had considered it a risky venture and reluctantly spent over $2,000 for 5,000 issues to be printed. As it turned out, all the available copies of the handbook sold out at a dollar a piece almost immediately and the rest of the amateurs who didn't have one were clamoring for a copy of their own. The second edition of 5,000 copies of the handbook was published just three months later. Even this did not meet the needs of the amateurs, as a lot of experimenters and other radio enthusiasts who did not yet have their licenses realized the usefulness of "Handy's handy handbook," and purchased a copy. A second printing of 10,000 issues of the second edition of the handbook was published the following April.

After reading his copy of the ARRL handbook Don wrote a long letter to Handy with suggestions on how to improve it. He and Don passed message traffic frequently and they often exchanged letters about league activities. Don's suggestions included comments on antennas, transmitters, receivers, and noise filtering systems. Many of the comments were incorporated in the third edition of the handbook, of which 10,000 copies were printed when it was published in October

1927. In the foreword of the second printing of the third edition (April 1928), Handy acknowledged the contributions of a number of people, including D.C. Wallace. About this time Don developed the habit of purchasing several copies of the publications that he found to be particularly valuable, especially the ARRL handbook, the government call book, and the latest issue of *QST*. One copy of each would travel with him on his business trips, another was kept at his operating position at home, and a third was put in his bookcase should one of the other copies become misplaced. In this manner a number of early editions of these publications have survived in essentially unblemished condition. Indeed, some of Don's file copies of the handbook and call book had spent 55 years in the same bookcase and were never opened.

By the end of 1926 Don had completely revamped his transmitter. His UV-206 was getting a bit soft and would no longer run the kind of power that it had in Minneapolis. His last couple of years there he had been tagged with the nickname of Don "Wallop," presumably due to the strength of his transmitted signal. But if he wanted to work Europeans regularly from out in California Don would need at least as much power as he had had in Minneapolis, if not more. One of the fellows in San Francisco with whom Don did business, G. Lewis, 6EX, manufactured tubes and Don purchased a water-cooled tube for his station, a Federal 328, rated at 1-kilowatt dissipation. The Southern California Edison Company in Long Beach would bring a high-current 220-volt power line down to a residence for the nominal fee of $1, so there was adequate line power available for his transmitter. The next spring Don described his station in the following manner:

"Station 6AM originally was located in Minneapolis using call 9ZT. The present station consists of one of 6EX's water-cooled tubes together with a 6EX mercury arc. The mercury arc is automatic and starts as soon as the transmitter power is thrown on and the key touched. If for any reason the arc goes out, it automatically starts up once more as keying progresses.

"The wavelengths for which the set is adjusted are 39 meters, 79 meters, and 21 meters. All of these wavelengths are regularly shifted to (depending on which was best at a given time).

"The antenna used on 39 meters is a 63-foot-long flattop zeppelin. A smaller zeppelin is used on 21 meters and a fundamental antenna is used on 79 meters. The water supply for the water cooled tube is supplied through a 25-foot rubber hose directly from the city water supply. The water is turned on automatically at the same time the 220-volt power switch is thrown. The water required is only about one quart per minute and this is drained out through the stationary tubs on the back porch. As it starts and stops automatically with the connection of power, very little water is used and no trouble occurs.

"The mercury arc equipment has purposely been made large enough so that two or three transmitters may be run from it if necessary. For instance, a 5-meter transmitter is now in process of being built and it may be found feasible to connect it in the same manner as the 5-meter transmitter previously in use at 9ZT. This transmitter was connected at all times in parallel with the main transmitter, so that anyone desiring to experiment with 5 meters could find this other set on 5 meters any time the main set was running. This same arrangement was also carried through at 9ZT during the early days of 20 meters.

"The circuit in use is the customary Hartley circuit, using a National transmitting condenser as a variable grid condenser, a large homemade special one inch spaced condenser for a closed circuit variable condenser, and two National transmitting condensers in series with the radio frequency feed line to the zepp antenna.

"The receiver is a modified short wave National Browning-Drake receiver, which has been described in Radio, *February 1927. "The station is headquarters for Official Wavelength Stations and Section Communications Manager. A number of certificates, such as the WAC (Worked All Continents), RCC (Rag Chewers Club), ORS (Official Relay Station) and others are held by the station.*

The first edition of the ARRL handbook was published in 1926.

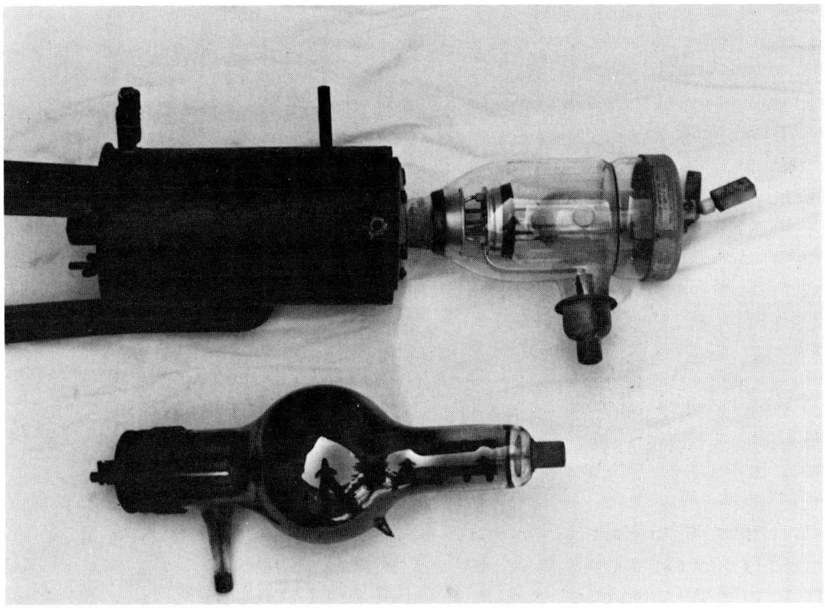

The 500-watt UV-206 and a Federal F-238 1-kW water-cooled bottle.

"As the writer is a traveling factory representative for the National Company, Inc., Arthur H. Lynch, Inc., and Day-Fan Electric Company, the station is in operation at irregular intervals. However, when the station is in operation, it is generally in operation pretty steadily for the limited amount of time."

Operating Don's station really was an exciting experience. When he stepped on the power switch and pressed the key the power transformers would emit a noticeable groan, the lights would dim, and steaming water would come hissing out of the water-cooled tube's drain hose and go gurgling down the drain in the sink near the transmitter. It was almost as much fun as the spark transmitter Don had used 15 years earlier. Considering the advancements that had been made in high voltage rectifiers, Don was probably running in the vicinity of 2 kilowatts of DC power input. This was not unusual among the handful of high-powered stations in the country. The Hoover Cup winner for 1924, 6AWT in San Francisco, was running in excess of 5 kilowatts. The typical ham station in the Unites States still used just a single 5-watt tube, running perhaps 20 watts of input power, although more and more amateurs were putting 50-watters on the air. In the Long Beach and Los Angeles area Don soon became the "Final Authority". As was usual, Don's station was drawing lots of attention. That spring *The Ham Meter* carried an article entitled "Description of Radio Station 6AM." The photo that accompanied the story was taken while Don was still 9ZT in Minneapolis.

Shortly after that article appeared Don received a letter from Mr. Williamson, (G)2ACI in Bedford, England. He wrote: "If you have one of your former QSL cards (9ZT) left I would very much appreciate one as a souvenir of the passing of the premier 9th District station."

On December 24, 1926 Don received an SWL from the wireless station at Mackenzie City in British Guiana. M. Solomon reported that Don was only the second sixth district ham that he had heard as early as 9 PM (6AWT was undoubtedly the first), and remarked that their transmitting licenses had not been granted as yet.

Early in the morning of New Year's Day, 1927, Don got on the air for a few minutes after he and Bertha had returned from a New Year's Eve party. Tuning across the band he found a clear spot and called CQ. A station signing CHA1CRS answered Don's call. The operator was located on the coast of China, in Shanghai. Don had received SWL reports from China before and he had worked ships off of its coast, but this was his first two-way contact with the mainland. The first contacts between China and the Unites States had occurred just a few months earlier. As in the case with Japan, Don received one of the first confirmations of a two-way contact with China. A few months later Don received an SWL from the *USS Simpson* in a destroyer squadron of the Asiatic fleet. Allen Potter noted: "Thot you would like to know that you were heard up the Yangtze Kiang (River). We have a modified Schnell hook-up. Navy regulations prevent our working u fellows, so all we can do is listen." Quite a number of the radio operators in the military would get their amateur licenses after monitor-

Don with his water-cooled tube transmitter at 279 Molino in Long Beach.

ing the ham bands and listening to all the fun that the ham operators were having.

New Year's Day, 1927, also officially marked the end of spark transmission. In the past year the number of amateurs still using spark had dwindled to just a handful of the diehards. Anyone who read of the achievements of the CW stations in *QST* and *Radio* had long since decided to convert to CW transmission. About the only cases of spark transmission after the first of the year would be the occasional experimenter who was unlicensed and using a Ford spark coil and who was not yet familiar with the current amateur regulations. Local amateurs would soon locate and visit the offender, then help him get his license and clean up the offending transmitter.

Even with his writing numerous articles for *Radio* Don retained a favored position with the ARRL staff. The January 1927 issue of *QST* carried the following comment in "Strays":

> Don Wallace of 9ZT fame has moved to the West Coast, being now located in Long Beach, California. His new call is 6AM, about which Wallace says "6AM is local standard time, the best time to work 6AM is 6AM LST regardless of whether you are in Singapore, Africa, or other foreign countries such as the East Coast."

Although contacts were tougher to get with Europe from California, even with his water-cooled bottle, Don found certain areas of the world easier to work from Long Beach. Asia, the south Pacific, and South Africa fairly roared into his new location. That spring Don worked all the continents. The November 1927 issue of *QST* noted,

> Don Wallace is the only one to have WAC (Worked All Continents) membership for stations located in two Unites States districts. First with 9ZT-9XAX and then later on the Pacific coast with 6AM.

By early 1927 the escrow had closed on Don's lot and the building contractor that he had retained began construction on his house. Don was still out of town for a week each month, but found time for plenty of operating when he was at home.

The first ARRL DX contest was held that spring, commencing May 9 and ending 13 days later on the 22nd. As this was the first competitive test of its nature, the rules were somewhat different than in the modern radio contests. Each contact with a foreign country counted one point. The station in the Unites States or Canada that had the most contacts with a certain country would be designated as the best contact point in the

The Associated Radio Amateurs of Long Beach came by to see 6AM and check out the new "Big Gun" in town. Bill Adams, 6ANN (W6BA) is seated second from left.

Unites States for that country and an Official Foreign Contact Station Certificate would be issued. If two or more stations in the Unites States tied for the number of contacts made with a particular country no certificate would be issued, as no one station was considered the best contact point. Naturally, this brought out the competitive spirit of all the "Big Guns" on the air. Conditions were good over those two weeks and when a DX station turned up, the pile up of stations calling him were ferocious. The results were published in the October 1927 issue of *QST*. An even dozen certificates were handed out to 11 stations, of which Don received two certificates. 6AM made the one and only contact with China and Don had a total of 19 contacts with New Zealand. Ten other stations received one certificate each. 6AM had a total of 72 contacts and tied for the most contacts with six other countries. The station closest to Don in overall performance was Bill Jackson, 2AHM in New York, who made a total of 90 contacts, received the certificate for Great Britain, and also tied for the most contacts with several other countries. Another station that made a good showing in the DX contest was Windy Windom, 8GZ. Although he didn't qualify for a certificate, Windy tied for the most contacts with several countries and his contact total of 82 was the second highest of the contest.

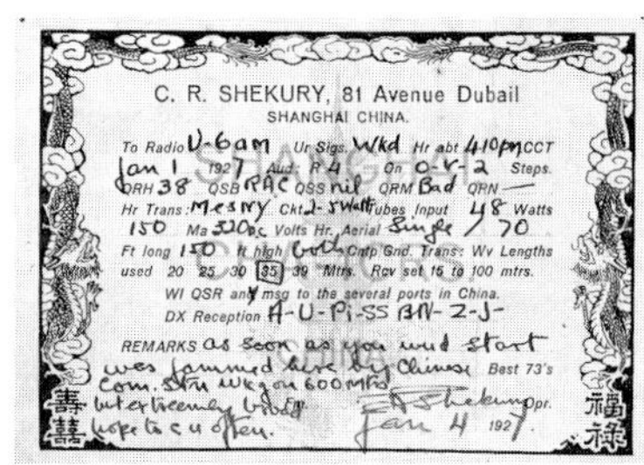

One of the first QSLs from China, CHA1CRS.

DX contacts were improving all the time. More and more stations were coming on all around the world. Inevitably, more prefix conflicts would start to crop up. Stations located in the USSR were frequently being worked by United States stations that spring and they too used the U prefix. At that time many of the stations that regularly made DX contacts adapted a system of placing a continent designator ahead of their call sign. In this manner Don's prefix became N (North America) U (United States), or NU6AM. Canadian stations began to use the NC prefix. Many of China's stations followed suit, using an AC (Asia, China) prefix. Australians began using OA (Oceania, Australia), and the English adapted EG (Europe, Great Britain), and so forth.

In May Don received an SWL report from Gordon Peve in England. He remarked, "I was very glad to receive you as NU6 is the most difficult DX in the world from here. I have logged 200 NU stations since January and you are the first 6 I have ever received." Of course Gordon included a request for a QSL from Don confirming his signal report.

Catalina Island is situated directly off the coast of Long Beach. The Wrigley family of chewing gum fame owned virtually the entire island, nearly 20 miles in length. There was an annual Catalina swimming contest, which started at the Isthmus Harbor in the western portion of the island, proceeded across the San Pedro Channel, and finished at the closest point of land, the Palos Verdes Peninsula 22 miles away. Over 100 stalwart souls would dive into the ocean at sunrise on New Year's Day and try to cross the channel. The race, held each January, drew national attention and the *Associated Press (AP)* wanted to be first with the race results and related information. They approached Don Wallace and asked him to arrange the communications to relay the broadcast. Don's location in the rented house at 279 Molino, just three blocks off the ocean, with his 750-foot long antenna from Minneapolis that had just been re-installed, was the ideal receiving station. A 50-watt transmitter that had been built by the Federal Telegraph Company was placed aboard the *AP* tugboat, operated by Howard Beckwith, 6BOL and Harry Pearce, 6DDO. The information from Don's station was relayed by telephone to the *AP* headquarters on the dock in Wilmington, just three miles away, where they put it on the "wire" to their offices across the United States. Lindley Winsor, 6OU, ran the receiving set at Don's for most of the 22 hours of the race. A 17 year old Canadian, George Young, came ashore at 3:05 AM at Point Vincente and won the race. Within one minute of the finish the news was broadcast by *AP* on their wire service. The radio station aboard the *SS Avalon*, 6XA, was not in position to report the finish till some time later, which they relayed for broadcast by KNX in Hollywood. An article by NU6AM "How Amateur Radio Scooped," appeared in the March 1927 issue of *Radio*.

The American Radio Relay League, Inc.
COMMUNICATIONS DEPARTMENT
Hartford, Conn., U. S. A.

CERTIFICATE OF MERIT
INTERNATIONAL RELAY CONTEST, MAY, 1927

This is to Certify that until further notice Amateur Radio Station 6AK, owned by Don C. Wallace, is appointed an Official Foreign Contact Station by The American Radio Relay League, Inc., in recognition of its exceptional performance in the International Relay Contest. Obtaining a score of 19 points in its work with New Zealand has entitled the above station to be considered the most reliable point in the United States and Canada for contact with that country.

Dated August 23, 1927.

Secretary. Communications Manager.

Don won the first ARRL International DX Contest in 1927.

That March Don received another letter from F.E. Handy. Handy reported that six amateurs in the Los Angeles area had nominated Don for the position of ARRL Section Communications Manager. The election was held later in the spring and Don was elected the SCM of the ARRL's Los Angeles section.

Toward the end of March Don attended a meeting of the Northern California Music and Radio Trade Association. A number of notables were present, including S.J. Podeyn of the Pacific division of the National Broadcasting Association, Jim Smiley, Assistant General Manager of Atwater Kent Company and G. Harold Porter, Pacific Division Manager of RCA. The main topic of the evening dealt with radio sales in the summer months,

Don at a radio show as a manufacturer's representative.

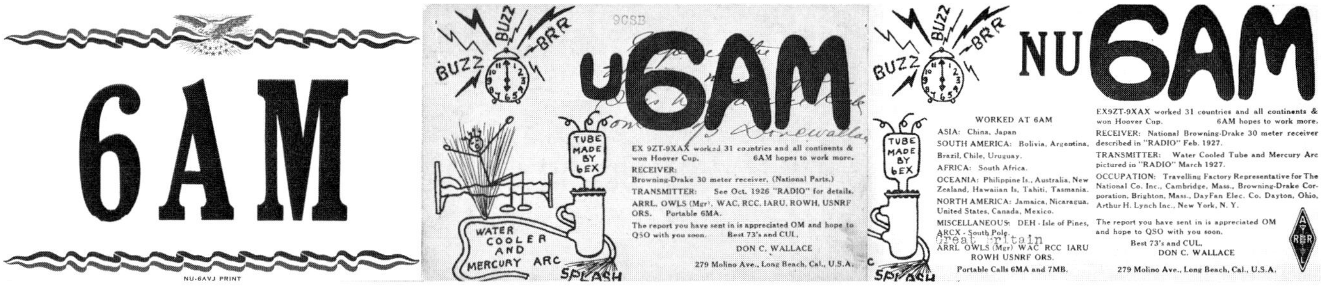

Don's QSLs in 1926 and 1927.

which were usually not as good as sales in the wintertime. Don wrote a 2-page description of the discussions and sent it to *The Jobbers Salesman*, where it appeared in the May 1927 issue, "Improvements in Summer Radio Business."

The income from Don's business was increasing and he had a steady source of money for his amateur radio activities coming in from his magazine articles and newspaper columns. Things were so good that he began taking the extra income and speculating in the stock market, which at the time was providing an enormous return on investments. The family was healthy, Don was on the way to becoming moderately wealthy, and all in all, life looked rosy for the Wallace family.

The American Radio Relay League, Inc.
COMMUNICATIONS DEPARTMENT
Hartford, Conn., U. S. A.
SECTION COMMUNICATIONS MANAGER

This certifies that D. C. Wallace has been elected to the office of Section Communications Manager for two years from this date for the Los Angeles Section of the Pacific Division. Until further notice he is invested with all authorities of that office and is subject to the rules and regulations of THE AMERICAN RADIO RELAY LEAGUE, INC., in the performance of the duties of the office which he holds.

It is his duty to co-operate with all radio amateurs in his territory in such a manner as will best further the work of THE AMERICAN RADIO RELAY LEAGUE, INC.

Dated this Twenty-seventh day of April, 1927.

F. E. Handy
Communications Manager

Don's SCM certificate.

Official Station Log

Station Call W6ZZA Sheet No.

William Taylor Hotel

McAllister and Leavenworth Streets
AT CIVIC CENTER
San Francisco

DATE OF CALL	TIME OF CALL	STATION CALLED	MY OSA	MY R	HIS OSA	HIS R	WKD	HRD	INPUT POWER TO LAST STAGE	FREQUENCY BAND USED	REMARKS
	William Taylor Hotel San Francisco Cal										
Oct 26/31	11 PM	cq									
		off									
Oct 27	540a	cq DX									
✓	550a	VK5PK	5	5	4	6	✓				AUSTRALIA
	Single "Solid Copy"										
	635a	cq DX									
	640a	VZXYX	RAC 5	7	4	6	✓				So. of Japan
	730a	cq									
	732a	W7UM	5	7	4	4	✓				
	745a	cq	85 steady								
	748	W6EYC	5	9	5	9	✓				
	800	cq W6LB									
	820	W6MA		9	5	9	✓				

Everything for the Amateur

Compliments of

Radio Manufacturers' Supply Co.
1000 South Broadway
Los Angeles, Calif.

Phone Tucker 2823 FREE PARKING

Don always pasted a picture of the hotel with the antenna location marked in his W6ZZA log.

The Crash

Often the local amateurs in Long Beach would stop in to see 6AM and check out the latest developments with Don's station. For all the time Don spent traveling, and the amount of time he spent operating his station when he was home, he still found time to modify it. Many operators didn't have the expensive meters to measure current and voltage in their equipment, so Don decided to do an article on the subject. He measured all of his station power parameters with a method as old as the hills: using the watt-hour meter installed on the house by the electric company. First he turned off everything in the house except a single 100-watt light bulb. The disc of the meter took 50 seconds to complete one revolution and this was his baseline. Using this calibration point and the fact that the disc completed one revolution in 12 seconds when the filament of his water-cooled tube was energized, Don calculated that the filament drew 417 watts. When the high voltage was applied to the mercury arc rectifier an additional 151 watts were consumed. With his transmitter adjusted to 1,000 watts input, the total power consumed from the 220 VAC line was 1,668 watts. Don spent a Monday evening writing the results and sent it in to the ARRL, where it appeared in the June 1927 issue of *QST*, "What is the Input to Your Set?"

Periodically Don would write up a description of his station and send it to one of the radio magazines. The September 1927 issue of *Radio* magazine contained an article by 6AM titled: "Clickless High Power DC Transmitter." In it Don noted that with a small transmitter it is quite simple to emit a clean signal. With the higher-powered stations a number of precautions must be taken to insure that the transmitted signal is relatively pure. This was accomplished by placing RF chokes in the primary of the power transformer and keying the primary voltage to the transmitter power supply. When these modifications were implemented Don also found an improvement in the output power efficiency of his transmitter (for the same amount of DC power input, an increase in the RF output power).

During these years Don's radio station was in a state of constant change. As he found modifications to improve the performance of his station they were implemented in short order. His water-cooled tube could not run at its full power capability due to the current limitations of the 220-volt single-phase line and the mercury vapor rectifier in the high-voltage power supply. When Don found that the city would bring three-phase power down to his house from the power pole for a $10 deposit and $3 per month, he had the power company install it.

Don also learned that the Edison Company replaced the power transformers mounted on telephone poles when their efficiency dropped below a nominal value, usually around 95 percent. However, they were still quite serviceable for amateur radio use. For $7.50 the old 3-kilowatt (KVA) transformers could be purchased. Don bought several and then set about rebuilding his power supply. He constructed a wooden rack two feet square and seven feet high. Shelves were installed in the rack about 20 inches apart and the power supply components were placed on various levels. The now familiar 19-inch wide steel rack cabinet would not be developed for several years. The new rectobulbs had just become available and they had comparatively low loss and high efficiency as high voltage rectifiers. Also, they didn't

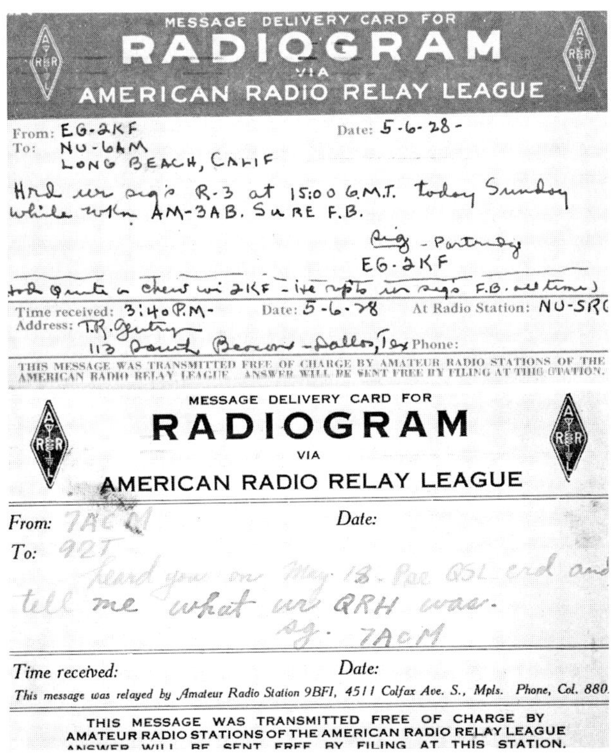
MESSAGE DELIVERY CARD FOR
RADIOGRAM
VIA
AMERICAN RADIO RELAY LEAGUE

From: EG-2KF Date: 5-6-28-
To: NU-6AM
LONG BEACH, CALIF

Hrd ur sig's R-3 at 15:00 G.M.T. today Sunday while wkg AM-3AB. Sure F.B.

Sig - Partridge
EG-2KF

Time received: 3:40 P.M.- Date: 5-6-28 At Radio Station: NU-5RG
Address: T.R. Gentry
113 South Beacon - Dallas, Tex Phone:

THIS MESSAGE WAS TRANSMITTED FREE OF CHARGE BY AMATEUR RADIO STATIONS OF THE AMERICAN RADIO RELAY LEAGUE. ANSWER WILL BE SENT FREE BY FILING AT THIS STATION.

MESSAGE DELIVERY CARD FOR
RADIOGRAM
VIA
AMERICAN RADIO RELAY LEAGUE

From: 7ACM Date:
To: 9ZT
Heard you on May 18. Pse QSL crd and tell me what ur QRH was.
sg. 7ACM

Time received: Date:
This message was relayed by Amateur Radio Station 9BFI, 4511 Colfax Ave. S., Mpls. Phone, Col. 880.

THIS MESSAGE WAS TRANSMITTED FREE OF CHARGE BY AMATEUR RADIO STATIONS OF THE AMERICAN RADIO RELAY LEAGUE

Typical Radiograms

The 3-phase power supply at W6AM.

need to be tipped and then again placed vertical as the mercury vapor rectifier had.

When Don finished rebuilding his set he had a power supply that would meet the maximum power requirements of his water-cooled transmitting tube. A total of 6 pole line transformers rated at 3-KVA each provided high voltage exclusively to the rectobulbs and the transitting tube. This enabled 6AM to run upward of 3 kilowatts of input power, if he pushed the transmitter to the limit. The AC line power to the city was 50 cycles, so the 3-phase full-wave rectified DC high voltage had a 300-cycle ripple frequency on it. As the high voltage was unfiltered, it imprinted a distinguishable 300-cycle note on Don's transmitted signal.

On October 4 Don made his first contact of the season with WNP, the current MacMillan expedition to the Arctic. Their operator's name was C.E. Himoe and he wanted a message passed to his mother in Glendale. After accepting the traffic Don chatted with Himoe at length. Afterwards he called the *Long Beach Press Telegram* to relay the information from WNP and the next-day their headlines read: "Long Beach Radio Amateur Talks With Friend in Arctic." Contact with the *Bowdoin* was big news and Don passed the word along to the newspapers whenever there was further information to report. The front page of the *Long Beach Press Telegram* in the January 9, 1928 edition had the headline: "Beach Amateur Talks With Arctic. Don Wallace, Radio Expert, Delivers Message to MacMillan Ship." The article went on to describe Don's 20-minute contact with the *Bowdoin*. The ship was ice-locked in Labrador and the outside temperature was 60 degrees below zero. Two weeks later 6AM was in the news again. The *Los Angeles Times* ran the headline: "Southland Operator Talks with Byrd." The story was syndicated and it was published across the nation. The *Chicago Tribune* reported: "Byrd Confident of Leaving Ice Pack Soon-Radio." A friend of Don's had cut out the clipping and sent it along with the enclosed note: "When it comes to getting publicity you have the world beat!"

At this time the *Press Telegram*, as well as the *Times*, still carried Don's weekly radio columns. Not wanting to be left out, the *Register*, an Orange County newspaper located 20 miles east of Long Beach, started publishing Don's weekly radio articles. It is not known how many newspapers outside of the Los Angeles area reprinted his articles, but they would periodically appear even in East Coast publications such as the *New York Sun*.

Due to the big signal that he put out, Don worked quite a few stations in the Orient. In October he received a letter from J. Mack, Chief Petty Officer on the H.M. Submarine *L27*. Mack noted: "I am AJ1MK (Asia, Japan) who worked you a few days ago. I am No. 1 radio guy of the above British submarine and like to occasionally do a bit of DX work with amateurs. You are, however, the first American I've raised. Say, you were pretty slick on operating. I wonder if you are a pro like I am? Congrats to you old boy on your set. Sigs very good here. Cheerio."

A few weeks later Don received a QSL card from Sukehide Kabayama. Hide reported: "Almost every night your steady, easy pickup signals come to my receiver. I have been on the air once, but now am kept QRT (no transmitting) by the authority, and all AJ stations are silent." A limited number of Japanese stations did continue to operate transmitters undercover, at great risk to the operators.

The Eighth Annual Convention of the Pacific Division of the ARRL was held in San Diego on October 14 and 15, 1927. Don was among the featured speakers and the convention activities included trips to local amateur stations, as well as the receiving station at Point Loma, NPL. The 32 page program and handbook included a 3 page article reprinted from The *Christian Science Monitor*: "Improving Transmitter Steadiness" by D.C. Wallace, 6AM.

In late December Don received a postcard mailed from Hawaii. The *SS Wilhelmina* had just docked and L.M. Hart, 6EM, was the operator of their wireless station aboard, WMO. Hart remarked "Yr sigs rec R5-R6 hr at sea 800 mls w SF at 647 AM. Are U Don Wallace? C somethin abt U in the Dec. *QST*. U are certainly a 'hamming man!" The CW operators had developed their own language over the years. It started with shortening words when sending the code to each other. With the passing of time the abbreviations naturally gravitated to QSL cards and letters and in some cases even voice conversation.

The technical editor of *QST*, R.S. Kruse, was impressed with Don's three-phase high voltage power supply and devoted over seven pages of the ARRL's magazine describing it. The February 1928 issue of *QST* carried the article labeled simply "6AM." Don's power supply had many unique aspects. For those stations with the higher voltage power supplies for high-power transmitters, the condenser was always the weak point. At this point of its development the filter condensors were not yet reliable and they were expensive to replace. A three-phase full-wave power supply created a ripple frequency on the high voltage that was three times higher than a conventional single-phase power supply. The higher frequency was much easier to filter and required a smaller condenser. Additionally, with all of the iron in the multiple power transformers, the voltage drop under full current load was quite small.

Each of the six rectobulbs (the high-voltage rectifiers) required 6 amps at 10 volts for their filament power. During operation their filaments were also raised to the level of the high voltage and had to be isolated from ground. Rather than wind a filament transformer rated at 6,600 volts insulation, Don reverse-connected another pole transformer, putting 220 volts into the secondary winding and getting 22 volts center-tapped out of the primary side. With a rheostat (adjustable resistor) he adjusted the voltage from 11 volts down to 10 volts and ran the filaments of all 6 rectobulbs from it. Don put a knife switch in series with each of the 6 high voltage outputs between each pole transformer and its corresponding rectobulb rectifier. In this fashion, when the power supply developed a problem, the trouble could be easily isolated and repaired. The 6AM station certainly had an impressive power supply. At that time the "Amateur's page of *Radio Doings* also carried a picture of Don sitting at the operating position of his station.

Sixty years later Jerry Faas, W6KMI recalled "I have a QSL from Don that I got in 1928 when I was in high school in Colton, California. I also have an Official Relay Station certificate signed by him when he was the SCM and by F.E. Handy, who was the ARRL Communications Manager. Don was using a water-cooled tube then and a 6-phase power supply. Most of us had just a 5-watter."

Mike Hiehle, W6FAU (now W6RZ), visited Don about the same time. "I met Don at an his open house on Molino. I belonged to the ARALB and he invited the club over once a year. He was using a vacant lot next door as well as his backyard for antennas. The poles were some 60 or 70 feet high. They were guyed every 10 feet or so and snaked back and forth. They also had crosspieces of wood nailed to them to permit climbing." One would have to be rather dedicated, or daring, to climb Don's antenna masts!

Dr. Hidetsugu Yagi, one of Japan's prominent radio/-electronics engineers, presented a paper to the IRE in January 1928. Entitled "Beam Transmission of Ultra Short Waves," he disclosed recent experiments that he had performed with a "wave canal" using microwaves. Dr. Yagi had performed his post-graduate studies in England, where he also worked with John Fleming, inventor of the vacuum valve. J. Dellinger, of the Radio Division of the National Bureau of Standards, reviewed Yagi's paper and characterized the work as exceptionally fundamental and was sure that it was destined to become a classic in its field. The "wave canal," consisting of a parasitic element, a reflector, and a director, would become widely known in the years to come as the "Yagi" antenna. Materials capable of supporting their own weight in the element size required for the lower frequencies would not be available for two decades. However, the antenna found immediate acceptance on the shorter waves, particularly 5 and 2-1/2 meters.

Due to his inordinate amount of station activity, Don continued to receive a lot of reports from various DX stations. A letter dated January 10, 1928 was received from N. Miyake, Minamihoribata, Tsu City, Japan. Mr. Miyake wrote: "NU6AM, I have been hrd ur sigs. I am one of the Japanese amateur, but I am sorry that can not QSO with you. Our government dose not allow xmit with me, but I have a xmitter and a receiver. CUL 73's OM. P.S. Kindly send me undercover, hi." Actually, a great number of amateurs around the world did not have government permission for their radio stations in their respective countries. Foreign post offices in these countries soon learned to recognize QSLs and one inadvertently sent without a cover envelope often resulted with the intended recipient disappearing from the air, sometimes never to be heard from again.

Don received a note in mid-April from AC2CK in Tientsin, China. The sender mildly scolded him: "Saw your letter in *QST* some two years back stating that you were such a busy stn wid so mani QSOs that cards were often unintentionally forgotten, so OB I'd appreciate it if you'd shoot one over for our numerous fb QSOs. Please OB es under cover. 73. A.W. Travers Smith." Don spent a great amount of time both working on his station and operating it. In another 20 years QSLs would become a lot more important to him. For the moment, however, QSLing was one of his lower priorities. Don had Bertha answer many of the cards, but she was not particularly enthusiastic about the task. Sixty years later numerous bundles of cards would be found marked "need to answer"!

The second annual ARRL Relay Competition (DX Contest) was held from February 6-19, 1928. When the results were published Don had the highest score in the Western United States, nearly triple that of the closest station. There were few stations out west that could give 6AM any serious competition. One evening that spring Don worked OH6ADH (Oceania, Hawaii). Donald deNeuf worked for RCA and was at one of their facilities in Hawaii. Five years earlier, when deNeuf had been in Petaluma, California, as 6ACZ running a Ford spark coil, he had first worked Don as 9DR. Two years later, when deNeuf had put together a 5-watt CW station at the same location he worked Don as 9ZT. Their chance meetings on the air, and future business contacts, would continue for over 60 years.

On February 23, 1927 the Federal Radio Commission (FRC) was created. One of their first acts was to call an International Radio Conference, which convened in Washington D.C. the following October. The number of amateurs had dramatically increased in the years immediately following the Great War and had leveled off for the past five years. Over 15,000 amateurs were now licensed. The FRC felt that it was necessary to redefine the radio amateur's frequency allocations. Amateur radio lost nearly 40 percent of the frequencies that they had, but gained a 10-meter phone band effective March 7, 1928.

Bill Adams, 6ANN (W6BA), had recently spent a lot of time operating on the shorter waves. Progressing from 20 to 10 meters, the 28 MC band seemed to have its own distinct characteristics. On Sunday, April 29, 1928 Bill worked 2JN in New Jersey. In his letter to Bill, Mr. Atwater remarked: "As far as I know OM this transmission of ours is the premier from coast-to-coast (on 10 meters) and we can claim the Transcon laurels. Shortly after I worked you I worked 6UF." (Bill Eitel, 6UF, would also be active for the next 50 years.)

In relating his operations on the shorter wavelengths Bill Adams recalled "I was licensed in 1923. There were five other hams who lived close to me. I grew up on East 7th Street in Long Beach. We all operated on 80 meter CW, and the interference to each other was tremendous. Our early home-built receivers were not very selective, you know. The six of us got together and decided to draw straws. I was one of the three who got a short straw; we had to go up to forty meters. That way there would be less interference between us local stations. We thought we were getting gypped at first, because everybody figured 80 meters was the better band. The trans-Atlantic contact had been made on 110 meters just about then. Well, 40 meters turned out to be a great DX band. We worked stations a lot farther away than the other fellows on 80 meters. Pretty soon I was trying out 20 meters and then on up to 10 meters when I worked 2JN. Later that year we started working European and Pacific stations on 10 meters."

The last couple of years Don realized that propagation between Europe and California was particularly good in the late spring. Consequently he spent a lot of time on the air during those months, making the most of it. At the end of May 1928 Don received a letter from J.H. Partridge, 22 Park Road, Collier's Wood, London S.W.19. He wrote: "You will have no doubt received my report on reception of your signals yesterday afternoon . . . I heard you work AM3AB . . . It is very interesting to learn that 20 metre signals travel from the 6th district when the 1st (district) and East Coast stations are not audible and I take it that your signals were not coming direct across the Atlantic but the other way round. What do you think? It is a very unusual happening to hear California during daylight hours and moreso at 1500 GMT in May is about the worst possible time one would expect it. I would have called you there and then but for the fact that I had just set the transmitter for 10 metres . . . EG2KF." 2KF was well known in America; he had made the first Anglo-American contact with 1MO less than five years earlier. Partridge was also among the first to note the radio propagation that would be the salvation of California DXers in the years to come, the long-path. Except for unusual conditions, signals from East Coast stations are typically a lot louder in Europe and the Middle East than those from the West Coast.

Don Wallace seemed to turn up in almost every publication of the day. In the "Research and Review in Radio" section of the May 1928 issue of *Radio Entertainment* Don's photograph was in the center of the page. The text notes that 6AM was the present Chairman of the Interference Committee of the Radio Trades Association of Southern California. Among the other items noted on the page was a brief note that Don Lee, owner of KHJ in Los Angeles, had his yacht completely equipped with receiving apparatus. In another 12 years Don would do a 2-week broadcast from a similar yacht, over the Don Lee radio network.

During this period Don was continually writing stories for various magazines on Monday nights. One of his articles appeared in the July issue of *QST*. It got him in some hot water with W.D. Terrell, the Supervisor of the Radio Department of the Federal Radio Commission. Don's article, "Becoming an Operator in 15 Minutes," described how, in 1926, he had got Bertha to learn the code (which included her transmitting to Don with his three-phase transmitter). In those days it was not unusual for amateurs to transmit prior to getting a license. Many an old-timer will privately tell you that they borrowed a call sign for a short period of time prior to receiving their own. Both Don and the ARRL were put in an embarrassing position after these details were published in *QST* and Don was fortunate to get off with only a warning from the FRC. He offered his resignation as Section Control Manager for the Los Angeles section, but the ARRL refused to accept it. However, he was advised to keep a low profile for a while. In the meantime, Don had successfully managed to coach Bertha to the point where she passed the code and theory test and received her amateur ticket, 6EN, on May 25, 1928.

Keeping a low profile was always a difficult task for Don. He was always experimenting with one idea or another and writing the results for publication. The past spring he had been reworking his receiver again. The preselector that Don had developed earlier was incorporated into his receiver, along with some other changes. The UX-222 tetrode (four elements) tube had been developed, which had the added shield, or screen, grid and the resulting higher gain on the short waves. The receiver's first amplifier stage was shielded from the rest of the stages by a copper partition. Don had also found that, by putting a small capacitor with a vernier dial in parallel with the large main tuning capacitor, easy receiver tuning with the band spread was accomplished. Plug-in coils were used to select the different wavelength bands, as that was still the most efficient method of changing frequency. The base of the UX-210 detector was removed to keep the stray capacitance to a minimum. The output audio stage used a UX-201A tube, driving a 1,700 cycle peaked transformer and the headphones. That reduced the unwanted background noise and increased the signal's readability. The batteries were internal to the shielded receiver. No ground was used on it as Don had discovered that with a sizable antenna the receiver's ground connection added more noise than signal. The latest version of Don's receiver appeared in the May 1928 issue of *Radio* magazine, titled "Snappy Reception on Low Waves," and his article ended with the comment "Receivers such as this make us feel enthusiastic about the possibilities of the future." Don's receiver modifications would continue for some years to come.

On October 1, 1928 the Federal Radio Commission, as a result of the International Conference, assigned the W prefix to amateurs in the United States. The unofficial prefix of NU was discarded and Don's call sign became the now familiar W6AM. That same year Don transferred his portable call sign, W6MA, to his wife. So Bertha had W6EN assigned as her portable call, and carried the call sign W6MA for the next 43 years. A general class licensee in California was usually assigned a 3-letter call at that time. For Bertha to have two preferred 2-letter call signs indicated that Don must have had considerable influence with the radio inspector in San Francisco. At the same time Don was issued a new portable call sign, W6ZZA, for when he was traveling. The two Zs in the beginning of his suffix indicated that Don had FRC permission to operate portably from any United States state or territory. This system also created some of the first 4-

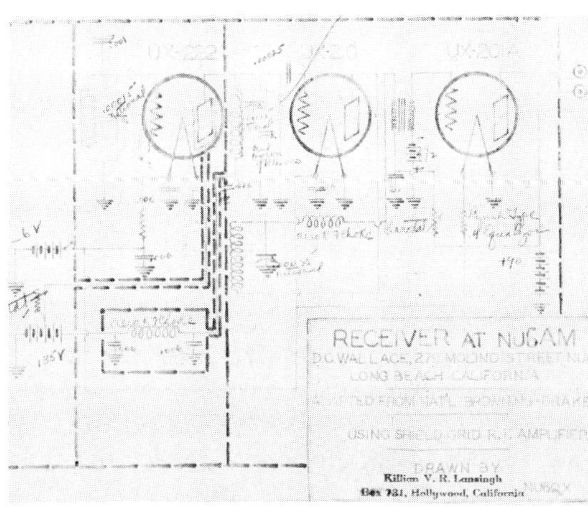

Don was continually redesigning and rebuilding his receiver. This is a schematic off his state-of-the-art receiver in 1928.

letter suffixes in the United States (the portable call sign of Larry Lynde, W6DEP, was W6ZZCE).

An enthusiastic young amateur whom Don had worked from to time to time was Eric Palmer, 2ATZ, located in New York. The young man would get so wrapped up in his amateur radio activity that he would stay up all night working stations and consequently fall asleep in school the next day. His parents felt this was detrimental to his school work and repeatedly tried to impress upon him the importance of going to bed at a reasonable hour. Their efforts proved to be fruitless, however, and as a last resort they eventually wrote the Department of Commerce and asked that his amateur radio license be revoked. The Department of Commerce complied with this request and this developed into a unique incident — a federal license being revoked when the license holder had not violated any federal regulations. In 1930 Eric would write a book, *Riding the Air Waves with Eric Palmer*, and relate the story of his license revocation. In the front of the book there is a picture of Don standing by his water-cooled transmitter with the caption: "One of the finest experimental stations in the world, W6AM, Don Wallace, famed pioneer in high frequency transmission."

Don occasionally worked Harold Graham, AC9GH, who was located in the interior of Northern China. Graham was a missionary and, although licensed, operated his station undercover. A Caucasian living that far in the interior of China was wise to keep a low profile. In the summer of 1928 Harold was to embark on a rather unusual journey which he recounted to Don when he returned to the United States three years later.

In the northeastern part of Tibet the Chinese were fighting the Tibetans. An American company in Shanghai had sold a radio set to the Chinese in the spring of 1928 and sent an engineer along to install it for them. After connecting the equipment and stringing an antenna, this fellow was unable to get the transmitter to operate, so he returned to Shanghai. Another engineer who was considered to be an expert rushed to Tibet to fix the transmitter. After an arduous journey the second engineer arrived in Koko-Nor, Tibet, and began work on the transmitter. Several weeks passed and he was still unable to get the set to function. At that point the patience of the Chinese military chieftain, a Mohammedan general, had worn thin. He placed the engineer under house arrest. The ultimatum was, "When machine works, you come out." The engineer-turned-hostage sent a letter back to his company in Shanghai describing the situation with a plea, "Don't send an expert, send a practical man."

The American company held some hurried consultations and contacted Harold Graham. Graham, a well known radio amateur and a man of peace, considered the situation. Knowing full well that if he was not successful he would end up sharing a cell with the engineer, he departed on this errand of war. It was during the first week of July 1928. He traveled four days by train, one day by plane, and another two days in an automobile over the most primitive Camel paths to Koko-Nor. Time was of the essence, for at those altitudes when the first snow fell, the mountain passes usually closed for the winter.

On his arrival, Graham set to work on the transmitter. The problem was that the power leads were reversed and the power transformer had burned out. He set about rewinding it and rebuilding the entire transmiter, as well as station installation. Three days later the rig was completely operational. Harold assigned himself the call sign AC4AA and fired up the transmitter on 20 meters in mid-July. Over the next week stations across China and India were worked, as well as a handful of Europeans, including ON4AU. One day Graham heard a signal from Long Beach, which was NU6AM with his water-cooled transmitter. That was the only time Graham copied signals from North America, but he couldn't get Don's attention. Don was passing traffic with another stateside station and couldn't hear AC4AA calling him.

Graham also spent that week showing the Chinese radio operator how to use the radio set on military frequencies. He hoped the fellow would take an interest in amateur radio and perhaps continue the operation on the ham bands after he left. Once everything was in proper working order and the Chinese general was satisfied, the engineer was released. He and Harold left Tibet and returned to their respective homes in China by the same route. Back in Changsha, Harold resumed opera-

A map of Tibet and China

tion as AC9GH, but did not hear AC4AA on the amateur bands again. The engineer, secure back in Shanghai, never again ventured out to the Tibetan frontier.

W6AM spent a lot of his time passing traffic and received a note from Victor Wilson in April, 1929. Wilson noted: "Many thanx for the quick service on the radiogram to my brother in Shanghai. I certainly think that amateur radio is the thing for me. I have been in the broadcast receiver construction game since 1922 and have not yet ventured into the realms of "ham." Don Clegg (6AKB and former op at 6BRS) informs me that you were once 9ZT in Minnesota where you established a wonderful record and I guess you are still making and breaking records, hi."

Meanwhile, Don was also accumulating more awards to put on the wall. When he was in town, Don continued to enter the various contests that the ARRL sponsored. In 1929 there was the ARRL traffic contest, in which Don won a certificate, and the Sweepstakes (SS) contest, in which he achieved the highest score in the ARRL's Los Angeles section.

The year 1929 recorded a number of changes in the radio field. The commercial radio manufacturers were putting out tremendous quantities of broadcast radios. The concept of the automobile assembly line that Henry Ford developed in 1913 now extended to the building of tubes, as well as radios. Atwater Kent alone was producing nearly 50,000 radios per month. Between mass production and the competition in the marketplace the price of a radio was dramatically reduced from just a few years earlier. Their internal design was better and the quality of signal reception had improved. The number of homemade radios that the general public used for broadcast reception was down to 20 percent of the radios in use, and rapidly diminishing. As a result Don's radio columns for the local newspapers tapered off to perhaps once a month by year's end.

Due to Don's travel schedule he spent a lot of time aboard trains. He decided to try using his receiver during the long rides, but found it to be incredibly noisy. Apparently the noise was created by the iron wheels rolling on the track, putting a static electricity on the ground system of the whole train. During a trip to Washington D.C. in January 1929 he was finally successful in reducing the noise to a level that was tolerable. Don recalled: " *I finally got my receiver to work on the train. I eventually hooked up my portable transmitter and I even worked home from the train. I had a letter from the chief electrical engineer and vice president of Southern Pacific Company. He had addressed it to all railroad conductors: 'Please extend to Don Wallace every courtesy that doesn't interfere with train operation, to operate his radio set.' So I would disconnect*

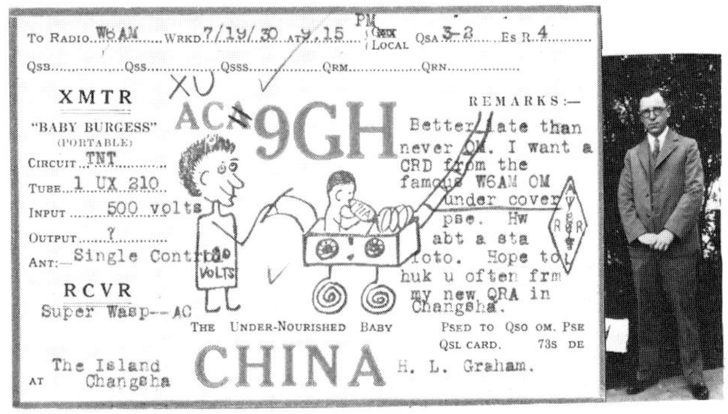

The first amateur radio operation from Tibet was by Harold Graham in 1928, operating as AC4AA.

their big Majestic radio that they had in the car and plug in my radio in its place. I soon found that you can't hook on to, or have anything touch, the railroad car. The electrical racket is just beyond comprehension. So what I did, I used the overhang on the observation car and the railing. I would put two ropes of light string down from each corner at the back of the car and run a grid of wire between them, just enough for a 40-meter antenna. Then I ran the other wire for a counterpoise — I stuck it on the inside of the windows with tape. The receiver never touched any metal, except for the AC line into which it was plugged. The counterpoise was a floating ground inside the car. I could hear stations on my receiver with no problem after that." Don wrote up the story and sent it in to the ARRL, where it appeared in the July 1929 issue of *QST*, "High Frequency Reception on Trains."

Don often drove down to San Diego to visit his various customers. On the long trips back and forth he would listen to assorted CW stations with his car radio. "*I got tired of listening to the broadcast stations,*" Don continued, "*they were rather boring. Sometimes I could hear high speed CW transmissions when I drove between San Diego and Long Beach. I had an RF converter that worked with my car radio. It was built by Waylan Horth. You flipped the switch to one position and you had the regular broadcast stations on the car radio. When you flipped the switch to the other position you had the RF converter. Mobile transmitters were not allowed in those days. The first authorization for mobile transmitters was in 1935. Then it was for 5 meters only. A few years later 10 meters was added. So I started with a receiver, put in a 5-meter rig when that was allowed, and when the authorization changed I soon had a big 10-meter rig in the car.*

In the meantime I had put in a receiver to listen to

the other hams. That was more interesting than the stuff that they put out on the broadcast band. I found that on some days I could tune Manila. They were running a high-speed tape, maybe 75 or 80 words a minute. At first I couldn't make out anything. It was like a purr it was so fast. After I had listened to it for a half hour I could recognize an occasional word that I knew. In another half-hour I could make out a few more words. By the time I got to my destination I could even distinguish the letters, which helped me in my code speed. It was a 4-hour drive back home from San Diego in those days, so I got lots of practice. When I got in a code contest, the code letters almost sounded slow. It is just a question of getting them written down. In those days I could write fast. I could write longhand at 49 words a minute; the fellows timed me. They couldn't believe it. So when I would get in a longhand CW contest I would most always win. The cups and plaques are over there (lined up along one wall). I won a few contests by typewriter. That was at about 55 words a minute. There was usually somebody that worked for some commercial company who would win those contests because they were still copying high-speed code over the air with a typewriter."

One day a year earlier a young ham, Frank Cuevas, 6AOA, knocked on Don's door and introduced himself. Frank had heard Don on the air and traveled down from Bakersfield to meet him and see the famous 6AM station. A friendly rivalry developed between them and 30 years later Frank would be the one who finally beat Don in a longhand code copying contest.

In the spring of 1929 the Wallace family moved into their newhome at 4214 Country Club Drive. It was two stories high, modeled in the New England fashion. Don had the radio room especially constructed to house his radio station and wired with three-phase power. Within

Don and his wife Bertha, in front of their new house at 4214 Country Club drive in Long Beach.

Don working on the rear pole, by the garage.

a week the transmitter was reinstalled and a zepp antenna strung outside. There were two 95-foot high telephone poles for antenna supports, one in front of the house and one in back. Don's trusty climbing spikes and safety belt would see service on a weekly basis in the years to come. At the back of Don's lot was the garage, next to the rear 95-foot pole. It had an alley entrance and was two story also, modeled after the house. The second story would soon be finished and Don planned to use it as his office. Some months later he moved his manufacturer's representative business from the McLaughlin building in downtown Long Beach to the upstairs above his garage.

Antennas, and testing them, were one of Don's favorite projects and most Saturday or Sunday afternoons would find him perched on top of one of his 95-foot poles, adding another piece of wire to those already fanning out from his lot. He had purchased the two lots adjacent to his house and the 25 acres immediately around him were vacant. Don put up temporary masts at the edges of the vacant property and no one seemed to mind. They were 75-feet high and were made from 2-by-4s bolted together. Don guyed them at three levels and they would support one end one of the assorted antennas that he was stringing in various directions from the poles next to his house. Don would climb up one pole, spend an hour working on the antenna, feed line, and its support rigging, climb down the pole and climb up the other pole and spend an hour up there. With antennas coming in from several directions, sometimes he would do this all day long. Don recalled, "*When I moved into the house, I was doing lots of pole*

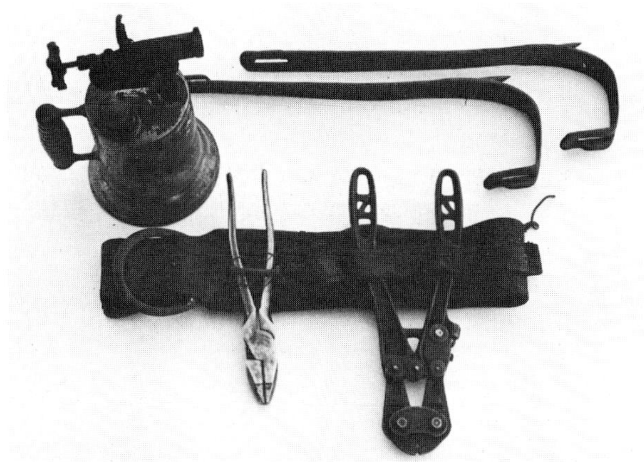

An antenna installation kit.

climbing. I found that I could only do six poles a day. I would go up, work for an hour or so, and then come down. I was pretty strong then, in good shape. I had played football. Going up the pole and spending all of that time working on the rigging and so on, it was very tiring. I found that after doing six trips up and down I was exhausted. That's when you get hurt, you know. You're so tired you start making mistakes. I brought the antenna feed lines down the pole and tied them off at a Redwood crossarm outside the hamshack. Eventually I had up to 14 transmitting antennas to choose from.."

In June 1929 Don worked quite a number of Europeans. Conditions always seemed to be pretty good that time of year. The first week of July he received an SWL report from Donald Sloper at Frying Pan Farm, Challeymead, Melksham, Wilts, England. Mr. Sloper wrote: "You have easily been the strongest W6 I've heard for a long time. As you know, it is unusual to hear W6 at any great strength. I use a 1V1 (one tube receiver) and aerial 75 feet long." The signal reports were for phone transmissions, which was unusual between the West Coast of

Don went to work for General Motors Radio in 1929.

the United States and Europe except for all but the high-powered stations.

The September issue of *QST* had an article by Windy Windom, W8GZ, "Notes on Ethereal Adornments." The antenna Windy described was the one that he had used when he contacted Australia using slightly over a half-watt three years earlier. Windy had performed additional experiments and found that when a single wire was used to feed the zepp antenna it had to be fed off center for a proper impedance match. Otherwise part of the output power would be reflected back to the transmitter, rather than being radiated by the antenna. Windy included a detailed description of his antenna and a chart that gave the proper antenna dimensions for each amateur radio band. In the years to come his modified zepp antenna would become known as the "Windom."

In early 1929 Don was offered a job as Zone Manager for the General Motors Radio Corporation. They had noticed how well he was doing with the National Company and decided to try to lure him away. The salary offer was somewhat better than his current job and would require less travelling than before. This job had a very promising future, especially with the potential size of his annual bonus figured in. Don accepted the position and parted with the National Radio Company and George Hill on good terms.

Don sold the National lines he represented to Bill Cook. Cook would take Don's position with National Radio, and they worked out an agreement that Bill would pay Don on an installment basis over a period several of years. Since the National Radio distributors were well-established they were practically a guaranteed source of income. Unfortunately, Bill died in an automobile accident a few months later.

When Don started selling radios for General Motors, he refused to compete directly against Day Fan. He felt that it was not good business practice; a former employer shouldn't be treated that way. So when Don visited various cities to represent the GM radio line, he would sign up what he considered to be the second-best radio distributor in town. As his staff grew and he traveled less, Don imparted the same wisdom on them. Day Fan distributors were not to be approached by GM salesmen.

Don continued to do well with his stock market investments until late in 1929. Some cautious investors had grown wary of the continued spiral of stock prices and eased their way out of the market. In September 1929 the great "bull market" began to turn. Prices fell and margin calls went out. On October 24 stock prices plummeted to all-time lows. President Hoover issued a statement which said, "The fundamental business of the country . . . was on a sound and prosperous basis." On October 29, "Black Tuesday," 16 million shares were sold and the New York stock market crashed.

Don was wiped out, counting his financial losses at upward of one quarter of a million dollars. Some of Don's stock holdings had been on margin. One would put down only a small percentage of the stock value, usually 10 percent. If the price of the stock went up, a tremendous gain was realized. If the price went down, the difference of cash value of the stock had to be paid. The nest egg the Wallaces had been accumulating for their children's college fund was depleted and the forthcoming bank failures took care of the rest of their savings. Here was Don with a wife and three children, a $10,000 mortgage at the bank, and a job that would barely pay his current expenses.

Saving the House

After the crash, Don's family tightened their belt and made ends meet. Their financial situation was not as bad as others. Don still had a regular job. Many people would be jobless for much of the great depression. Things were going reasonably well and he was holding his own until a bank officer at the local bank, to remain unnamed, took a liking to Don's house and decided that he wanted it for his own. Don explained "*The bank called in my loan. They could do that in those days, you know. There was no protection like there is now. They foreclosed, even though I had been making my payments on time. The bank demanded that I come up with the entire $10,000 of my mortgage — I had 30 days. They didn't do it to anyone else around there, just me. One of the fellows at the bank wanted my house for himself. He lived in the neighborhood, I know who he was.*

"I tried to get a loan from another bank, but there was no way. Banks were failing then and those that survived the stock market crash were not giving loans. My Dad had also been hard hit when the market crashed and he didn't have any money to spare. Well, one day I was talking to a fellow I had shared office space with, Bill Gaede, trying to come up with some ideas. Bill said, 'I will tell you what we can do. I work for the Metropolitan Life Insurance Company. I can write you a life insurance policy for $10,000 because you have a good job. Now, you can make the first payment on the policy to the insurance company and then borrow that $10,000 value of the policy from the insurance company, using your house as collateral. With that money you can pay off the mortgage that the bank holds and make the payments to the insurance company instead.' So we tried that and it worked, and that's what I did to keep my house." The local bank that had held the mortgage to the Wallace's house survived the depression and prospered. However Don declined to do business with it in the future.

The next several years were lean times for the Wallace family. On January 16, 1930 Don received the loan from the Metropolitan Life Insurance Company. The payments Don made to the insurance company were $180 a month for five years, which left very little money for other family expenses. However, they still had their home.

The tenth annual Pacific Division convention was held at the Hotel Alexandria in Los Angeles on November 29 and 30, 1929. A message of greeting from Commander Richard Byrd was received by radio just in time to be read by ARRL Director Allen Babcock at his opening address. The banquet, under direction of W6AM as toastmaster, represented the apex of the entire convention and was declared to be the largest convention ever held at a similar function in the history of the Pacific Division. In the convention program an ad from National Radio Tube Company noted that for those amateurs who used the UV-204A (250 watts dissipation) and had poured too much coal to the tube, it could be repaired for the nominal sum of $75. Also included with the 28-page convention program was a 12-page insert of *The Oscillator* (forerunner to *R/9* magazine), an amateur radio publication that had started the previous year.

Two months later Admiral Byrd was in the headlines again. The front page of the January 27, 1930 edition of the *Boston Daily Globe* carried the headline:

Byrd Not Alarmed Over Ice Conditions. In radio chat with coast man, he says that party expects to leave Antarctic soon. Don Wallace, operator of radio station W6AM, was in touch with the Byrd expedition near the South Pole

The article had been syndicated and appeared in newspapers across the nation. The next day Don received a letter from Louis Chik, W8CQJ. He wrote "Congratulations OB (Old Boy). In the *Detroit Free Press* I read about U working Byrd Expedition. Would you please tell me what meter wavelength Byrd Expedition located on? I'm just a beginner OB."

The third annual ARRL Relay Competition (DX Contest) was held on February 15-28, 1930. Up to 90 hours of operation were permitted during the nearly 2-week long marathon. Don turned in a score but did not spend too much time in the contest. His priority in the early 1930s was to try to keep enough money coming in to pay the bills; he had little time to spare for a marathon DX Contest.

In Southern California there is a wind condition known as the Santa Ana, or Santana, depending on which of the natives you talk to. It is a wind that blows the desert air from the east, down through Santa Ana Canyon and across the Los Angeles Basin to the west, then out to sea. It occurs periodically in the winter months and plays havoc with boats along the coast. The wind is channeled by coastal canyons and can be very strong, at times reaching upward of 60 miles an hour. Don and Bertha had accompanied a group of college students on a weekend trip to Santa Cruz Island off Santa Barbara. About halfway across the channel the Santa Ana winds struck unexpectedly. They soon became fierce and the in the midst of the turmoil the boat's engine quit. Due to the severity of the winds, the families ashore became quite concerned with the safety of their children. After safely reaching Valdez harbor at the island, Don got on the air. He had brought his portable transmitter along and worked John Lewis, W6ZBJ, across the channel in Santa Barbara. Through Lewis, Don passed a message to each of the families of the students aboard. The headlines of the May 31, 1930 edition of the *Santa Barbara Daily News* stated: "Noted Radio Man Tells Of Isle Party's Safety."

Amateur radio activity was a newsworthy item and the public seemed interested in this unusual hobby. The Sunday, August 17, 1930 edition of the *Long Beach Press Telegram* carried a photo section of the better stations that belonged to the Associated Radio Amateurs of Long Beach, which now had over 70 members. Pictured were Foster Young, W6MK, M. James, W6DQI, V. Trowbridge, W6EWK, Hal Nehmans, W6HT, George Weller, W6CZZ, Bob Fisher, W6DYJ, Larry Lynde, W6DEP, who took the photos of all of the stations, and of course Don Wallace, W6AM. A separate photo of Don's 95-foot unguyed mast (telephone pole) was also pictured.

The ARALB had been formed two years earlier in May 1928. Their initial meetings were held at Recreation Park, in the recreation room which was located among the eucalyptus trees, in Long Beach.

In his day-to-day activities of getting on the air and passing traffic, which comprised the majority of his amateur activity, Don was often called by DX stations across the Pacific and Asia. During the 1930s he would work a hundred stations in China and a similar number in the Philippines. Don worked virtually all of the members of the Philippine Island Amateur Radio Association, PARA, and the president of the club sent him a photo of the membership that had been taken at a meeting in 1930.

The fall of 1930 marked the first open house at the Wallace's new location on Country Club Drive. Now that their home was again secure, Bertha and Don had time to plan a proper housewarming. They had quite a turnout and their open house would remain an annual event for the next 55 years. Hams from all across the Los Angeles Basin stopped in to see the new W6AM QTH (location), as well as a lot of the local fellows from Long Beach. A total of 50 people turned out for the event. Naturally, the photo taken of the group at the end of the afternoon appeared on the front page of the October 25, 1930 edition of the *Long Beach Press Telegram*, along with the headline: "New Short Wave Radio Station Tested."

Don's open house had become an annual event in 1923 and his guest books go back to 1927, when Bertha first suggested the idea. Ed Mariner, W6BLZ (W6XM), recalled: "Back in 1931 I visited Don in Long Beach. The tubes in his receiver had the bases removed; he said that he got better reception. He also had a tube in a jar of oil for transmitting. (This may have been the mercury rectifier.) My brother was a signalman on the *USS George Washington* in World War I when Don was the chief radioman. A few years ago (1982) at his open house I saw my name in the log book from 1931."

By 1931 Don had progressed to the position of Regional Manager for GM Radio Corporation, which was located in the Bendix Building at 12th and Maple in Los Angeles. A decision was made that year by the United States government that had a tremendous impact on Don's career. They decided to bring an antitrust suit against General Motors. The government felt that GM was creating a monopoly. Rather than fight the suit, GM closed down their radio division and sold the assets, and the bonuses for the executives that year were canceled. Don's bonus, had he received it, would have exceeded his regular salary for the entire year. GM offered Don a job as Manager of their Fridgidaire Refrigerator Division. Feeling that his future was not working for a large company in refrigeration, Don quit his job at GM and went back into the manufacturer's representative business.

For the first five years of the great depression money was extremely tight in the Wallace household. Don's oldest son Bill remembers these days well. "We had gone for a drive to the Pike one Sunday afternoon. It was an amusement park located in downtown Long Beach by Rainbow Pier. That was something you could do that didn't cost any money, walk around the amusement park. We children were looking in a store window at the chewing gum— it was a nickel for five sticks. Dad bought a package and everyone got a piece of gum. That was our treat for the day."

Whether it was a consequence of the hard times of the depression, or the fact that ever since his first rig was thrown out in 1916, Don didn't like to toss anything out. The author discovered boxes upon boxes of tubes, totaling several hundred, in Don's basement from this era. Each one would have an individual note tied to it:

The first open house at W6AM was held in 1930.

"Hot spots, save for emergency; some filament strands broken; emission about 50 percent; or NG as doubler, may be ok as amplifier." Any tube Don used in his equipment was in no danger of ending up at the dump.

Broadcast radios had changed dramatically in the past few years. Most of them now had internal power supplies and no longer required external batteries. By 1930 the homemade radio market had virtually disappeared. Only the exceptionally well-made set would outperform the better commercial radios. Due to the increased competition among manufacturers there were no cost savings realized in a home-built radio. Thousands of these broadcast enthusiasts began looking for a new challenge. As a result the ranks of radio amateurs would swell to double its size in the next three years. At this time there were still just 18,000 licensed amateur stations in the United States.

Money was so tight that Bertha's engagement ring was sold so Don could buy a new suit for his upcoming business contacts as a supplier's representative. Don's first call was to an old friend who ran the Radio-Television Supply Company in Los Angeles, Mitchel Hirsch. Don stopped in to say hello and mentioned that he was reentering the representative business. Don reminisced, *"Mitchel said, 'Don, glad to see you again. What lines do you handle. I'll buy them,' and I replied, 'Well, I just got started again and I don't have any products to represent just yet.' 'Oh, wait right here.' Mitchell went into his office and made six calls to New York. He came out and said, 'You now have six lines; Vibroplex, Conant, Amperite, JFD, Gordon Knobs, and The Instructrograph Company. Come on in and let's write up your orders.' "* As of that moment, Don was officially back in the manufacturer's representative business.

Now that he was established and had an income again, Don decided to look up an old friend from college that had started a company in 1922, Edgar F. Johnson, 9ALD. Johnson had started his machine shop in a small room at the back of a barber shop in Waseca, Minnesota, and had slowly built it up into a good business. Don wrote him and told him that he would like to represent the E.F. Johnson Company. Johnson replied with a letter and said that every sales rep he had ever met was a bum; no offense intended. He would handle his company's sales by advertising, thank you. Not to be discouraged, Don carried an ad containing E.F. Johnson Company products along while he was out visiting his various customers. When he was taking orders for his assorted product lines Don would write up any product from Johnson in which his customers were interested. In a short period of time he had accepted enough orders to literally fill a boxcar—the freight rates were most favorable when an entire car was filled. Don sent the stack of orders in to Johnson and said that this was for old times sake, no charge. Ed, realizing the effort involved in taking this large number of orders, sent back a thank you note and a commission check for 10 percent of the total order. So Don continued to take orders for the E.F. Johnson Company and pass them along when a boxcar was filled, and the commission checks kept rolling in from Ed. In this manner Don soon became the exclusive West Coast representative for the E.F. Johnson Company.

Don's relationship with Ed Johnson proved to be very beneficial for both of them. Aside from the business he was bringing in, Don also used the various products made by Johnson at his amateur station. As he saw improvements that could be made, or additional products that could be offered, Don would make a sketch of the modifications (or proposals) and send it to Ed. A great many of these were implemented by the E.F. Johnson Company and as their product line improved their sales improved. A number of these innovations probably could have been patented by Don, but that did not interest him. He felt that his larger profits from their increased sales was sufficient reward.

Don was traveling on trains again, representing the products of various manufacturers. When he left on a business trip, Don would always buy a regular ticket for a train that left in the evening from Los Angeles. Once aboard, he would tip the porter a couple of dollars and get put in the Pullman sleeper car. It was cheaper that way, money was so tight, and he would still get a full night's sleep. If the train arrived at its destination in the early morning hours, the Pullman cars were dropped off in the railroad yard so as not to wake the sleeping passengers. In mid-morning, after the passengers had freshened up and departed, the Pullman was hooked up to another train and was on its way again.

Don was a personable sort of fellow and not surprisingly his favorite subject was ham radio. When he was transferring trains, he would carry a copy of *QST* around in the train station, with the cover showing. Often another ham would come up and introduce himself and they would pass the time by talking radio. Don's son Bill chatted at length about his Dad's trips. "When Dad arrived at his destination he took a cab from the train station to his hotel. He would get out of the cab and start into the hotel carrying four bags; his portable transmitter in one suitcase, 10 days worth of suits and clothes in a Gladstone bag, all of his paperwork and literature for business in a briefcase under one arm, and a portable typewriter under the other arm. A bellboy would come running up and offer to take the bags. Dad would say, 'Oh, sure. Thank you,' give him a tip and walk on in. Well, the bellboy could never carry the bags. They had a combined weight of over 200 pounds. He would have to get another bellboy to help him, and split the tip, or go and get a cart in order to get the bags to Dad's room." Between his playing football in college and climbing poles to do antenna work, Don was in top physical condition and now tipped the scales at 240 pounds.

One week Don was visiting customers in the Bay area and was staying at the San Francisco Hotel. He set up his portable transmitter, with the antenna on the roof, and after a CQ that evening was answered by three DX stations simultaneously. He worked all three of them at the same time and found propagation conditions to be extremely good that night — a total of five different countries were contacted. One weekend Don, with his W6ZZA portable station, worked all continents. That was no mean feat, even for a fixed station.

When he wasn't traveling, Don continued to spend any spare time he had up on the poles outside his house, installing and testing various antenna configurations. It would be another 30 years until the material was available to put together and support large Yagi antennas that would approach the gain of fixed wire antenna arrays.

One particular antenna in this category that Don was fond of was the popular "Lazy H" or four half-waves in phase. It was easily adaptable for use on 10 or 20 meters because of its compactness. The antenna was often referred to as two-above-two and it resembled the letter "H" tilted on its side. Two of the half-waves are formed by the upper part of the antenna. The inner ends of these are connected to the inner ends of the lower pair of half waves by a transmission line that is one half-wave long at the operating frequency and is transposed once in order to bring all of the elements into the same phase. To the lower end of this connecting transmission line is attached a quarter-wave closed matching stub. The transmission line from the transmitter is attached

Don (seated, middle) with his portable transmitter set up at a ham fest in Oceanside, circa 1932. William Wallace is seated, left.

at the 600-ohm point on the stub. The feed line, phasing section, and stub are typically constructed of No. 12 wire spaced six inches and having a nominal impedance of 600 ohms.

To tune the Lazy H, its stub was resonated by parasitically exciting the antenna and running a flashlight bulb attached to a small loop of wire along the stub near its lower end to the point of maximum brilliance. This point was the location at which the stub should be shorted. The point on the stub to connect the transmission line was found typically from one-third to one-half the distance from the far end of the stub to the point it is attached to the antenna and phasing section. The point that the transmission line was attached was indicated by using the flashlight bulb to measure the standing waves along the transmission line. When the minimum indication of standing waves was found, indicating an impedance match between the feed line and antenna, and the corresponding maximum transfer of power, the line was at the optimum point to be attached. The gain of this antenna array was typically six dB over a half-wave antenna.

An aerial photo of Don's house shows the open space for antenna work.

Another directional wire antenna that Don put up was known as the bi-square. It can be pictured as a square resting on one corner. The sides of the square are each one-half wave long and this antenna, as with the Lazy H described above, was well suited for the higher bands, particularly 10 or 20 meter operation. A single pole served as the supporting structure for two of these antennas and the antennas themselves were used as part of the guying system for the pole. Four insulators were attached to the top of the pole and spaced 90 degrees apart. Four antenna wires, each two half-waves long, are tied to the four insulators. At a distance of one-half wavelength from the top of the pole another insulator was attached to each antenna wire and a guy wire attached to the opposite end of each insulator. The ends of the antenna wires were then tied to the lower portion of the pole through insulators at a point 23 feet from the top insulator for a meter antenna, and 46 feet down from the top for a 20-meter antenna. Two quarter-wave closed stubs were then connected to the opposing pairs of antenna wires at their lower attachment points. Once the antenna wire and insulators were prepared the pole was raised to the vertical position. Upon erection, the four guy wires previously attached through insulators at the mid-point of the antenna wires were secured to anchors in the ground at 90 degrees spacing, so that each antenna would be at right angles with the others to minimize the coupling between them. The guy anchors are oriented so that the broadside pattern of each antenna will be directed to the favored part of the world. As each antenna was bi-directional and had a wide aperture, the two antennas effectively covered the world. A gain approaching 5 dB over a dipole could be realized with this pair of antennas. The stub and feed line adjustment are the same as described for the Lazy H. By means of a low capacity double-pole double-throw ceramic knife switch connected to each pair of feed lines, the antenna giving the better receiver response from the received station could be selected and used for transmitting in that particular direction.

One of Don's favorite antennas that he installed in Long Beach was the rhombic. The rhombic antenna was diamond-shaped, mounted horizontal to the ground, and it was directional following a line drawn down the center of the diamond. The antenna feed line was connected to one end of the diamond and if the antenna was left unterminated at the opposite end, current radio theory said that the RF signal was radiated from each end, in two directions. Terminating the opposite end of the rhombic in its characteristic impedance, typically 600 to 750 ohms, resulted in the antenna radiating a narrow beam of RF in just that one direction. Don found the rhombic to be superior to the other antennas that he had installed, but due to the enormous amount of space that it required—it was 260 feet from end to end—he only had room to put up one. Don wrote a brief description of some his findings on antennas and sent it off to *Radio* magazine. The April 1932 issue carried an article titled "Short-Long Wave Combination Sets Can Be Made to Perform 100% Better If You Follow the Advice of Don Wallace."

When Herbert Hoover was inaugurated as President in the spring of 1929, a number of changes took place in Washington, D.C. His son, Herbert Hoover, Jr. became licensed as W4SR in late 1930 and was quite active from Asheville, North Carolina. He kept a weekly schedule with Don for passing traffic to local friends and family for most of the next two years. In 1933 Herbert Hoover, Jr. returned to Southern California and was assigned the call W6ZH, along with the portable call of W6ZZK.

One afternoon that year Don got a call from a friend who wanted to stop in and see his station. Harold Graham, AC9GH, had just returned from China and wanted to meet the "big signal" from stateside. He and Don swapped stories all evening and Harold related his experience on the Tibetan frontier during July 1928.

The ARRL Trans-Continental Relay contest took place in December 1931. Two months later Don received a letter from George Grammer, the assistant technical editor of *QST*. George noted:

> Now that the smoke of battle has cleared away, (referring to the Transcon Relay) certain stations, of which yours is one, stand out. April *QST* will have a bang-up story on the Relay and we want to have station descriptions in that issue of the stations that did the best work.

Don described his station for *QST* as follows:

W6AM is a 1000-watt crystal controlled transmitter. On the right is the power panel with 6-phase plate supply. Six rectobulbs are seen above the 10 old pole line transformers which juggle voltage and phases. The 36 heater element resistances allow primary keying which is clickless, and "safe" for operator, tubes, and the children of the household.

The center rack is still being completed. On top is the power supply for the oscillators and amplifiers. A separate 3720 KC crystal controlled amplifier equipment is just below, so it can be switched directly into the 1-kW water-cooled tube built especially for the station by the National Radio Tube Company. Below this are other pole line transformers for more 6-phase expansion later on. On the floor is the W6ZZA portable, which travels with W6AM on trips so that a daily QSO may be had with Mrs. W6AM (W6MA). It is crystal-controlled 7004 KC, a completely self-contained transmitter and receiver, delivering 150-watts output to its own portable zepp.

The left-hand rack is the 3600 KC crystal, thermo-controlled, doubled to 7200 KC, amplified and doubled with a 50-watter, and then amplified with a 75-watt screen grid tube. This feeds the water-cooled tube directly below at 7200 KC, 14,400 KC, or 28,800 KC. The panel has a 6-pole switch for main line (two power lines came into the radio room: one a 220 volt with the usual three wires, the other a private 3-phase line). The left-hand 6-pole switch has a rope on it, which automatically turns on the water, with the big filament. The 220-volt 3-wire line serves all steady voltages, (filaments, C power pack, amplifier power packs, automatic keying, portable, etc.) The 3-phase serves the final amplifier stage and is keyed with two Leach break-in relays — 3 of the contacts for 3-phase, the 4th contact for the portable (when the portable is home). The keying devices consist of a light key, an electro bug, Omnigraph, and Heterograph (tape transmitter with assorted tapes for calling).

The receiver has 6 tuned stages and so is very selective. The built-in cabinets over the receiver house the monitor, frequency meters, call books, message blanks, and other ham lore.

W6AM "just grew" for the past 20 years (under different calls) and is still growing. It is WAC in two different districts (9th and 6th), has the 1926 Trans-Pacific Test "Class A" certificate, the Section certificate in the 1929 sweepstakes contest, and the 1927 International Relay best contact point for New Zealand. The cup is the 1923 Hoover Radio Cup."

Crystals had been introduced in 1924 and by the late 1920s the advantages of crystal-control transmitters had become apparent. The transmitted signal took less space in the radio spectrum because the transmitter didn't drift around, and stayed on a single frequency. The crystals were rather expensive, however, typically ten dollars each. Consequently, those stations that were crystal-controlled usually had just a few crystals of different frequencies. Sometimes a ham owned only one crystal for a low-frequency band and would devise circuits to operate on harmonics of the fundamental frequency, to generate a signal on a higher band. Don's transmitters had predominantly been crystal-controlled the past five years. His ARRL designation was OWLSc, indicating crystal control. The April 1931 issue of *QST* carried the description and current pictures of Don's rig.

Wayne Cooper remembered a visit to Don's station. He recalled, "Don had a telephone pole with maybe half a dozen sloping doublets for 20 and 40 meters. He demonstrated to me the beam effect of these antennas. His transmitter was in the breakfast nook. The final was a special water-cooled tube with '6AM' formed with wire inside the tube. The water cooling was from a rubber tube to the kitchen faucet." Bertha must have been very understanding, to allow this intrusion into her kitchen.

Los Angeles celebrated it's 150th anniversary by having La Fiesta de Los Angeles that summer. Many of the 800 amateurs in the Los Angeles area sent messages to the mayors and governors across the United States and to dignitaries in foreign lands. More than 5,000 invitations were sent by radio and this was a prelude to the amateur effort for the 1932 Olympics. The resulting story was on the front page of the Sunday July 19, 1931 edition of the *Los Angeles Times*, along with a picture of Don at his station.

The American Radio Relay League, Inc.
Hartford, Conn., U. S. A.
COMMUNICATIONS DEPARTMENT APPOINTMENT

OFFICIAL RELAY STATION APPOINTMENT

This certifies that Amateur Radio Station W6AM, located at LONG BEACH, in the LOS ANGELES Section of the PACIFIC Division is an OFFICIAL RELAY STATION of the AMERICAN RADIO RELAY LEAGUE.

This appointment shall remain in effect for the term of one year from the last date of endorsement by the S.C.M. which appears hereon, unless sooner suspended or revoked at the discretion of the S.C.M. for failure to live up to the established qualifications (set forth in the Rules and Regulations) for Official Relay Station appointees, or automatically cancelled by lapse of membership in the A.R.R.L.

In Witness Whereof, we have hereunto affixed our signatures.

#225# 11/1/30
Number and Date of Appointment — Communications Manager — Section Communications Manager

In view of creditable performance, regular reports, and continuing activity, this appointment is continued in effect by endorsements as follows:

Date	S.C.M.'s Signature	Date	S.C.M.'s Signature
11/1/30		11/1/31	
11/1/32		11/19/40	

I, the owner of the amateur station designated above, do faithfully promise to obey the communication laws and regulations, especially to observe our frequency allocations carefully, to send monthly reports of station activities to the Section Communications Manager in whose jurisdiction this station comes, to handle messages in accordance with the recommended operating procedure, to deliver or mail messages within 48 hours when impractical to relay nearer to destination, to participate in every A.R.R.L. communication activity to the best of my ability, and to try to live up to all those ideals set forth in "The Amateur's Code."

Station Owner's Signature

Don's ORS certificate.

The 12th annual Pacific Division ARRL Convention was held in San Francisco that year, on September 5 and 6. Among the speakers present was Professor F.E. Terman (ex-6AE) of Stanford University. In the years to come Terman would publish a number of books that were very valuable to engineers and amateurs alike. In 1932 he came out with *Radio Engineering* and 11 years later followed with *The Radio Engineer's Handbook*.

Don read Professor Terman's first book with great interest. Here was some of the first theoretical information on diamond (rhombic) antennas. One characteristic that Terman mentioned didn't ring true, however — that an unterminated diamond antenna was bi-directional. Don reflected, "*In my work with a rhombic antenna I found that an unterminated one always favored one direction. It was never equally bi-directional. I had also found that if the antenna was sloped in the desired direction, it radiated better in that direction. One day at the convention Terman and I had lunch together and he was quite interested in my observations. So when I got back I wrote him a long letter and told him the results that I had found experimentally. He had a facility up at Stanford with the antennas set up. I believe that he went up in a plane trailing an antenna and actually flew past the rhombics at some distance away and took field strength measurements. Anyway, I noticed that my corrections appeared in a later edition of his book.*"

In what would have far-reaching implications in Asia and the Pacific for the next decade and a half, Japan embarked on a policy of colonizing the Asian mainland. Due to their historical conflict, Japan believed that it was destined to battle the Soviet Union for domination of northeast Asia. The Japanese archipelago was woefully short of natural resources, while the neighboring territories on the continent were not. Japan invaded

Manchuria in September 1931 and had complete control of the country by the spring of 1932. China was quite concerned by this incursion and news stories around the United States carried much speculation. The January 9, 1932 edition of the *Long Beach Press Telegram* carried the headline:

> Long Beach Operator Talks To Radio Friend In China. Don C. Wallace, W6AM, talked for forty-five minutes this morning with Earle Chang, AC8GO, the Minster of Communication, at Shanghai. He was told that Shanghai is seeing no fighting.

The situation in China would remain fluid for the next six years.

Now that he was back in the representative business, Don was also sending articles in to the various trade publications. The April 1932 issue of *Western Radio and Refrigeration Journal* carried one of these entitled: "Are You Ready For the Big Push with Shortwave this Summer?" The article noted that in the summer months, when broadcast reception was relatively poor, shortwave reception was much better. A picture of Don with his portable rig set up in a hotel room accompanied the article. The caption stated: "For many years Don Wallace, a pioneer in radio in the west, has been an authority on long wave, short wave, and television reception." Television was still in the experimental stages, with the receivers quite primitive. However, TV was frequently mentioned during the thirties.

Another article by Don appeared in *Western Radio and Refrigeration Journal* at this time, "How Short Waves Travel." He described how radio signals usually follow a straight line on a great circle chart, which can be quite different than they would appear on a flat map. After observing a world map many people seemed to think that England was east of the United States. Of course, its true direction of north is apparent on thegreat circle maps. With directional antennas, one has to be careful to be sure that their antenna is pointed in the desired direction.

At this time a station was being assembled in Ventura that would be the other "big noise" from California during the 1930s. Charles "Doc" Stuart, W6GRL, had grown up in Ventura and was licensed before the war as 6CW. He had graduated from the University of Southern California School of Dentistry in 1924. He purchased a lot on Pierpoint Boulevard, in a housing development by the ocean. Only a few homes had been built there when the stock market crash occurred in 1929. Due to his dental business Doc was relatively secure from a financial standpoint and he purchased several lots adjacent to his in the coming years. The 200 acre housing development had just a handful of homes scattered about and no new ones were being put up, so Doc set about erecting a number of 80-foot-high poles at various locations on the property. No who lived there seemed to mind and the owners of the various vacant lots never came around to object. They were probably too busy trying to survive than worrying about an empty lot. By 1932 Doc had 7 rhombic and vee antennas spread across 60 acres, pointed to various directions around the world. These masts were located among the sand dunes on the beach and, indeed, the high tide seriously encroached on two of them. It was one of the best receiving and transmitting locations on the California coast, combined with one of the best amateur radio antenna and station installations in the country at that time.

Doc Stuart was one of the premier DXers in the 1930s. By the end of the decade he would share the top of the pre-World War II ARRL DX Century Club with just one other operator, A. Hopper, W2GT. The drive to attain the top position was perhaps best described in a speech he delivered in 1946. Doc stated: "As a radio amateur, I have always been most interested in working DX. The highly competitive nature of that fascinating phase of amateur radio is the most thrilling and satisfactory field I have explored. It may be compared to fishing — you scan the bands for rare distant countries and I might add, elusive ones. It sometimes requires many hours and days of careful watching and observation to hook the particular rare one. Should you not succeed the first time more effort must be expended, a better antenna for stronger signals in both receiving and

Doc Stuart putting up a rhombic in 1934.

transmitting, increased efficiency in the transmitter for more power output, and all the thousand and one improvements that can be instituted to accomplish a difficult feat. It spurs the soul to call forth the best that is in one."

Doc Stuart noted that he had been most fortunate in having sufficient space for vee antennas and rhombics pointed in various directions around the globe. Indeed, by 1934 his station was nearly on a par with Chung-ju, the CGRA station outside of Shanghai. In multi-antenna stations such as his, switches were provided for selecting the best antenna for the received signal and using this same antenna for transmitting in establishing contacts.

In Doc's early days he initially put up a vee antenna. This antenna enjoyed quite wide use at the time, for those people located in a rural location with sufficient space around them. However, it is really a single-band antenna because the included angle between the legs of the vee must be correct to produce the addition of radiated lobes upon which such an antenna depends for its signal gain. High gains upward of 16 dB can be accomplished with a vee and it requires only three poles for support. The antenna could be made uni-directional by terminating each leg at the far end to ground in its characteristic impedance. The vee, however, requires quite an expanse of real estate and can only give good performance on the one frequency for which it was designed.

Doc found that the most satisfactory antenna for overall use was the rhombic or diamond antenna. It was usually erected over flat terrain and required four poles for support. Due to its characteristic of acting as a radiating portion of a transmission line it could be used over a frequency range of two-to-one with power gains of from 20 to 30 (13 to 15 dB). That is, a relatively moderate size rhombic, 270 feet on a side, will give a power gain of 20 on 7 megacycles (40 meters) and proportionally more on 14 and 28 megacycles (20 and 10 meters), where the wavelengths are shorter. Height was also an important factor in developing maximum gain, but compromises in the rhombic angles could somewhat compensate for this and the maximum lobe in the vertical plane can be steered by altering its shape. Lengthening the antenna would also help compensate for insufficient height.

Although a single-wire flattop rhombic gave good results, it was difficult to properly match the feed line to the antenna and eliminate standing waves, except by stubbing or the use of tuning networks. That, of course, makes the antenna narrow in bandwidth and it could not be used on more than a narrow frequency band without retuning. However, placing one or two wires directly beneath the top wire of the rhombic flattop will

The station of Doc Stuart, W6GRL.

accomplish the desired result of broadbanding the antenna. These additional wires are connected together with the top wire at either end of the diamond and fanned out along its length, so that the wires were from three to five feet below the top wire at the side poles of the diamond and attached with separate insulators. This lowered the feedpoint impedance of the antenna and broadened its bandwidth.

Single-wire rhombic antennas had impedances in the neighborhood of 750 to 1,000 ohms, but by the addition of the curtain wires to the antenna flattop the impedance could be brought down to the region of 600 ohms. This allowed the 600-ohm open-wire transmission line to be connected directly to the antenna without the use of stubs or matching networks and permitted its use over a wide range of frequencies without any serious amount of standing waves, provided the antenna was terminated in its characteristic impedance.

Initially Doc found that the best means of terminating his rhombics was to attach a line made of resistance wire, such as nichrome wire (used in electric heaters) to the far end of the antenna, that is, to the end of the antenna toward which the radiated signal was desired. This line must be the same impedance as the antenna and must have sufficient DC resistance to dissipate at least one half the transmitter power output without excessive heating. A length of the nichrome wire that had 300 ohms resistance was connected to one side of the end of the rhombic, and the wire was brought down the pole until its end and tied off with an insulator. The adjacent side of the end of the rhombic was similarly connected to another nichrome wire, which was also tied off down the pole. The ends of the two pieces of nichrome were then connected together and formed a 600-ohm termination for the antenna. At that time the large resistors that would dissipate upward of a kilowatt of power and exhibit the low inductance necessary for a 600-ohm resistive termination were not available. The termination served to make the antenna uni-directional, bcause the power which would ordinarily be reflected back on the antenna and radiated in the opposite direction was dissipated in the resistance wire that terminated the end of the rhombic.

With very little effort the direction of radiation of the rhombic antenna could be quickly reversed when it was desired to transmit or receive in the opposite direction. This was accomplished by bringing feed lines from either end of the rhombic in to the station and the resistance termination line was extended out from the station, instead of being attached to the antenna at the pole. A suitable large four-pole double-throw ceramic switch was so connected that the receiver or transmitter could be coupled to either end of the antenna at will. At the same time this switch attached the termination line to the opposite end of the antenna, reversing its directivity. W6GRL had seven rhombic and vee antennas connected to a bank of seven four-pole double-throw switches so that the antenna for the desired direction could be selected. Two additional banks of double-pole six-throw switches were used to change between the rhombic antennas and another eight phased wire arrays, along with a rotary curtain antenna, in order to accomplish complete world coverage on 7, 14 and 28 megacycles. The rhombics had such high gain that if a particular station around the world was between the major lobes of the two closest antennas, it was sometimes necessary to switch to a lower-gain antenna that had a broad lobe in the required direction and encompassed the desired station.

The antennas that Doc used were all of the horizontally polarized type. While high gains could be realized from some vertically polarized antennas, they were not used for receiving in ordinary cases because of their high sensitivity to man-made interference such as automobile ignition noise and domestic and industrial appliances, which were vertically polarized. Therefore, Doc's choice of horizontally polarized antennas provided the quietest reception. The noises were not be eliminated but their amplitude was considerably less on the horizontally polarized antennas.

By 1932 there was only one call book being published in the United States. Perhaps it was a result of the government's financial position, but the call book that it had published annually since 1913 was last issued in 1931. The privately owned *Radio Amateur Call Book Magagazine*, which had come out with a quarterly edition for the past twelve years, was now the only one available.

The summer Olympics were held in Los Angeles in July and August 1932. Most of the events took place at

The familar "Flying Horse" Call Book.

the Coliseum, with certain specialized ones, such as sailing, at Long Beach or other sites. An amateur radio station was set up at the Olympic Village, in what is now known as Baldwin Hills. W6USA was the authorized operator for the months of June, July, and August. The station was on the air for a total of 60 days, until the closing ceremonies on August 14. The main station personnel were Bill Lippman, W6SN, C.A. Chaetham, W6CUU, F.C. Martin, W6AAN, Don Shugg, W6ETJ, Charlie Perine, W6CUH, and N.L. Madsen, W6FGQ.

Many of the local amateurs turned out to operate over the period and W6AM also spent a day operating the station. In the first weeks of operation the fellows worked a lot of stations around the United States and did some DXing. After the athletes arrived in the Olympic Village, W6USA spent most of the time passing traffic back to their home countries. A total of 1,450 hours of operation were logged and over 5,000 messages were passed, most of those destined for foreign countries.

The Thirteenth Annual Pacific Division ARRL Convention was held at the Breakers Hotel in Long Beach on September 3 and 4, 1932. Convention activities included visits to the ham shacks of A. Mutter, W6BCK, Harold Nahmens, W6HT, Larry Lynde, W6DEP, and W6AM. The speakers included F.E. Handy, W1BDI, and Colonel Claire Foster, W6HM, as well as a talk by Don Wallace on "Practical Transmitting and Receiving Antennae."

Don continued to experiment with various antenna configurations for a project that he had in mind. He wanted to publish a short-wave manual and evaluating every conceivable type of antenna was part of the necessary groundwork. Don and his family were just about to sit down to dinner on the evening of March 10, 1933, when the city of Long Beach, and most of Southern California, was shaken to its very foundations.

Earthquake

When the earthquake rocked Long Beach at 5:54 PM on March 10, 1933 Don was on the air. He noticed a fast QSB (signal fading) on his received signals a moment before it hit, perhaps from the antenna feed lines vibrating. He gathered his family members and rushed them outside. Bill had been listening to the big General Motors console radio in the living room. Don scooped him up and carried him out front and hurried back inside for Bertha, who was in the wash room holding on for dear life. The washing machine, being on casters, was rolling back and forth. He took her outside and rushed back upstairs for his daughter, Betty Jean. The hardest part was trying to hold on to something solid. Bill Wallace recalled the day vividly: "When dad dropped me outside and went back in for Mother, I grabbed hold of the tree out front. It was over a foot across but I couldn't hang on to it, the shaking was so severe. I fell to the ground, but it was heaving up and down so much that I really took a pounding then. I got back up in a crouch and finally got my balance. The shaking lasted for over a minute."

Don, Jr. was on a bicycle a couple of blocks away from home and was thrown off when the quake hit. He hit his head on the curb, but was otherwise unhurt. Everyone breathed a sigh of relief when he arrived at the house.

Once the shaking stopped Don took a look in the house. Several walls were cracked, making it unsafe to stay inside. He set up a tent from the garage in the field next to the backyard. The family was afraid that one of the 95-foot poles would fall over, so Don paced off 150 feet from the pole and erected the tent at that spot. There, his family lived for the next several weeks. His 200-watt portable transmitter was placed on a card table in the backyard. He immediately contacted a station in Sacramento, the state capital, and notified Governor Rolph of the quake. Over the next four days Don handled traffic, taking cat naps whenever a break in the traffic and message handling would occur.

"I think that I swept up rubbish in my dreams for weeks afterward," Don reflected. *"The excitement was intense. We couldn't imagine how anything could still be standing. Oil refineries in the valley below (San Pedro) blew up, oil fields behind us (Signal Hill) caught on fire, smoke poured in, and a very cold wind hit us standing out there in the pitch black.*

"I used the 200-watt portable set out in the field next to the house. I cut off the 40-meter zepp antenna feed line where it entered the radio room and connected it to my portable set outside. Almost immediately I made contact with stations in Sacramento and San Francisco and passed the news of the earthquake. I stayed up all night and in the morning at 6 AM when there was light, I inspected the transmitter in the house and found that the water-cooled tube had been cracked by the earthquake.

"After a couple of days a Navy first class radio operator came up to me and showed me his furlough. He said, 'Read the reason Don.' Reason: 'Help Don Wallace run his radio set.' So I put him to work. The Navy sent more fellows and then I had several operators. Pretty soon some officers who were hams came around, Lieutenant Commanders, and they wanted to help. I said, 'Well, I tried to deliver some messages, but it took me 20 minutes to get past every corner. The MPs didn't want to let me get through without a speech on what I was doing. You officers can just go on by and they will salute you.' I needed some messages delivered so I had a whole chain of commanders as delivery boys.

"The fleet sent ashore 5,000 MPs, and put one on

Don's portable transmitter set up in the field after the 1933 earthquake. Don's family stayed in the tent in the background.

every corner in Long Beach. It was a wonderful thing for the town because it could have been bedlam otherwise. The message traffic continued for three weeks. No telephones in town were operating at that time. I had an hourly schedule with the governor's office in Sacramento. We would exchange messages, get blankets and food sent down here, and so on. The neighbors would see me out in the field and come over to ask me to send a message to their relatives, cousin Joe back in Minnesota or wherever. I said OK.

"We were afraid to go in the house. You couldn't see across the room from all of the plaster dust. In the kitchen you couldn't walk on the floor, it was full of canned goods. Downtown Long Beach said they had the worst earthquake damage and Compton said that they had the worst earthquake damage and we were right between, exactly. All of the houses in back of me slid off of their foundations and were tipped up. Mine didn't, because I had built the house with a great big foundation with bolts in it — I paid extra for that. As a boy I was out there and saw this adobe soil, we used to slip around on it, and when I built the house I thought, 'Well, the house might slip.' So we put in an enormous foundation, along with a large basement. It sure paid off. So my house was still on its foundation. The chimney came down. I had a great big chimney and it shattered. The shock from the earthquake had been tremendous. Friends asked us to Santa Barbara, so I sent the family there for a few days, as they could not be induced to sleep in the home yet. Five days after the quake the frequent earth slips were still bad enough to knock down plaster from the walls and ceiling. We were all okay except for a few minor bruises on my youngest son, Don, Jr."

Gordon Elliott, W6CIT, remembers those days well. He recalled, "I was a radio man aboard the Battleship *USS New York*. Our homeport was Long Beach. Just prior to the earthquake the *New York* proceeded to the Bremerton Navy yard for routine maintenance. Many of the crew aboard the *New York* had families in the Long Beach area. Since outages of power and telephone facilities existed in Long Beach just after the earthquake, it was difficult to receive any news concerning damage. The Communications Division Officer aboard the *New York* assigned licensed hams to proceed to Lieutenant Commander Lowell Cooper's QTH (location),W7LQ, for further duty. Our instructions were to QSO (contact) any hams in Long Beach for the purpose of handling traffic. We each took four-hour shifts until commercial traffic communications were reestablished at Long Beach.

"One of the first stations I heard was W6AM, and of course I knew his QTH was Long Beach. We used 40 meters during the daytime and at night shifted to 80. In addition to handling traffic for members of the crew who had families in Long Beach, we also handled traffic for the Seattle newspapers and advised them of the general conditions in the Long Beach area as reported by Don Wallace. Don handled a great deal of traffic for us during that period. He was a great CW operator, and in later years when I would hear him working DX on CW, my memory would go back to the aftermath of the Long Beach earthquake and recall the great service he provided. Nearly a half-century later I was President of the Northern California Chapter of the Quarter Century Wireless Association about the same time Don was President of the Southern California Chapter. Like Don, I am also a member of the Society of Wireless Pioneers. Don was a great guy and did so much to foster ham radio and to help make it such a wonderful hobby for so many."

Three days after the earthquake, on Monday, March 13, the second page of one of the San Francisco newspapers carried the headline: "Radio amateur among heroes of earthquake." The story went on to describe Don's passing traffic from his portable set to a local amateur in San Francisco, Richard Barrett. A number of other amateurs in Long Beach were on the air immediately following the earthquake. Many of these also handled on the order of 700 messages daily. Charles Gilchrest, the Radio editor of the *Chicago Daily News* also printed a story that Monday. He noted: "Most of the messages the hams were sending from the West Coast were from people who wanted relatives and friends in other parts of the country to know they were all right. Sometime, some how, these amateurs are going to get the credit due their work."

Official messages from Mayor Harris of Long Beach and Mayor Porter of Los Angeles were relayed by many stations. Dozens of stations in Long Beach and the Los Angeles area came on the air through the night and weekend as rigs were repaired and power restored. KFOX, the main broadcast station in the Long Beach area, was knocked off the air for four hours immediately after the earthquake. Long Beach stations that relayed traffic included W6ANN, W6DEP and W6RO. Dozens of stations received a letter from the Long Beach city manager thanking them for their efforts. Some time later the ARRL would issue public service certificates to these stations. In the community's time of need, amateur radio rose to meet the challenge. Over the years Don would accumulate 17 such public service awards.

At the time of the earthquake the population of Long Beach was almost 160,000. A total of 21,000 homes were badly damaged and over 1,800 had been destroyed. Five years later when the Richter Scale was adopted, the strength of the earthquake was estimated to be 6.3. Bill Wallace felt that it was much stronger. "If you had spent one minute of not being able to stand due to the strong shaking, you would understand. Having lived in Long Beach most of my life and experienced many earthquakes, I personally think that the one in 1933 was in the neighborhood of a 7-1/2-to-8 strength earthquake."

In those days most buildings in the downtown area were not reinforced with steel and many collapsed from the shaking. Additionally, some contractors did not put an adequate amount of cement in the mortar when classrooms were being built and most schools were virtually flattened. Had the earthquake occurred earlier in the day when school was in session the casualties would have been much worse. In the aftermath of the disaster, the City of Long Beach adapted strict building codes for new construction projects.

The chimney at Don's house had shattered when the earthquake hit and was now in pieces spread along the side of his house. Fortunately, no one had been near it when the shaking started. It was rebuilt with different materials than the original one, and re-inforced. Don used the rocks from the shattered chimney to build a barbeque, which came to be known as his "earthquake-barbeque."

Not only Long Beach experienced a disaster that year. According to a story on the front page of the December 3 edition of the *Long Beach Press Telegram*:

> Fishermen Flee Homes at Volcano. Long Beach Short Wave Radio Operator Talks With Honolulu friend. Don Wallace, who has achieved international fame as an amateur radio operator, was in communications with Clifton Evans Jr. (K6HLP) of Honolulu, also a short wave enthusiast, shortly after 8 o'clock last evening . . . Just prior to the lava eruption an earthquake had shaken the island. After passing along the status of Mauna Loa, a 14,000 foot volcanic crater, the operators discussed football scores as well as the earthquake situation.

In the following weeks Hawaiian tourists would charter planes and fly over the volcano to obtain a better view of the fiery pit.

Once his house was repaired and life returned to a more normal pace, Don concentrated on the business of making a living. His livelihood depended almost completely on income from self-employment and he had spent much of the previous year working on a new project, the *Short Wave Manual*. Don wanted to appeal to the newcomer to radio and wrote his manual in such a manner that it could be understood by the novice. Don had always had a good relationship with the ARRL, but

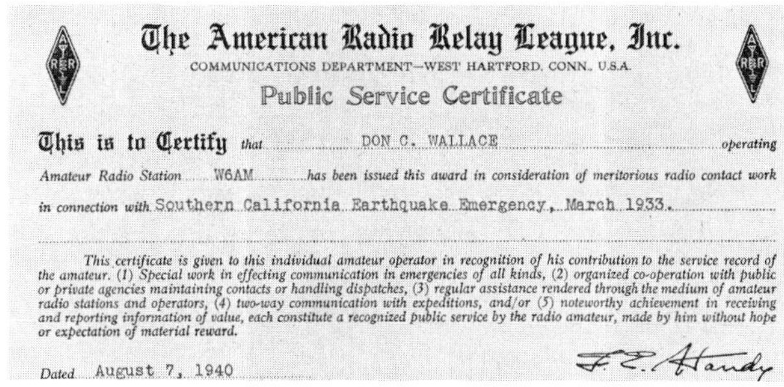
The American Radio Relay League, Inc.
COMMUNICATIONS DEPARTMENT—WEST HARTFORD, CONN., U.S.A.
Public Service Certificate

This is to Certify that DON C. WALLACE operating Amateur Radio Station W6AM has been issued this award in consideration of meritorious radio contact work in connection with Southern California Earthquake Emergency, March 1933.

This certificate is given to this individual amateur operator in recognition of his contribution to the service record of the amateur. (1) Special work in effecting communication in emergencies of all kinds, (2) organized co-operation with public or private agencies maintaining contacts or handling dispatches, (3) regular assistance rendered through the medium of amateur radio stations and operators, (4) two-way communication with expeditions, and/or (5) noteworthy achievement in receiving and reporting information of value, each constitute a recognized public service by the radio amateur, made by him without hope or expectation of material reward.

Dated August 7, 1940 F. E. Handy

The first of many Public Service Certificates that Don received.

their feathers were ruffled over the publication of his *Short Wave Manual* because they considered it to be competitive with the ARRL handbook. Don recalled, "QST *had always been the big ham magazine and* Radio *magazine started up. Dick Howe, the editor, thought he ought to have a handbook and he asked me to write it. I knew him pretty well. I was writing for his magazine all of the time in previous years.*

"The Radio Handbook *was named for* Radio *magazine. I actually wrote three books. I wrote the first* Radio Handbook. *It wasn't called that yet. It was called the* Short Wave Manual, *a beginner's book and an antenna book based on my 25 acres of experimenting in Long Beach. There wasn't any antenna book available at that time. I wrote from about 5 to 6 o'clock every night for about six months, then I would eat dinner. Afterward I would operate my radio set, get up in the morning and work somebody, and then go to my office. Dick Howe, being the crack editor that he was, edited the book. I didn't. I didn't have to put it in form. My wife, being an ex-English teacher, would correct it. I never had to read my material. My wife would always read it, correct it, then I would give it to a neighbor girl. She would type it up for a quarter a page and I never had to read it. If I had to read it I would have been so bored, having just written it.*

"Earlier I had been writing for a dozen magazines, a dozen newspapers, and The Christian Science Monitor. *I wrote about two or three articles a year for* QST *in those days. They welcomed external authors at that time. They don't any more because they have a big staff. Everybody feels like they have to get an article in the magazine every year to earn their pay. So now [1983] the magazine is practically written by the staff. They aren't on the air much. The articles they write aren't as interesting as the ones written by the fellows who are on the air. The enthusiasm is better. You can feel it in the articles. I reread my book the other day and thought, 'My God I was enthusiastic.' Well, that is the secret. The articles in* QST *by some of the other fel-*

Don published the *Shortwave Manual* in 1933.

lows like Frank Clement, W6KPC, they were enthusiastic.

"Frank Jones, W6AJF, took over publishing the handbook after me. I think he wrote it the next two or three years after I did. Then, some time in there, Bill Orr, W6SAI took over. When I wrote the book I made more income on that then I did in my business. When I got the book published I became so well known in the business that it was easy to make a living."

As they were getting close to publishing the manual at the end of 1932, Don received a letter from Dick Howe concerning the advertisers in it. Dick noted: "There are many manufacturers who make small parts that should be advertised in the manual. I am setting the advertising rate at $500 per page, so we can get a decent lot of manufacturers into the manual but the small fellow cannot pay this price. So I have struck upon the idea of giving these people a page at the same rate and turning around and giving them 1,000 copies of the manual with their ad, so they can sell these at wholesale price and get back the cost of their advertisement, or the retail price of one dollar each, and make a profit. They can pay 100 dollars a month for five months. They get 1,000 manuals, a page ad with their name, and more business. The first printing of the book will be for 250,000 (25,000?) copies." In that quantity the cost of publishing the manual, including shipping, would be less than 20 cents per copy.

The 1933 *Short Wave Manual* foreward read:

> Simple, non-technical information for building and operating the Wallace Hoover-Prize-Cup Short Wave Receivers and Transmitters. Instructions for building Noise-Reducing Short Wave Antennas. Log and Audibility Chart of Short Wave Stations. Data on Ultra High Frequency Communication, Vacuum Tubes, Experimental Equipment. Instructions for learning the code. How to become a licensed Radio Amateur. How to pass the U.S. Government Examination. How to build a modern Short Wave Receiving and Transmitting Station.

In the previous three years Don had experimented with over 700 different antennas at his house in Long Beach. In the manual he described the best one to use for all-around general short-wave reception: a half-wave dipole cut for 40 meters with a tuned feed line that was one wavelength long and transposed periodically at 2-foot intervals. This balanced feed line has a high immunity to noise and noise is usually a big problem for most beginners in short wave. The antenna should be strung using glass Pyrex insulators, and as high as possible. If reception was to be optimized for one direction only, sloping the antenna in that direction improved reception.

The modifications that Don continually made to his receiver the past 10 years resulted in a good design for the new enthusiast. Coupling coils were connected between the input of the antenna to the receiver's first stage of amplification. That way the maximum amount of energy at the desired frequency is coupled from the antenna to the receiver. That coupling circuit reduced interference from nearby transmitters operating on adjacent wavelengths. Don also provided a chart of coil winding data and described how to make your own coils or how to locate sources to purchase them. A factory-built Powertone-Wallace receiver kit was available or the reader could build the receiver described from scratch. A simple one tube receiver using headphones could be built and then a second tube could be added for increased volume. An audio amplifier tube could be added to the receiver so a loud speaker could be used in place of headphones. The receiver was designed to operate from batteries, eliminating noise from an AC operated power supply in the receiver output. Additional circuits were described for those who wished to try various hookups. Don included an explanation of how AC and

DC tubes differ and how to bias them properly. For those who wished to operate the radio directly off of the AC mains rather than a battery, a circuit for a battery eliminator was included. A brief description of how to properly tune a short-wave receiver rounded out the receiver section.

The transmitter section of Don's *Short Wave Manual* began with a brief overview of amateur radio and how to learn the radio telegraph code. Don described the typical transmitting antennas in use, including the end-fed zepp and the doublet. A brief explanation followed on crystals and their use in transmitters to control frequency. A simple CW transmitter was then illustrated, along with a description on how to properly build and tune it. Additional transmitter circuits were shown, as well as one with the tubes connected in a push-pull configuration. The advantages of each type of circuit were discussed. For those experimenters who were more advanced, Don went on to describe his 200-watt portable transmitter and how to build it. After evaluating several types of transmitters that would run from 10-watts to 200-watts input, Don launched into a discussion of radio telephone operation. A simple circuit for a low-power modulator to permit voice operation followed.

The last portion of his book contained information on 5-meter operation and he described how to build a receiver and transmitter to operate on that wavelength. Don's manual contained all the information needed for a newcomer to radio to build and operate a station. The manual was 168 pages long, including 37 pages of manufacturers ads for the various components listed. The price of most radio items was at a bare minimum. Manufacturing methods had improved in recent years and the reduced costs were passed along. Equivalents of the UV-202 5-watt transmitting tube that had sold for eight dollars a decade earlier, such as the 10, 45, 47, or 65, were widely available for under $2. Over 30,000 amateurs were licensed and the number was growing daily. Fifty-five years later Rick Ferranti wrote: "I have a copy of Don's Short Wave Manual. Don signed it twice; once in 1933 (before I got it) and once 44 years later, in 1977."

At this time the United States was in the depths of the great depression. Franklin Delano Roosevelt was inaugurated as President that spring. One of the first programs he initiated was his Fireside Chats. Broadcast weekly across America, his intention was to help the population deal with their economic problems. Upward of 25 percent of the workforce in the country did not have jobs. Many of those that did received the wage of 25 cents per hour.

Don and *Radio* Magazine were doing a good business with the *Short Wave Manual.* Sales were brisk and the manual's advertisers had increased sales of their products also. The January 1933 issue of *Radio* magazine reviewed it in their "Radiotorial" section and carried a photo of an aerial view of Don's house, showing his two poles and the adjacent 25 acres of land on which he had evaluated antennas. The same issue also carried an article by Don: "Short-wave Antenna Design and Construction." Don's basic position was that an antenna required proper installation and maintenance. By taking care in the numerous improvements he suggested concerning wire size and type, connection methods, insulating material, antenna location, and the use of a transposed feed line, a substantial improvement in antenna performance could be realized.

That spring the pages of *Radio* were filled with ads for the *Short Wave Manual* and the Wallace Short-Wave Receiver. The receiver was available in kit form for $14.70 and wired for an additional $3.70. In the June issue Don was listed on the masthead of *Radio* magazine as Short-Wave Consultant. Don also sponsored a number of products at this time. The July issue of *Radio* had an ad for "Don C. Wallace Transmitting Transformers; Engineered and produced under the supervision of Don C. Wallace" by the Universal Microphone Company. Fifty-five years later Dan Bathker, K6BLG, furnished the author a copy of the August edition of *Short Wave Craft*, which carried an ad for the *Short Wave Manual*. Bathker noted: "I hope, in the interest of history, the enclosed is of some value to your book." Submissions like Dan's were quite valuable, as copies of those early magazines are very difficult to locate.

Another amateur who wrote the author was Jerry Tannembaum, W9EY, who was first licensed in 1936. He recalled, "Don Wallace was something of a legend. A local ham in Chicago had built the Hoover Trophy receiver that Don had designed as a young man. It was a neat, elegant design, and high performance for its day. About 12 years later I picked up one at a ham fest and cannibalized it for a receiver of my own design. The parts were first rate. I wish I had it now."

Don's receiver became very popular for a number of reasons. One was that it worked very well, better than most other home-built receivers. Another was that a commercial receiver for amateur radio use only had yet to be marketed. Practically all of the receivers that covered the amateur frequencies were also general coverage with broadcast, government and military bands. Seeing a market that had yet to be tapped, that year the National Company came out with the SW-3, a receiver designed specifically for the amateur bands.

Since he had graduated from the University of Minnesota Don always made it a point to take at least one college class each semester. John Roberts, N6FXU, recalled, "I was in my senior year at the University of California at Berkeley in 1933, working for a BSEE with a communications major. Don was in some of my classes and always seemed short of sleep. The San Francisco-Oakland Bay Bridge was under construction and Don had a contract to provide communications between the work stations and mobile equipment such as tugs and barges, so that equipment and material deliveries could be coordinated with requirements.

"Don was invited to speak to our student AIEE group about the radio system and despite the demands on his time for system maintenance and his studies he graciously accepted. The meeting was held on the first floor of a three-story concrete building. Don brought a compact, low-power transceiver and a dipole antenna which was suspended from an overhead light fixture and fed by a short ladder line from the set. It was a spare from the system that was in use on the bay.

"He started his talk by telling us how he had convinced the bridge authority and the contractors of the saving of money and time that could be achieved with reliable communications. Next, he made a trip to Washington D.C. to get the frequencies assigned that he needed, around five meters. He implied that he had some political help in getting the assignment expedited.

"With the frequencies established, Don and some associates drew up plans and completed the necessary sets and spares in 60 days. They were designed around available receiving type vacuum tubes. To maintain high dependability and reduce service calls with reasonable economy, all tubes were labeled with an installation date. If a tube tested good at the 1,000-hour check, it was returned to service; at 2,000 hours it was removed from service and discarded regardless of how it tested. To most of us this was a new concept wherein expected tube life was balanced against reliability and cost of servicing. Once the sets were installed their value in coordination and time-saving became apparent to all.

"Next Don went on to the equipment and turned on the set. Expressing doubt that it would work from that location he picked up the microphone and gave a short call. Almost at once he was in communication with a tugboat captain out in the bay, loud and clear despite our location. This demonstration was most impressive to our group. Don was probably 15 years older than the class average and had demonstrated a knowledge and practical skill that far surpassed that of even the hams among us. Our professors included such dignitaries as Dr. Fuller, Dr. Reukema, and Dr. Abe Tilles. They too were somewhat awed by Don's demonstrated knowledge of radio in all its phases."

Don had left San Francisco and was visiting customers when the first annual ARRL Field Day was held in June 1933. The purpose of a field day was for amateurs to set up portable equipment for simulated emergency conditions. The portable stations would contact as many fixed stations as possible from Saturday morning until mid-day Sunday. Initially, just a few portable stations participated in Field Day; there were only a couple of dozen portable transmitters in the whole country at that time. Since Don was traveling that weekend he operated as W6ZZA and worked a couple dozen stations with his 200-watt portable transmitter. In the years to come this event became more and more popular as additional amateurs built portable radio sets. In another 20 years it would develop the largest annual participation of any radio event and has remained that way ever since.

In July, the ARRL announced the formation of the A-1 Operator Club. Requirements were good general keying, correct procedure, accuracy of copying, judgment and courtesy. Don scanned the accompanying material that the ARRL had provided to the League members and noticed that he had been chosen as a charter member.

The November 1933 issue of *Radio* magazine had a picture of some members of the Squirrel Club; W6AM, Dwight Williams, W6RO, Larry Lynde, W6DEP, and Leroy Trowbridge, W6EWK. Most were wearing jackets with an 852 transmitting tube on the back. There were no particular requirements for membership, but all of the members seemed to live in Long Beach and had 1-kilowatt rigs. Other members not shown in the photo were Foster Strong, W6MK, Joe Huff, W6FYF, George Weller, W6CZZ and A. Drake, W6EZN.

Don's contacts in the publishing industry ran far and wide. The February 1933 issue of *CQ*, a journal devoted entirely to the commercial radio man, carried an announcement: "Don C. Wallace, who won the 1923 Hoover Cup, is publishing a short wave manual . . ."

In 1933 a new company started up by the name of Radio Manufacturing Engineers, or RME. Don became their representative and their sales picked up as he pitched their product to his various customers. They would develop a receiver line for amateur use that was very competitive over the next 30 years. As RME developed new receivers, Don would try one for a while and see how it stacked up to his home built one. When a new receiver became available, the old one was shifted down the operating table. Once two or three receivers were on the table when a new model came out, the one at the end was put downstairs in his basement. Fifty years later the author discovered several mint condition early RME receivers in Don's basement, still on the same shelf where they had been placed back in the 1930s and 1940s.

A number of magazines were interested in Don's radio articles and occasionally they contacted him about writing a particular one for their publication. The April 1933 issue of *Short Wave Craft* had an article by Don entitled "Some Things You Don't Know About S-W Aerials." It was well received by their readers and when *Short Wave Craft* published the *Short Wave Beginners Book* in 1935, Don's article was included in it. Forty years later Don was sent a copy of the book by George Carson, W0ZUY, with the comment "Things sure have changed."

With his assortment of antennas and the water-cooled tube to put lots of power to them, Don's signal into the Orient was tremendous year after year. In November he received a QSL from Yasuo Tsukamura, J3-CW, in Kobe. Yasuo reported "Vy plsed to wk wid you. I have not been on the air so long, but I know that you are famous in J." Later the same month while Don was portable as W6ZZA, he worked Kyozo Asamura, J3CR. Kyozo noted on his QSL, "Your signals were pounding in here about R-8 and very fb (fine business) to copy." Don's 200-watt portable transmitter with an antenna on top of a hotel also put out a good signal.

The World Wide Amateur Radio Convention was held in Chicago on the first weekend of August 1933. The speakers included Hiram Percy Maxim, W1AW, Boyd Phelps, W9BP, John Rheinartz, W1QP, Robert S. Kruse, W1FG/W1DYO, Clinton B. DeSoto, W1CBD/-W1ZZF, Fred Schnell, W9UZ (ex-1MO), Arthur A. Collins, W9CXX, McMurdo Silver, ex-8VM, James Millen, W1HRX, and, of course, Don Wallace. The Old-Timers dinner consisted of those amateurs licensed prior to 1925, just eight years earlier. Don had the convention committee reserve a particular room for him, near the top of the hotel. During the convention it was the highlight of the evening to come by and see the portable transmitter of W6ZZA in action.

One of the rising stars at the convention was Art Collins, who had come on the air as 9CXX in 1925. After Collins graduated from college he started Collins Radio Company in 1931 in Des Moines, Iowa. He decided to manufacture a line of amateur radio transmitters. It was quite a gamble, as every one figured that most amateurs preferred to build their own. In the years to come Collins would become the Cadillac of amateur radio equipment.

In 1934 the Federal Communications Commission (FCC) was formed. It was now the organization that licensed radio amateurs and administered the regulations that guided them. The Federal Radio Commission, formed just seven years earlier, was dissolved. Radio amateurs themselves noticed little difference between the two organizations; most of the staff of the FRC transferred to the FCC.

In his spare time between business engagements and getting on the air, Don was again busy making addi-

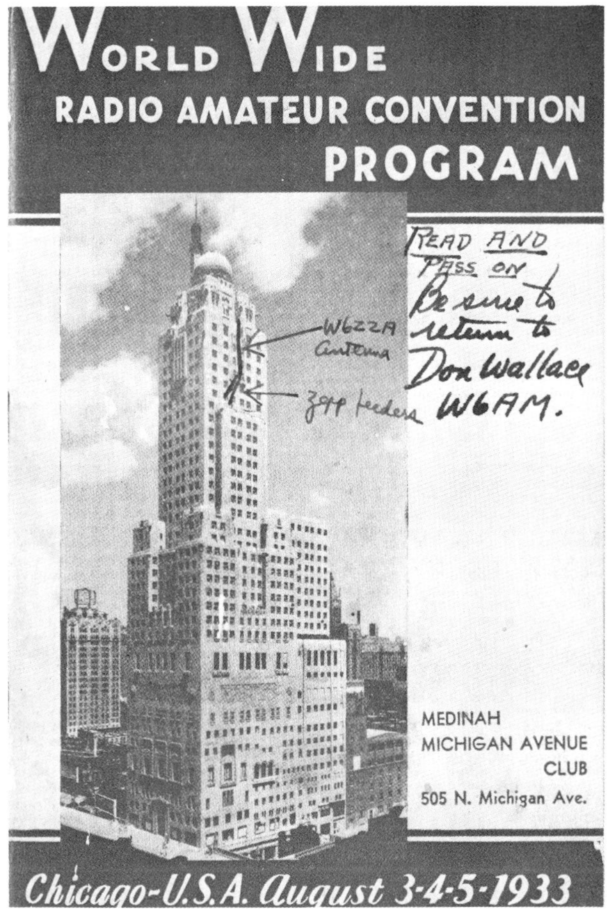

Don always marked in the convention manual where he put up the antenna for W6ZZA.

tional modifications to his station equipment. He had found that the 10-meter band is often open to other parts of the world. The problem was that no one was operating up there when the band sounded quiet. Don would often work a station on 20 meters and ask if they had equipment to go on 10 meters. Band changing took quite a bit of time, what with changing plug-in coils and retuning the transmitter. After much experimentation he found that he could perform many of these operations by carefully installing large ceramic switches in his transmitter. Don also built a separate power output stage for 10 meters only, using an 852 at a kilowatt input. A new amateur magazine had started up that year, *R/9*, and an article describing the modifications, "Rapid 14-28 Mc.Changeover," appeared in their July 1934 issue.

"A Century of Progress" was the theme of the World's Fair exposition, held in Chicago in 1933 and 1934. Their call sign was W9USA and the Chairman, Fred Hinds, W9APY/W9WR, wrote Don and asked for advice on the use of a diamond (rhombic) antenna. Don responded with the details for the necessary feed lines, coupling information, various dimensions, and the proper angle between the legs of the diamond for each of these installations. At the end of his letter Don noted: "It also receives very well and is the most quiet antenna I have used for receiving. However, outside of the line of the beam a half-wave antenna works better. I use the Johnson 'Q' antenna here for directions not actually covered by the different beam antennas I have."

Inspired by the success of Don's *Short Wave Manual*, the technical staff of *Radio* magazine spent much of 1934 developing the *Radio Handbook*. It was a combined effort on the part of Clayton Bane, W6WB, Frank Jones, W6AJF, W.W. Smith, W6BCX, C.C. Anderson, J.N.A. Hawkins, D.B. McGown, and Chas. Perrine, Jr., W6CUH. The book was virtually a complete rewrite from the manual that Don published in 1933. The final result was introduced in 1935. It was 296 pages long and had just 10 pages of advertisements at the end. This edition was very complete and took the reader from basic theory to the most sophisticated amateur radio projects to which the technology had advanced. In short order the *Radio Handbook* became widely read by amateurs and experimenters alike.

The ARRL felt the competition from the revised *Radio Handbook*. The ARRL handbook had been moving forward technically each year, but its total length had remained basically unchanged since its introduction nine years earlier, about 250 pages long with 10 pages of advertising. The ARRL staff spent much of 1935 rewriting its handbook and the next year came out with an entirely new and revised edition. It was almost double in length, 470 pages long with 90 pages of advertising, and over 73,000 copies were printed, up from 48,000 in 1935. The technical articles were beefed up and there was a wider variety of construction articles. The competition from *Radio* had an overall positive effect on the quality of technical information available to amateurs.

Don's articles on antennas proved quite popular and the June 1934 issue of *Radio* carried another called "Directive Antennas." In it Don notes, *"My hobby is antenna tinkering. The reason for this is my good fortune in being situated in a wide-open location where I can erect new antennas 'till the cows come home, without having to take down the old ones. Thus I am able to make direct comparisons under actual operating conditions."* Don goes on to state that 20 kilowatts of effective radiated power (ERP) was achieved with just 1-kilowatt input with his directive antennas, due to their beam effect. He also felt that it was equally important to tune the feed line of the antenna.

Between Don's activity on the bands and the good signal he put out, SWL reports continued to roll in. Morris Meyers of Sydney, Australia, wrote, "U are absolutely the strongest W sig I have ever hrd. QRM (interference) was bad but u pushed everything out of the way hi! At times peaked to R max on speaker—u must be fairly QRO (high power)!" The extra kilowatt or so Don put into his antennas certainly didn't hurt his signal strength.

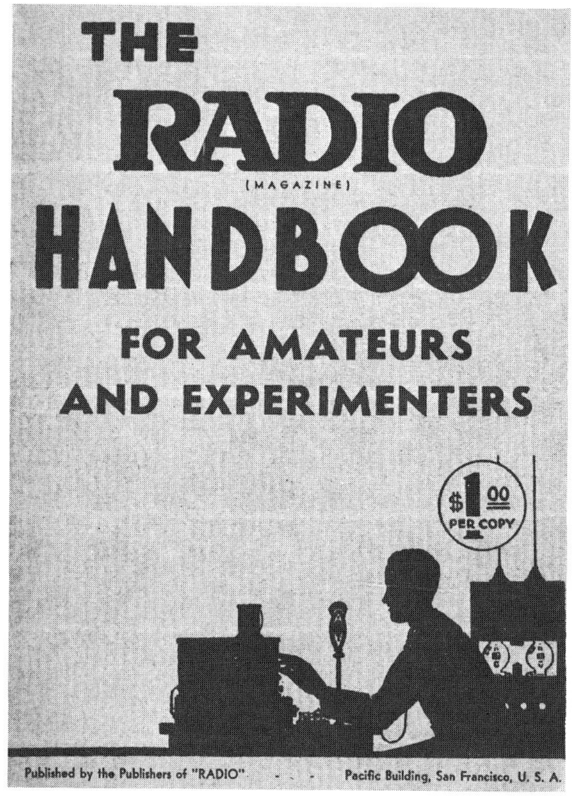

The *Radio Handbook* in 1935.

The ARRL 15th annual Pacific Division Convention was held on November 10 and 11, 1934, at the Fresno Hotel in Fresno, California. The mayor opened the convention and technical talks followed by Frank Jones, W6AJF, Jack McCullough, W6CHE, and Clayton Bane, W6WB. The next day the speakers included Fred Schnell, W9UZ, Charles Perrine, W6CUH, A.L. Budlong, W1JFN, and Don Wallace. Don delivered a talk on "Directive Antennae." After the convention Bert Ayers, W6CL, and Bill Driml, W6NAT, drove up to Eimac, where most of the substantial power tubes would be manufactured. Bill Eitel and Jack McCullough had started the company a couple of years earlier. Bill Eitel, W6UF, gave Ayers and Driml a couple of big bottles, 750Ts, a prototype tube with 750 watts of dissipation that would run several kilowatts of input power. Bert and Bill brought the tubes back south for Herb Becker, W6QD. Herb and Charlie Perrine, W6CUH, were combining their assets to put a big rig on the air. Herb lived in Manhattan Beach, an excellent DX location. They built the high-voltage power supply in the basement, using a large pole transformer. The amplifier was built breadboard style on a large piece of varnished wood. There was lots of open space around the big tubes for heat dissipation. When they lit off the big amplifier the power input was calculated at just over 5 kilowatts. Herb christened it his "corn-fed kilowatt." It was an amplifier that would give W6AM a run for his money.

One of the first commercially made amateur receivers, the National SW-3.

The Sky Hook

In 1934 Don Wallace became determined to improve his transmitted signal strength. More and more radio amateurs were putting 1-kilowatt transmitters on the air and a few stations around could run upward of 5 kilowatts. Don's transmitters were optimized for their present location in his house and it was not feasible to try to boost their power output. He decided the improvement in his signal would have to come from raising his antennas. For the rear 95-foot pole on his lot, he fashioned a 24-foot wooden tower, and by a means yet to be discovered, attached it to the top with the help of Forrest Wright, W6LFC. The final height of the rear pole was now 119 feet. While Forrest and Don were enjoying the view of the countryside from the 95-foot level of the pole, they began speculating on how far they could see from twice that height. Back on terra firma, Don expressed a desire for such a sky hook. The seed, thus planted, and began to grow. Forrest sketched out a design for a 181-foot tower and built much of it while Don was on a trip to Canada. It was constructed in 20-foot sections and then hauled to Don's house and assembled. The tower was made entirely of wood: 2 1/4-inch square pieces 20 feet long for the bottom sections, 2-inch square pieces 20 feet long for the middle sections, and 1 3/4-inch square pieces for the top sections. Cross bats for the center were 1-by-2, and for the top and bottom 60 feet the cross bats were 1/2-by-2. The center 50 feet of the tower was 4 feet square and tapered toward each end. Cement coated nails were used throughout, as they had a tendency to hold tight after being driven in, even if the head rusted off. The tower was constructed of ordinary clear pine, with no knots, and given a coat of grey paint. At 181 feet in length, this was a tower worthy of the effort.

Such a tower was normally guyed in the middle. But it was to stand next to a 95-foot high telephone pole and be tied to it, which would provide adequate support. Don decided to attach antennas for 4 directions at the 166-foot level, giving additional support. Before the tower was erected, the correct length for all of the antennas and feed lines at that level had to be computed. The calculations were verified by building a 2-foot high scale model of the tower. Lengths, angles, and anchor points were all figured before the tower ever started up. As strain insulators tend to be somewhat leaky for high-frequency signals, Don made insulators from oak strips of 1 1/2-by-2 that were 18 inches long. He boiled them in paraffin wax.

The first try at putting up the "big stick" was less than successful. While it was being raised, the tower broke in the middle. The next week was spent removing 10 feet of the middle section, making it 171 feet long, and reinforcing the entire middle portion of the tower. Don reworked his calculations, as the antennas that acted as guy wires would now be at the 156-foot level.

Two sets of block-and-tackle and 3 automobiles were used to raise the finished masterpiece. The 95-foot pole in front of Don's house was used as a gin pole and the 171-foot tower was tied to it when it was erected. On the morning of November 4, 1934 the big stick was raised in less than two hours.

Don went up to the 110-foot level to rig the halyard lines and clear the excess tackle, which took an additional 7 1/2 hours. Lunch was lifted to him up on the pole at the 95-foot point, along with a half gallon of water. When Don came down the tower, he and Forrest built a box around its base and poured concrete 12 inches thick, so the sky hook would have an adequate footing. Once it was finished, as well as the antennas at 156 feet, there were 4 large brass pulleys up on top at 171 feet, with a total of 1,400 feet of half-inch tarred manila rope connected to them. These would be used to raise various antennas, which would effectively be the only guy wires on the top of the tower. Even with the increased competition from the kilowatts coming on around the city, Don was firmly in place as the Final Authority in the area.

The antenna details were reported in the January 1935 issue of *R/9*: "Wooden Masts PFD Hit New High." Thirty-six years later Don was sent a copy of the article with a note from Woody Smith, W6BCX:

> Thought you would get a kick out of the enclosed. I was co-Editor of *R/9* after going to work for them three months earlier. About a year later we bought out the old *Radio* magazine and the *Jones Radio Handbook* [renamed *Radio Handbook*] of which I was editor until World War II when I joined the Marine Corps. I wrote the article from info obtained from you and Forrest Wright.

Don's 171-foot sky hook was thought to be the highest amateur radio tower on the West Coast, perhaps even in the entire United States.

Don's tower and antennas were the favorite subject of a number of the magazines of the day. A two-page spread appeared in the June 16-July 1, 1935 issue of *Radio Doings All Wave Magazine*.

Among his 13 transmitting antennas were a vee beam and a rhombic. At the end of the story Don remarked, "*There are several minor towers on the edge of the countryside, used to support the (other ends of) various antennas described. It is a quarter mile from the extreme of one antenna to the extreme of some of the others. Any antenna not valuable for some specific purpose is taken down to make room for others. The apparatus and antennas undergo a change almost every Saturday, as that day we try to switch things around to get some different effect or different results. The change is tried over the weekend.*" Just two months earlier Don had an article in the same magazine: "Getting the Most from Radio." Apparently he followed his own advice.

On top of the big stick Don placed a Johnson "5Q" antenna for the 5-meter band. Don still had the 5-meter kilowatt that he had built ten years earlier in Minneapolis. That spring John Rheinartz, W1QP, stopped in one day and they got on the air with the 5-meter rig. John was rather surprised to be able to easily work stations over 40 miles distant, as 5 meters was primarily a line-of-sight band. However, since the antenna was 171 feet high, the horizon was probably over 50 miles away. Rheinartz was suitably impressed and that afternoon's activity was published in the "DX news" of the September 1935 issue of *Radio.*

Don recalled, *"I had a 5-meter mobile rig within 30 days after they were allowed, in 1935. I always wanted something pretty big; a 50-watter was enormous for a 5-meter mobile rig in its day. That was the only band that was allowed mobile yet. That set was probably the most efficient 5-meter rig that I ever put in the car. I had an aerial mounted on a screen door hinge on top of the car. It was made out of a steel fish pole. The tuning network was in the ceiling of the car. There were no leads to reduce efficiency. I had a little light mounted on the tuning network and I tuned the capacitors for maximum brilliance. They were shunted across a larger capacitor, to make tuning the transmitter easier. It was fed by the amplifier on the floor and the power pack (a 500-volt generator) was under the hood. The power pack was built by a motor company up there and it was a walloper. I could talk to anybody in the area that had an aerial just 50 feet high all the way home from my business appointments in Orange county; 5-meter signals would boom in, there were lots of people with home-built sets you could talk to.*"

Don made the most of his 5-meter mobile rig. While he was out visiting customers during the day he would chat with the local hams between calls. At the end of the day he had a schedule with his oldest son Bill, now 12 years old. Don would report his location and estimate when he would arrive at home. Bill Wallace could legally operate the equipment of W6AM because Bertha was licensed as W6MA and technically was in control of the station. The *Long Beach Press Telegram* thought that this made an interesting story and published it on the front page of their March 5, 1935 edition. The article was also printed in the newspapers of a number of other cities and three months later it appeared on the front page of a newspaper in Buenos Aires. Don's old friend from Minneapolis, Boyd Phelps, sent him a copy of the Buenos Aires paper with the note: "W6AM everywhere! Even the foreign newspapers. Have not read the Bible lately, but presume it now has a few pages on W6AM! 73 'Beep', W9BP." The article seemed to become more popular daily and it even appeared in the June issue of *Popular Mechanics*. Fifty years later the author was sent a copy of article by Jerry Geiger, N0BDK.

Don with his 5-meter mobile rig in 1935.

Don drove up north one day on the new section of the Pacific Coast Highway, State Highway 1. When he passed a sound truck the driver honked "HI" in code. The driver of the truck turned out to be Bob Carter, W6NTU. Bob had a 5-meter receiver in his car, so Don used his 50-watt rig to transmit over the few car lengths between them, and Bob used his car horn to communicate with Don. The optimum speed for the horn to be most audible over all the road noise turned out to be at 50 miles per hour. They maintained a conversation for the better part of 1-1/2 hours, to the amazement of the residents that they passed along the way. Some time later the story was related in the September 1937 issue of *Amateur Radio News*: "Seventy Mile QSO."

"Then we decided," Don recalled, *"to go up to Mount Whitney for our vacation. So I took a 5-watt transmitter for 5 meters, with extra batteries, so it would be more power than 5 watts, maybe 10 watts. I sent a letter out to everybody in my log that I had worked on 5 meters. Told them that I would be on from Mount Whitney. When I got on the air it was the biggest bedlam you ever heard. I heard them all transmitting but they couldn't hear me. They were all talking to each other about my trip and I was a lot weaker than the local stations down south. I finally got through to one in Lake Arrowhead, near L.A., and we had a good QSO. I never heard any of the fellows in San Francisco the first day. I had worked an awful lot of the fellows in San Francisco previous to that time. I used to sign marine mobile on the ferry as I went over across the bay. I would try to get on last, so I wasn't under the steel superstructure. I was either the first car or the last car, so my antenna was out in the open. I had a mule that carried up these batteries, and the radio set, and blankets for all of us. The whole family went, three kids, my wife, and myself. I had to carry Betty Jean the last half of the way up the mountain, she was just nine years old.*

"The guide I had hired got sick during the trip on the way up the mountain and had to go back. At several places on the mountain the trail made a sharp switchback, and the Forest Service had cut a little wedge about 3 feet wide through the snowbank for pedestrians and mules. At times we had to hoist our equipment on our heads to go through the snowbank and then reload the mule on the other side. Once we had arrived on the top of the mountain they sent another guide up to get the mule, as the mule couldn't stay up on the summit overnight due to the cold and the exposure.

The junior op, Bill Wallace, on the 5-meter rig at W6AM.

"I had written the Smithsonian Institute before we left. I heard that they had a cabin up there on the summit. They said, 'Yes, but vandals had destroyed it.' I was welcome to use what was left of it. The doors had been knocked in and the cabin was full of snow. I set up everything inside, thank goodness for that, because the wind was howling. We had overtaken the Boy Scout troop that was up there. The leader was so sick he could hardly move. I said, 'If you can come over and hold a flashlight, I will put up an aerial.' He was terribly sick, but he held the flashlight while I put up the aerial. It was dark by then. I put up a Yagi beam antenna that I had made on the spot. The elements were vertical wires and strung on a string. I had carried a 20-foot mast up the mountain with me. My predecessors had never worked anybody with their radios on the previous trips up there. They felt that because you were on a mountain it was easy to work anybody. You have to get the antenna off of the ground and up in the air a ways, no matter how high the mountain is.

"Like my present location here in Rolling Hills. I'm on a hill, which helps a little, but the 140-foot poles are what make the antennas work well. So I had this 4-element vertical Yagi strung up with bell wire that I carried along. I worked a station in Los Angeles right away. I was on the air for four days at the end of July. When I left, I had to carry the 80 pounds of equipment back down myself, as the mule was brought back down the mountain by the guide the first day. I sent in the results of my trip in to QST, and they ran a big story. That's what made it such a success. That was the Mount Whitney DX-pedition." The story appeared in the October 1935 issue of *QST*. A week after the Wallaces returned to Long Beach a description of the trip appeared in the August 4, 1935 edition of the *Long Beach Press Telegram*. The story was headlined: "Alpine Climbers Ascend Slopes to Wintery Heights," and concluded with "So far a it is known, they are the first family ever to hike up Mount Whitney to the summit at 14,502 feet. Last year they hiked up San Jacinto and the previous year they climbed Mount Baldy."

Don liked to take his family on vacation each summer, where they could enjoy the outdoors and get some exercise. In August 1936 they spent two weeks in Yellowstone National Park. They had first visited the park 10 years earlier, when traveling from Minneapolis to their new home in Long Beach. Don had his portable radio along and he strung an antenna high in the trees next to the lodge at Old Faithful. Just one day before they left Don talked to Larry Lynde, W6DEP, in Long Beach, and found out that a summer storm had taken down his sky hook. The 171-foot tower had fallen to the east, across the roof of Don's house. Larry and a couple of fellows had patched the roof well enough to keep out the water until Don got back and could have it properly repaired. Within a week he had all 14 transmitting antennas back up, with the highest again at just 95 feet. The remains of the 171-foot tower was put behind the garage to use when a piece of wood was needed. The big stick was now just a fond memory.

The third annual ARRL Field Day was held in June 1935. Don thoroughly enjoyed operating in it because he was one of the few stations that were mobile in motion. Don had gotten the 5-meter rig installed in the car and working properly just before the field day. He found that it was a lot more fun operating mobile than using his portable 200-watt station in the contest, as he had the past two years.

One evening Don was on the air with his portable transmitter and Bertha relayed word that his father had passed away. Don immediately cut short his trip and returned to Long Beach in time for the funeral. William Wallace was eulogized by several prominent city officials. He had served on the Harbor and Police Commissions and had been president of the Long Beach Chamber of Commerce and Realty Board.

Thanks to Don's articles appearing practically every month in one or another of the amateur magazines, and human interest stories on his station periodically published in the local newspapers, which often turned up in other papers across the nation, he was one of the better known radio operators in the country. Even the United States post office, not known to be flexible, recognized his call sign. Dozens of postcards addressed as W6AM, Long Beach, California, were routinely delivered to Don in the 1930s.

Don reflected: *"One of the ARRL staff came out here on the train about that time. He had brought along a receiver with him and tried to use it while he was on the train. He remarked to me, 'Gee, I sure couldn't hear anything during the trip.' I said, 'Didn't you read my article in* QST *four years ago?' He was one of the staff engineers, you know. 'If you had read my article about receiving on a train, you would have heard a lot of stuff. I even worked home from the train while it was in motion.'"*

That year the ARRL advertised for donations of early wireless items for a museum display that they were preparing at their headquarters in West Hartford. They realized that the early days of radio were a bygone era that was slipping away and they wanted to preserve it. Don sent in one of the spherical audions that he had used while attending college at Hamline 19 years earlier. On June 21, 1935 he received a letter from the assistant secretary of the ARRL.

A penny postcard SWL.

> The (spherical) audion has arrived and is now occupying a particularly prominent position on one of the shelves in our exhibit; it is the first and only one of these really old types that we have received. I want to express to you our very deep appreciation of your kindness in donating this for our growing collection. We prize it highly and are deeply indebted to you for your thoughtfulness. 73, A.L. Budlong.

Even with all of the DXing and VHF activity on 5 meters, Don's main amateur radio activity was passing traffic on the lower bands. Over 50 years later Ted Burmeister, W8BSS, an early Quarter Century Wireless Association (QCWA) member (#292), would recall, "I worked Don twice on CW in 1935. I took a message from him for a relative who lived in a small community between Cleveland and Lorain, Ohio. I told him that I was going to a ham fest in Lorain on the following Saturday and would deliver the message in person and he was delighted. I still have the two QSL cards from him."

In April 1935 Don received a letter from Vayra Edwards, 35 Alexandra Terrace, Byculla, Bombay, British India. She wrote;

> I wish to report on the reception of your station W6AM. It came as a big surprise to me, as I never expected my 4-valve (tube) all wave set to reach out as far as California. I have heard you on two occasions, the morning of March 30, and again this morning. On both days the signal strength of your station was R5, with negligible fading. Your speech was very clear, the slow and deliberate style making a distant listener easy to follow. I am enclosing a reply coupon, please send me your QSL card.

The International Reply Coupon, introduced in recent years, could be purchased in the post office of those countries that belonged to the International Trade Union. By enclosing an IRC with your QSL card, the recipient in another country could present the IRC to his local post office and receive the stamps necessary to send a first class letter back to the country of origin.

In January 1936 the Worked All States (WAS) award was announced by the ARRL. The award soon became quite popular and numerous stations qualified for it. Don sent in his cards from 48 states some time later, and on June 13, 1938 he received award number 554. He had been one of the first to confirm all states in 1923, but the award was not yet available.

Since placing very high in the first ARRL DX contest in 1927 (an overall winner was not announced) and taking the high score in the second annual competition in the western states in 1928, Don had not seriously entered the contest since. It was held annually and in the early 30s the DX contest usually started on the second Friday in March, continuing through the week until the next Sunday night. Up to a total of 90 hours of operation were permitted. For several years after the stock market crash Don was occupied with supporting his family. Once he started to recover financially in 1934 he was interested in rejoining the annual competition. As much as he loved to operate, Bertha put her foot down when it came to the annual ARRL DX Marthon. The 90 hours of operation required over a week and a half by the serious competitors was simply too much to tolerate, especially with Don traveling more than a week each month. Don would work a few stations each year during the contest, but knew where to draw the line to keep peace in the family.

Herb Becker, W6QD, had taken high score in California in the ARRL DX Marathon for 1934. By the 1935 DX contest Doc Stuart, W6GRL, had perfected his rhombic and vee beam switching system. Even Herb's "corn-fed kilowatt" was no match for Doc's antennas. That year W6GRL achieved the high score in the western states and finished fourth in the country. This was a remarkable achievement, considering the geographic advantage that East Coast stations had in relation to contacting stations in Europe. The greatest number of overseas operators in a large number of countries were clustered there. Doc used a pair of 250Ts, running at just 1 kilowatt input. Another up and coming DXer did very well in this contest, Charlie Mellen, W1FH. In another 15 years, Mellen and Don Wallace would begin a DX competition in the DXCC awards program that spanned decades.

Don Wallace and Doc Stuart seldom competed directly. For all of the DX that Don worked, he tended to operate on just a few frequencies and let the overseas stations find him. Most of his operation was usually passing traffic. Doc Stuart was among the first of the true blue DXers. Passing a little traffic was all right once in a while, public service and all that. However, the aspect of amateur radio that interested Doc was working DX. Doc spent the majority of his time combing the bands, looking for unusual or rare DX stations. Consequently, W6GRL would accumulate QSLs from some of the most remote corners of the earth. Few other operators from this era can show a QSL from AC4YN in Tibet or VQ8AS on Chagos Island in the Indian Ocean.

Don Wallace continually worked to improve his station and antennas. Pictures of his station at this time show commercially-built receivers starting to appear on the operating position. His home-built receiver was still being periodically modified and would perform as well as the commercial ones for a few more years. The Federal 328A water-cooled tube was repaired after the earthquake three years earlier and Don's transmitters were still crystal controlled. His main transmitter was plate modulated by a pair of UV-204As during phone operation. There were a total of 7 transmitters and 14 transmitting antennas at W6AM. Don wrote a description of his current station and sent it in to *All Wave Radio*, where it appeared in the March 1936 issue: "W6AM, Like 6AM in the Morning — The Station of Don C. Wallace — A Ham of World Fame — at Long Beach, California."

In June, N.E. Thompson, KA1XA, who had recently returned from Manila in the Philippine Islands stopped in to see Don. They had chatted from time to time over the previous 18 years and he wanted to meet Don and see his station firsthand. While he was there Don fired up the rig and worked another station in Manila, who called Thompson's family and proceeded to patch them in to his equipment from the phone lines. A picture was taken of KA1XA and Don, which appeared on the front page of the June 11, 1936 issue of the *Long Beach Press Telegram*, with an accompanying story.

Don sent a letter to the ARRL in January and offered them the UV-206 transmitting tube from his Hoover Cup station. He noted, *"It used to make a lot of noise in the old days, as it is rated 1-KW."* Byron Goodman replied the next week:

> Thanks for your kind offer of a UV-206 for our museum. I regret that space limitations makes it impossible for us to display tubes that did not enjoy widespread usage in amateur radio, and we will have to decline your kind offer. Incidentally, the "round" audion (single wing) you so kindly donated occupies a distinctive place among our tubes and furnishes us with one of the earliest examples of tube manufacture.

Fifty years later the UV-206 was discovered by the author in a box in Don's basement marked "museum piece". The assistant secretary of the ARRL returned from a trip out west a short time later and dropped a letter to Don in September.

He noted:

> I enjoyed the trip thoroughly. Every place I went there was some tall fellow that kept talking about directive antennas! Ross (Hull) and I have a diamond (rhombic) antenna working into Australia now. Many thanks for the help you gave me on the trip. Byron Goodman.

In December Don received another letter from Byron:

> Many thanks for the Audiotron (tubular audion) you recently sent us for use in our museum. Yes, I think we could use one of those Recto-bulbs. As I recall, they were rather widely used.

The Hotel Leamington in Oakland was the site of the seventeenth ARRL Pacific Division Convention. One of the guest speakers was A.H. Holloran, of Farnsworth Television Laboratories. Most of the work in this field was still experimental, as practical, mass-produced television receivers were still over a decade away. Don gave 2 talks at the convention; "Curtain Beam Antennas" and "A Fifty-Watt Five-Meter Mobile Rig."

Don was to provide quite a bit of public service in the day-to-day operation of his station. He worked Max Hilliard, W9WEE, on 20-meter phone. Hilliard wanted to know if Don could arrange a phone patch to a relative in Long Beach who had been recently injured and was still hospitalized. Don arranged the telephone call and an account of it appeared in the September 20, 1936 edition of the *Long Beach Press Telegram*.

The first annual convention of the recently formed Southwestern Division Convention was held at the Hotel Biltmore in Los Angeles October 3 and 4, 1936. The program was contained in pages 3 to 10 of a ham magazine that had started up two years earlier called *73*. The editor was C.H. Haas, W6EAH. The magazine would not survive the decade, but it provided an inexpensive way for the convention to have its program printed.

Gene Kille, W6CRI, wrote the "Here's To DX" column in *73* magazine. Gene noted in the December 1936 issue that W6AM's antenna system at this time included five curtain beam antennas and that Don had maintained weekly phone skeds (schedules) with Manila and Chile for the past five months without a miss. Also, during a four-month period, Europe was raised every day on phone, and his schedule with Hugo Romander, W2NB on Tuesdays had not been missed in the past two years.

Don kept a weekly sked with an acquaintance in Chile, John Pyster. One day, while he was chatting with John, Mrs. Allen DuRee, who lived just a short distance away from Don, chanced to tune them in on her shortwave set. After listening a few minutes she called her husband, the chief of the Long Beach Fire Department. It seemed that Allen DuRee and John Pyster had been friends for over 25 years, when John was still chief engineer of the Southern California Power Plant in 1911. DuRee called Don and was pleasantly surprised to find that he could be patched in right over the phone. The two old friends hadn't talked for over 10 years. The event was reported in the February 11, 1937 issue of the *Long Beach Sun*, and the March 1937 issue of *73*.

A National Fire Prevention Week was held in 1937. Don acted as control station and the Los Angeles Fire Commissioner came to his house. Fifty years later Bill Savage, VE6EO, would recall "At that time Don was acting as a control station for a get-together of as many fire chiefs as possible during Fire Prevention Week. I had my ham station in the fire alarm switchboard room. I had our fire chief, Robert Lindsey, at my station and there were chiefs at Chicago, and several other cities, along with Don. They all had a talk with each other and it was a real good net. Later I was transferred to the Electric Department and installed the first 2-way radio for the Fire, Police, and Public Works. I have a large photo of Don at his rig with Red Wyatt, W6MYO, B.M. Blake, deputy chief of Los Angeles Fire Department, James Cairnes, fire commissioner, and O.J. Emery, captain of public relations." By approaching Don Wallace they certainly came to the right place for publicity, as the photo appeared in newspapers across the nation.

The Secretary of the Navy sent an annual message of greeting to radio amateurs in the United States each year on Navy Day. The ARRL participated in the Navy Day celebration by holding a competition among amateurs in copying the message. The twelfth annual greeting was broadcast at 25 WPM on October 27, 1936, which Don copied and forwarded to the ARRL. Afterward he received a letter from the Navy congratulating him. Don always participated in these annual events that the ARRL and the Armed Forces sponsored. Over the years he accumulated several dozen letters.

Meanwhile, the competition between *Radio* and *QST* became more serious. The editors of *Radio* published the *Radio Antenna Handbook* in 1936. The 80-page handbook was the first amateur radio publication devoted exclusively to antennas. It started with antenna fundamentals and contained design guidelines and construction information for the popular types of antennas. Then the ARRL got busy on their own version of an antenna book. They were quite meticulous in verifying all information in antenna design and they actually built and evaluated each type of antenna they described. Three years later the ARRL *Antenna Book* was published. It was 144 pages long and contained similar

Commercial receivers are starting tc appear at the W6AM operating position by 1936.

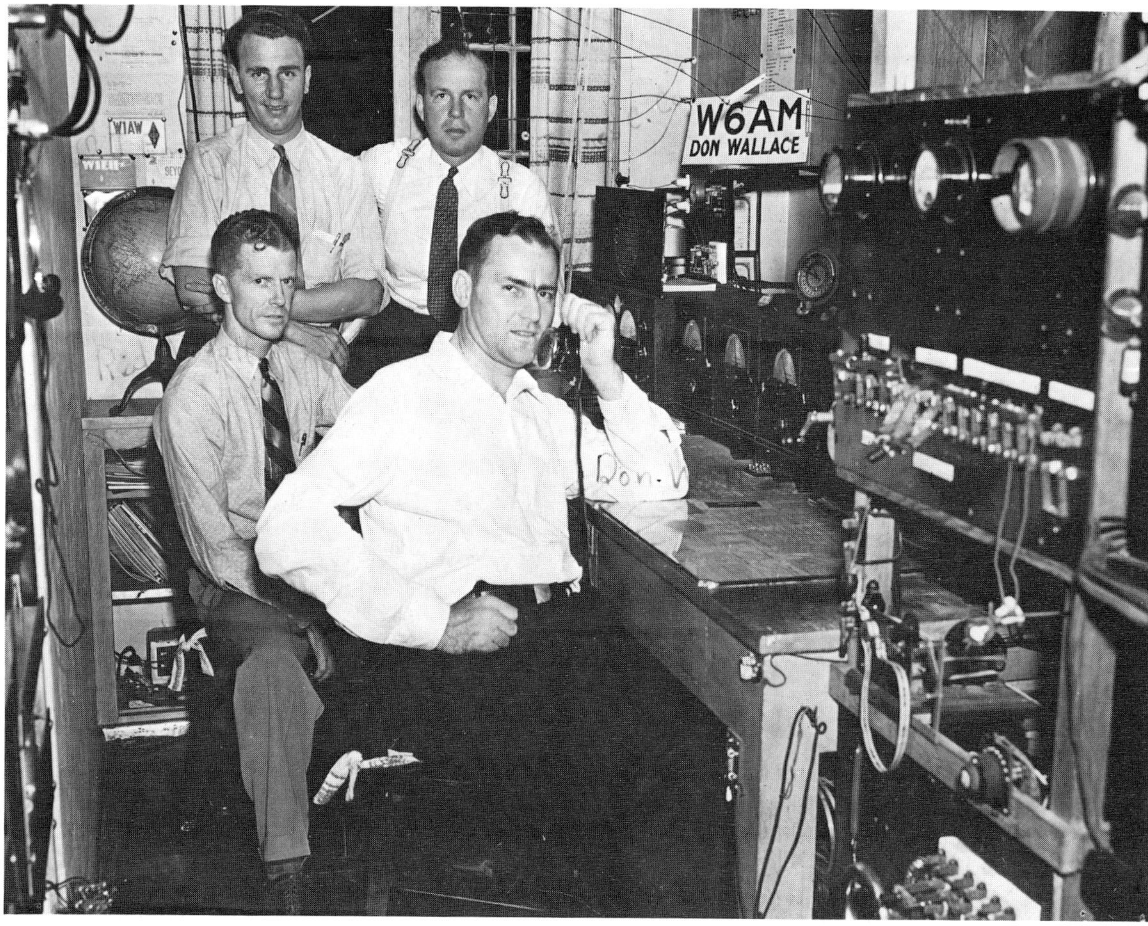

Visitors during Fire Prevention Week. Standing Left to right: O.J. Emory, Captain of the LA Public Relations, James C. Cairns, LA Fire Commissioner. Seated rear, Fred Wyatt, W6MYO, and Don Wallace, W6AM.

design information, with some varying antenna types and ideas for construction. The amateur fraternity was to generally benefit from the ongoing competition between *Radio* and *QST*.

All Wave Radio had a section called the "Ham Bands," which was written by George Hart, W8GCR. George reported that the new bi-directional beam at W6AM was measured at just 5 degrees wide. He noted that "Wallace is getting in here when all other 6s have faded. Don claims 36-KW of effective radiated power."

What with the high-voltage present on the final amplifier of his station, Don developed a safe way of making adjustments when the equipment was "hot." One safe approach was to have only one hand in the equipment making the adjustments, so the chance of a serious shock was reduced. Another idea of Don's was to have the other hand on the power switch, so if you got shocked the voltage would immediately be shut off. Don installed a spring-loaded bypass switch around his main power switch for just this situation. He held a yardstick that pressed against the spring loaded switch while reaching into his high power amplifier. That way, if he brushed the high voltage and received a shock, he would jump, the yardstick would go flying, and the high voltage would be shut off. No doubt there are many amateurs whose families wished they had taken adequate precautions, as the high voltage for final amplifiers was often lethal. A brief description of Don's precautions appeared in the September 1937 issue of *Amateur Radio News*. Sadly, just one year later a friend of Don's who was also the associate editor of *QST*, Ross Hull, was electrocuted by a high voltage power supply while trying to receive experimental television transmissions.

Changes were taking place on the amateur bands in 1937. The old call for Burma, VU, was dropped and the prefix of XZ implemented. The British had military men stationed there and they often put a good signal into the West Coast. As more stations came on the air around the world it became easier to work all continents.

The July 1937 issue of *Radio News* noted that W6AM

had worked WAC each of 5 consecutive nights. Although Don had worked WAC on phone shortly after coming to California he had never requested another WAC certificate, as the WAC program did not have separate phone and CW awards. Byron Goodman asked Don to send in the cards to the ARRL and they issued Don a special WAC award for phone only. After that, the ARRL started issuing both awards. The Worked All Zones (WAZ) award was announced in the November 1934 issue of *R/9.* The prior measure of a DXer's prowess had been the WAC award. As more and more DXers achieved WAC, a new system of gauging individual achievement was needed. The new WAZ soon became popular. The world was divided into 40 zones and the goal was to contact a radio amateur in each one. (See Appendix 6.) As the equipment for many amateur stations overseas was primitive, this was certainly a challenge for even the best stations.

The editor of *R/9* magazine, Killian Lansingh, W6QX, who had started the magazine in 1931, purchased *Radio* magazine in 1935 and merged it with *R/9*, starting with the January 1936 issue. The new journal was called *Radio*, and at 6 x 9 inches had considerably smaller dimensions than the old 9 x 12 inch *Radio* magazine. Incredibly, by late that year, 6 stations around the world had achieved the 39-zone mark. In March 1937, Jacques Mahieu, ON4AU, in Brussels, was the first station announced as achieving WAZ.

The rig of E.Y. Nepean while he was operating in Tibet as AC4YN.

In the middle 1930s there was amateur activity from 39 zones around the world. The one that had no activity and would become known as the rarest of the rare was zone 23. It consisted of Tibet, Mongolia, a region of

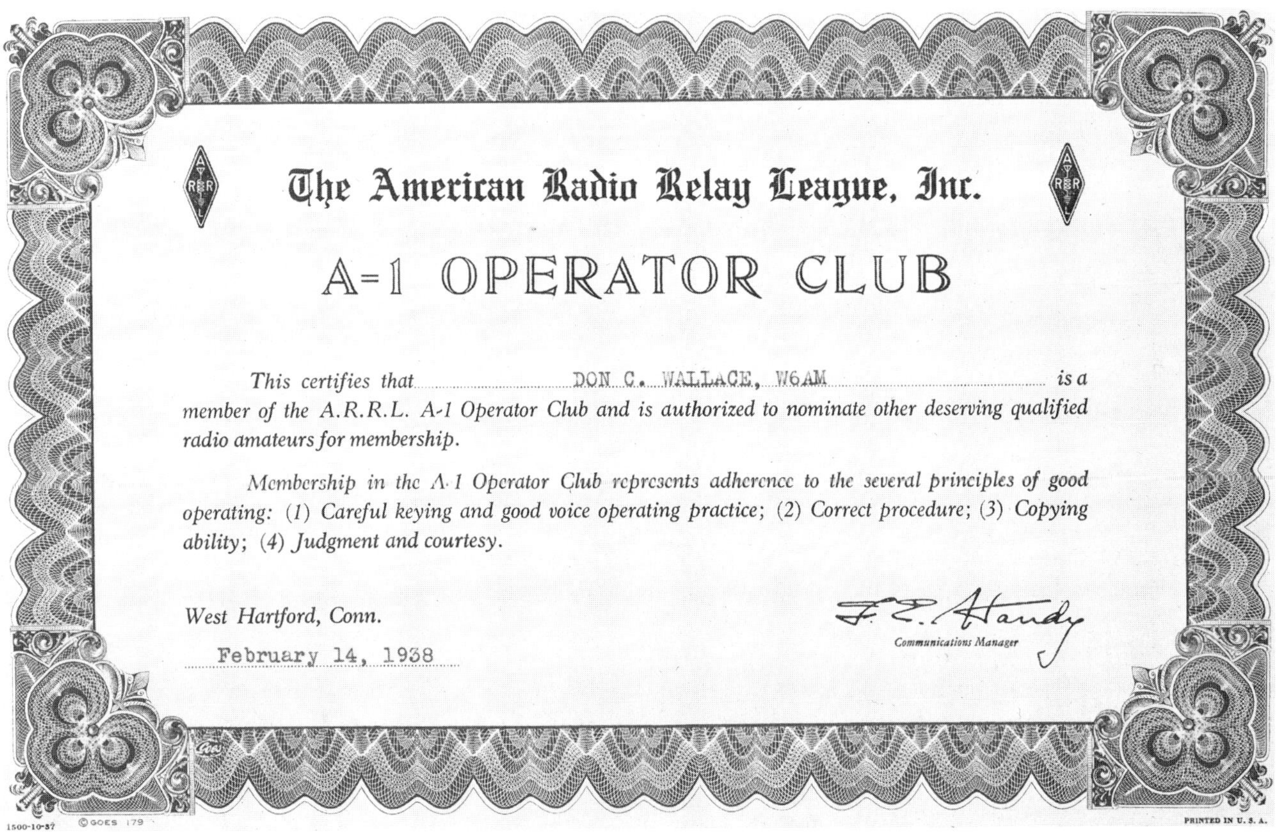

The American Radio Relay League, Inc.

A-1 OPERATOR CLUB

This certifies that DON C. WALLACE, W6AM *is a member of the A.R.R.L. A-1 Operator Club and is authorized to nominate other deserving qualified radio amateurs for membership.*

Membership in the A-1 Operator Club represents adherence to the several principles of good operating: (1) Careful keying and good voice operating practice; (2) Correct procedure; (3) Copying ability; (4) Judgment and courtesy.

West Hartford, Conn.

February 14, 1938

F. E. Handy
Communications Manager

1500-10-37 ©GOES 179 PRINTED IN U. S. A.

Don's A-1 operator club certificate.

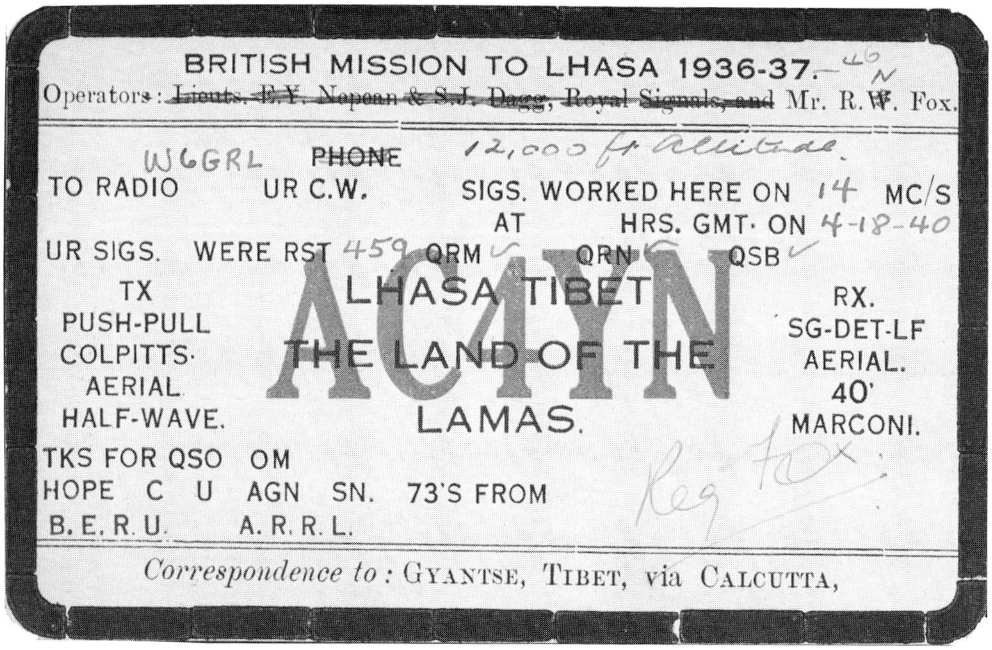

The QSL of AC4YN.

northwest China, and a small portion of the U.S.S.R. The only amateur radio operation in that area to date had been by Harold Graham, as AC4AA, in the summer of 1928. Jacques had worked Graham during the week he was in Koko-Nor, Tibet, eight years earlier and that had provided his margin of success for this award.

After Graham departed, the next amateur radio operation from Tibet would not take place until 1936. Lieutenant Evan Y. Nepean, G5YN, was stationed in the Peshawar District Signals, in what is now known as Pakistan. At that time the territory was part of British India. Located just a few miles from Afghanistan, Nepean was assigned the call sign VU2YN. Tibet invited the British to install a political mission in their capital city of Lhasa in 1936 and Nepean was selected signal officer. He was assigned a military receiver and a 100-watt transmitter to take with him. He later found his own home-built receiver to be superior to the one issued him by the military.

Nepean spent one month traveling from Gankok, Sikkim, where the train tracks ended, to Lhasa, Tibet. His primary mode of travel was on a pony, although one could walk if desired. Evan commenced operation in mid-1936 with a 50-foot-long dipole fed by 40 feet of open-wire line. His rig was set up in his permanent quarters, a tent in the garden. As there was no licensing authority in Tibet, Evan issued himself the call AC4YN and he operated for the next 18 months. Because of official duties, his operating time was quite limited, and no contacts with North America were made. In 1937 a frontier war started with Afghanistan and he was ordered back to his old unit in Peshawar. Upon his return Evan's

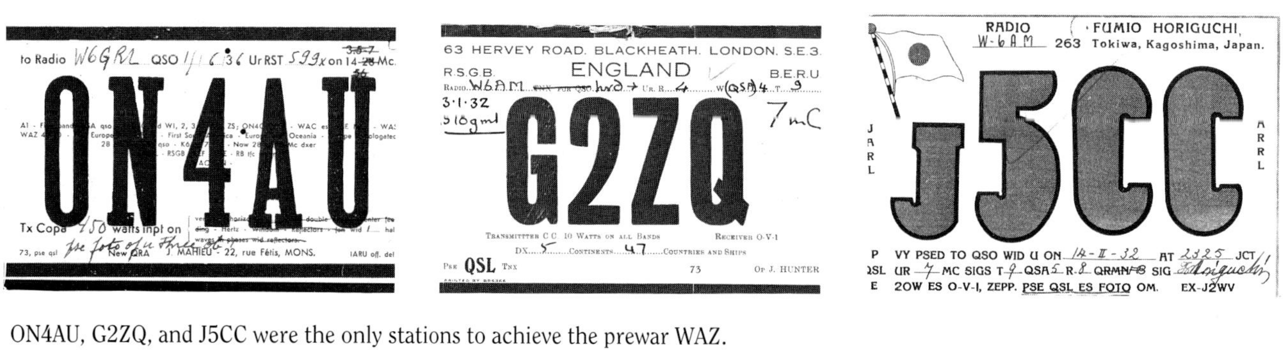

ON4AU, G2ZQ, and J5CC were the only stations to achieve the prewar WAZ.

amateur radio operation ceased. The world was in turmoil and it would not be until after World War II that he would be on the air again. Evan later inherited a Baronetcy, which was created on July 16, 1802, and he became the sixth Baronet.

Nepean's replacement arrived in Lhasa in 1937, a young ex-signal officer by the name of Reg Fox. Reg had been with the British mission in British India the past five years, with the call sign VU2DR. He continued operation of Evan's station using the same call sign, AC4YN. Evan had to take his military receiver and transmitter back to Peshawar with him, but left his personal receiver behind. Reg built a transmitter which functioned rather well, but ran just 7 watts input. Whenever official duties permitted, Reg fired up on the amateur bands. That year he worked John Hunter, G2ZQ, in England and three years later Fumio Horiguchi, J5CC, in Japan, both the only other amateurs to achieve a prewar WAZ. One day in 1939 the band opened to the United States and Reg worked 3 stations, including Doc Stuart, W6GRL.

Doc Stuart would confirm just 39 zones before amateur radio closed down for World War II. The QSL from his last zone, AC4YN, did not arrive until after the war. The only other amateur radio station active from Tibet before the war was J. Schultz, AC4JS, located in Choni. Schultz came on the air in late 1939 with a Jones exciter that used a 6L6 and ran 75 watts to a single-wire Hertz antenna. Jules Wenglare, W8OSL, now W6YO, was one of the few stations AC4JS worked prior to the war and Schultz was not heard from afterward.

The Institute of Radio Engineers held their Pacific Convention at Spokane, Washington in July 1937. Don attended and while he was there had dinner with H.H. Beverage, the national president of the IRE. Beverage had accompanied Don on the President's ship for the trip to Versailles just 18 years earlier. Beverage presented a paper on "Propagation at Ultra High Frequencies," which Don found quite interesting due to his extensive 5-meter operation.

While Don was on a visit to Hartford in late 1936, K.B. Warner asked him to write another article for *QST*. At that time Don had the amateur's dream—beam antennas for virtually every part of the world where amateurs are located in quantity. F.E. Handy visited his station shortly thereafter and was quite impressed with it. Don felt that an article on antennas would encourage more building, in that many amateurs didn't realize the number of antennas that can be put up in a small area. Don prepared a three-page article "Making the Most of Directive Antennas," and it appeared in the November 1937 issue of QST.

One evening in December 1937, Don received a phone call from a short-wave listener, R.C. Anderson, a retired jeweler who lived nearby. Anderson told Don he was clearly hearing a station on an Arctic expedition. Don fired up his rig. Captain C.J. McGregor, located at Reindeer Point, Greenland, was coming in loud and clear. Don had a 20-minute contact with the party, just 720 miles from the North Pole. Don also put Anderson in contact with them for a few minutes by means of a phone patch. Members of the party had messages to pass to friends in Long Beach. Don tried to call both Jack Dowling, a USN chief petty officer, and Lieutenant Commander C.D. Watson, retired USN, but neither was home. So Don wrote down the messages and delivered them the next day. The base where the expedition was located was one of those used by MacMillan when he had taken Admiral Byrd north for certain airplane tests. The story appeared on the front page of the *Long Beach Sun* on January 1, 1938, and was reprinted by the *Amateur Radio News* in their January edition.

John Koch, W6MND, remembered some of the camaraderie of the early days in Long Beach. "After attending an ARALB club meeting one evening in 1938, a bunch of the Gas House Gang went to a local pub for some refreshments. When the bar closed we were feeling pretty good and someone suggested that we go over and wake up W6AM. The group consisted of George Sinclair, W6GAL, Carl Louis, W6NRQ, Fred Warner, W6MHH, Lefty Harmon, W6GHU, Steve Stambuk, W6MEN, and myself. We knocked on his door at 2:30 AM and in a few moments he appeared in a robe. When he saw who it was he had a big grin and invited us in. We went to the radio room and he fired up the big rig. When he pressed the key the big tube glowed cherry red. We asked him how much power he was running and with a grin he said, 'Roughly a kilowatt.' He gave us a thorough tour of the shack and we of course, felt somewhat sheepish after his gracious hospitality. As we were leaving he bid us all goodbye with a QSL card and an invitation to return."

Steve Stambuck, W6MEN, added a few details. "The radio room seemed to be in an alcove under the stairs on the ground floor. When Don pressed the key the final tube turned very red. He then reached up with a yardstick and tapped the crystal in the oscillator which was mounted on the ceiling. This apparently started the oscillator, and the rig settled down. He then pulled an arc for us. This was the custom when showing off a final amplifier of any power and consisted of using a carpenter's pencil to touch the plate cap and pull an arc from the unloaded final amplifiers plate circuit. The length of the arc indicated its potency. As Don graciously showed us out after the visit, I remember George, W6GAL, whispering to me, 'I feel about 6 inches tall.'"

The "Gas House Gang" was the name that a group of young hams adopted in the late 20s. It was the nickname of some fellows who also belonged to the Southeast Radio Experimenters Association, which met every Tuesday evening at the Ebell Club, in the city of Bell. Afterward these fellows usually stopped in at a local pub, often staying until it closed. Consequently, they got into a bit of mischief from time to time.

Ward Atherton, W6MRU, now W4RVE, was a teenager living in Orange County in the late 1930s, and he remembered a visit to W6AM. "One Sunday afternoon several teenaged friends and myself drove in to Long Beach to see W6AM's setup. We arrived unannounced about 2 PM. Don and his wife, along with another couple were just leaving as we arrived. He invited us in and spent 30 minutes showing us the station and answering questions. Don was quite gracious and his wife appeared very tolerant about having their plans disrupted."

Don enjoyed amateurs dropping by to visit and early in their marriage Bertha had learned to accept the inevitable. A couple of local teenagers who were beach bums in the late 1930s were Carl Louis, now W6SR, and Keith Jones, now W6OB. Keith recalled: "Carl and I lived in Compton and hitchhiked to the beach almost every day in the summer. After a swim we would often stop at Don's on the way home. We were always amazed at the size of the power transformers that he had. Don always fed us cookies and milk and was a hell of a good guy."

The weather in Southern California is among the most moderate in the world. The southwest corner of each continent enjoys this region of temperate climate. The summertime high temperatures are not as high as the rest of the continent and the wintertime low temperatures are not as cold. The average annual rainfall in the Los Angeles Basin is just under 12 inches. Palm Springs, 90 miles to the east, receives less than 6 inches per year. The humidity is mid-range and typically the coastal weather is quite moderate. But in 1938 a tremendous storm struck Southern California with a fury the likes of which had never been seen, before or since.

The Philippine Island Amateur Radio Association. Back row (left to right)-KA1HR, KA1AN, KA1DL, KA1CS. Center row-KA1AR, J.H. McDonald, KA4GR, KA1OR, KA1EL, KA1GZ. Front row- KA1NA, KA1JR, KA1LG, KA1TS, KA1UP.

Prewar Philippine Islands QSLs.

Prewar Japanese QSLs.

Prewar Japanese QSLs.

Prewar South African QSLs.

Prewar South African QSLs.

Prewar DX QSLs from most countries.

Prewar DX QSLs from most countries.

Prewar DX QSLs from most countries.

Press Wireless

A tremendous storm swept in from the Pacific on March 1, 1938 and hit the coast. Heavy rain pounded Long Beach and the mountains surrounding the Los Angeles Basin. Over 11 inches of rain fell in five days, the amount the area normally receives over a 12-month period. Hour by hour the rivers rose. Bridges across the Los Angeles River in Long Beach and nearby cities began collapsing.

The Long Beach emergency committee of the ARRL met at the request of Lieutenant Peterson of the Long Beach Police Department. Hams with 5-meter mobile rigs in their cars parked near bridges where they could observe them safely and reported back to the base stations of W6AM and Al Edgerton, W6EDF. Don and Al then relayed the information to City Hall. Larry Lynde, W6DEP, was stationed at the Long Beach Edison plant, which had to supply power to portions of Los Angeles when the power lines from Boulder Dam went out in the early evening. The local National Guard unit had been unable to get through to Sacramento until W6AM contacted KADJ, the Sacramento office of the California Highway Patrol. Contact with the state capital was then maintained through the use of patch panels to Don's transmitter. For several days Long Beach would have been isolated from the outside world were it not for amateur radio. Hundreds of amateurs up and down the coast passed traffic in and out of the area and kept concerned relatives posted about late-breaking events.

Since all the telephone lines were down, Fred Reineking, Los Angeles manager of the *International News Association*, brought his facsimile equipment over to Don's late March 3. Contact was established early the next morning between W6AM and Victor Reubhausen, W9QDA, and the necessary receiving equipment was brought to Reubhausen's house from the *INA* office in Chicago. From this station the data was relayed over the telephone lines to the Chicago office of the *INA*. In this manner the first flood pictures showing destroyed bridges were transmitted by W6AM and appeared in newspapers across the nation later that day. A picture of Don operating his station also appeared, in front of the ever-present RME receivers.

The Long Beach city manager sent Don a letter the next day.

> Success of work carried on by all coordinated agencies in Long Beach through the Disaster Preparednes Council under trying conditions which were experienced at the height of the flood last night was due in no small measure to the prompt response and assistance given by the communications divisions. I particularly want to express the appreciation of the City Administration for the assistance which you rendered in radio communications during the hours of the night when you were asked to remain on duty.
>
> R.M. Dorton

The details of the amateurs' efforts were reported on the front page in the March 4, 1938 edition of the *Long Beach Sun* and in the March issue of The *Radio Amateur News*.

Interestingly enough, an SWL in Liverpool, England, heard all of Don's conversations with W9QDA in Chicago. Walter Morris sent Don a detailed SWL report, with the note that he was using a straight detector (1-tube receiver) and a 15 foot inverted-L antenna.

Steve Haberfeld, who lived in Bakersfield, remembers the storm well. "My dad had taken us fishing for the weekend. He had a 65-foot power boat, all mahogany,

Don Wallace and Fred Reineking of the INA preparing to send facsimile pictures of the 1938 flood.

and we had spent the afternoon fishing off of the Channel Islands. We had eaten dinner that night and everyone had gone to sleep. About midnight the skipper came around and awakened all of us. 'We gotta git,' he said. Jan Christensen was an old Norwegian who had skippered Dad's boat for many years. We were all very tired and didn't want to get up; I think that the adults may have had a few beers. Jan repeated 'We gotta git,' and we took the boat around to the north side of the island. The storm hit shortly after that. Jan had been-raised in Norway and had an instinct about the sea in him. Apparently it became so calm around midnight that he woke up, knowing that a big storm was about to hit. Weather fronts usually come in from the north Pacific, so everyone always anchored off the southeast side of the islands. This storm came in from the south, completely without warning. We spent the next two days in blue (deep) water, riding it out. Every couple of hours I would crawl down to the galley to get something for the men at the binnacle to eat; it took two of them to hold the wheel. I would reach in the food locker and the first can that I fished out, a can of beans or whatever, that's what they got to eat. After the storm passed we saw that many boats had blown ashore and had broken up in the surf. Jan said that it was the fiercest storm that he had ever seen on the West Coast."

Doc Stuart, W6GRL, was also to have a few anxious moments in that storm. His house was located adjacent to the sea wall at the beach in Ventura. The first day of the storm, when the tide rose and the waves got large, his house was in danger of being inundated. The waves were breaking over the sea wall and Doc knew that it was a matter of time before his home would be swept away. He called a local house mover, and within 24 hours it was jacked up, and at low tide moved back several hundred feet, to another lot that he had previously purchased. It was just in the nick of time, as during the next high tide a portion of the sea wall in front of where his house had stood was washed away. It was several weeks before his new location had the plumbing and electrical wiring installed and the antenna feed lines re-routed.

A few weeks before the storm hit, Lew Bellen of Providence, Rhode Island, had traveled to Pitcairn Island in the South Pacific. The island was the home of the 200 descendants of the H.M.S. Bounty mutineers. She was the great-great-granddaughter of Fletcher Christian, who led the mutiny against Captain Bligh a century and a half earlier. Pitcairn was located some 6,000 miles south of Long Beach, between the tip of South America and New Zealand. Lew strung an antenna between 2 palm trees and using the call sign VR6A her first contact was with W6AM on March 8. Since the local emergency with the storm had passed, Don had repaired the damage to his antennas.

One of the islanders, Andrew Young, had been using a crystal receiving set the past year. It was very similar in design to what United States hams had used before audions were developed 25 years earlier. Lew set about teaching Andrew ham radio and he got on the air using the call sign VR6AY. They were also preparing to install a rhombic antenna that would be pointed at San Francisco. The following week Don reported the developments on Pitcairn Island to the *Long Beach Press Telegram* and the story appeared on the front page the next day.

A few days later Don worked a station in Austria, OE3AH. Don had previously talked to him, only this time Anton Hapsburg, the Archduke of Austria, was a protected prisoner of German troopers. The previous day, on March 12, the successful German "putsch," when they occupied and annexed Austria, occurred. The

Doc Stuart's house was jacked up and moved in 24 hours.

Don, tuning receiver, with RME executives. Seated, Adolph Schwartz, W2CN. Standing left to right. Unknown, J.J. Cota, W4FAP, Bob Cooper, W9KNH, R.M. Planckk, W9RHG.

winds of war were again stirring in Europe, just a generation after the "war to end all wars" had been fought, the Great War. Don received a QSL from the Archduke a few weeks later and the conversation was reported the next month in the *Long Beach Press Telegram.*

Don still spent most Saturdays out in the fields adjacent to his house working on his antenna farm, putting up antennas of one type or another. One weekend he built an extended double zepp for 10 meters and added a reflector to it. The antenna worked well and Don devised a simple method of rotating it. One of the stations he worked while testing it was Frank Capellupo, W2JZQ. Fifty years later Frank sent the author a copy of the QSL card from the contact. Don wrote up the results of his latest antenna project and it appeared in the July 1938 issue of *QST*: "A Ten Meter Rotatable Alford Beam."

In May Don received an SWL report from F.G. Maynard, Isle of Sheppey, England. He wrote:

> The QRM gets bad at times on your frequency, but you drove right through it. Are you using a heap of power? Here you compared with W2BRO, W2ZC, and the other fellows I get so well. I have been a radio experimenter for seven years now, and I have never-logged a W6 station! California always seemed to elude me. I would be awful grateful for your card.

Don was traveling for a week that July, and he worked Bertha each night with the portable rig, W6ZZA. One evening Rod Newkirk, W9BRD, broke in on their conversation and asked for a signal report. Rod was running 25 watts to a 6L6 with a 66-foot-long Zepp antenna. Don gave him a signal report and turned it down to Bertha, who also supplied Rod with a report. Shortly afterward W9BRD received one of the few W6MA QSLs to be sent out. Rod misplaced the card in the intervening years, and 22 years later he requested another. Rod wrote the "How's DX" column in *QST* at that time and remarked: "You're as hard as the Aldabras to get a QSL card out of — W6MA-wise anyway. I'm refiling for a replacement. I remember the QSO well; clunky primary keying (smooth QRS fist, though). I was really clicking on that date; had just worked VK3XB for my first Australian."

The next month Don attended the dedication ceremony of the ARRL headquarters station, W1AW, and met the new president of the ARRL, Eugene Woodruff, W8CMP. The Maxim Memorial Station was dedicated on September 2, 1938. Hiram Percy Maxim had become a silent key 2 1/2 years earlier, on February 17, 1936, while traveling to the West Coast. He had been the youngest student in his class at the Massachusetts Institute of Technology, where he graduated in 1886. Maxim had been the first and only president of the ARRL and provided the organization with his firm guidance for its first 20 years of operation. Overtures were made to his family by the League after his death and they received permission to ask the FCC that a memorial call sign be issued. In what was the first license of its kind, the FCC assigned W1AW to the ARRL headquarters as the Maxim Memorial station.

The Oregon Amateur Radio Association (OARA) convention was held late in September. During the convention activities the usual code speed contest was held. As usual, Don took the cup for copying the code faster than anyone else there.

Don had reached the point where he no longer had to travel extensively for his job. His problem now was justifying to the FCC that he still needed a license for his portable transmitter, W6ZZA. In a letter to them on October 3, 1938 Don noted:

> *The property at W6AM is covered with 14 antennas, and any additional antennas installed would begin to interact. My mother's home, six miles away, furnishes an ideal spot for additional experiments, and several antennas are erected there. I have had a radio at this location on and off for the past 27 years. The portable transmitter made it possible to handle seven hundred messages in the aftermath of the earthquake of 1933. It was also in operation during the recent flood. Accordingly, please renew W6ZZA in the usual manner, as it is in the interest of the public.* W6ZZA was renewed.

Whenever the opportunity arose, Don would visit stations that had an outstanding signal. While he was there he would try to identify the particular item or concept that made that individual station stand out from others and try to adapt it to his own station. Peter Onnigian's (W6QEU) first remembrance of Don goes back to his high school days and 160 meters when he was a sophomore at Fowler Union High School. Peter recalled, "Coming out of the great depression, nickels and dimes for chores were collected to visit the Phil Davis (W6TO) radio store in Fresno for parts to build amateur receivers and transmitters. With my Breting receiver I had heard and worked in a contest or 2 the big voices of W6AM and W6ITH (D. Reginald Tibbetts). Much to my surprise one Saturday in the winter of 1938, while browsing into one of Phil Davis's glass show cases, I was introduced to Don Wallace, who was then a radio parts manufacturers' representative. Don was interested in my 'big signal' on 160 and wanted to see where it came from. After my invitation to come see, he arrived at the family grape ranch in Fowler in the late afternoon.

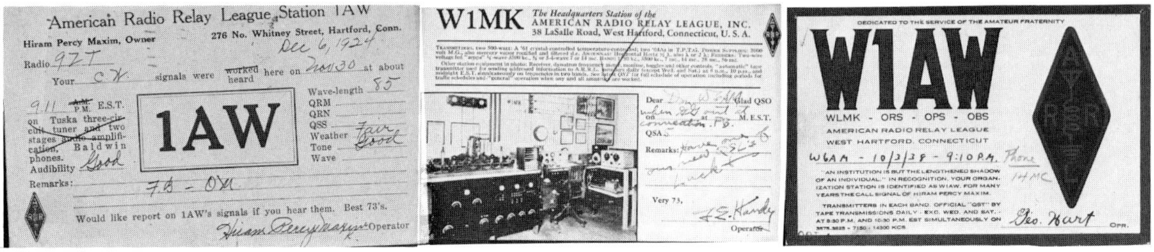

The QSLs of Maxim, 1AW, of the first ARRL headquarters station, W1MK, and the Maxim Memorial station, W1AW.

"We visited for a while and he said, 'Your big signal on 160 does not come from that pair of push/-pull Eimac 35Ts running 100 watts.' He wanted to see the antenna. It was 532 feet (162 meters) long, fed in the center with zepp feeders. And it was really two half-waves in phase for 160! Supported about 15 feet above a large poplar tree by a braced 2 X 4, its center was about 70 feet above the ground. The far ends, broken up with small porcelain egg insulators, were also held up with guyed 4 X 4s and the antenna ends were about 35 feet above ground, located in rows of grape vines. The grapes Dad was farming were Thompsons, picked in early September and laid out in the hot sun on paper trays to dry and turn into raisins. In the early spring these grape vines were trained with their pruned canes wrapped and arbored on 2 levels of #8 galvanized steel wires. The rows, each spaced about 10 feet apart, had two runs of continuous wire, 660 feet long. The antenna ran north-south while the grape rows with their wires ran east-west, taking up more than 5 acres. Upon seeing all this during the walk Don turned around and said 'Peter, your antenna has a ground system which none of us can ever hope to duplicate.' Now, after 50 years, I'm filled with nostalgia when I think back to my teenaged ham radio years. I'm grateful that Don encouraged me to 'go into radio.' Taking his advice, I went into AM, FM, and TV broadcasting, and managed an antenna manufacturing company for 20 years."

Each year in March and April the propagation is particularly good from the West Coast of the United States to Asia. Annually in the mid-30s Don worked MX2B in Manchuko. Takeshi Nagano was a radio operator in the Japanese armed forces that had invaded Manchuria in 1931, and Japan renamed the region Manchuko. The Japanese had occupied northern China from 1932 to 1936 and in 1937 the Nationalist Chinese and Communist Chinese joined forces to attempt to repel the invaders. In July they resisted Japan at the Marco Polo Bridge. Within a matter of weeks Japan launched a full-scale attack. This marked the beginning of World War II in Asia and the cessation of amateur radio operation by Japanese operators on the mainland. Don worked Takeshi for the last time in March 1938, when MX2B had still continued to operate undercover for a brief period. He was never heard after the war and his fate remains a mystery.

When Don visited his customers in San Diego he usually drove his car rather than take the train. *"After my last call I would drive up Pacific Coast Highway. Sometimes I would stop in and see my uncle on the way back. He was the mayor of Newport Beach. Uncle Lew was the one who created the harbor down there. I always figured that that was Uncle Lew's harbor. When he was mayor, Newport Beach was a river's delta; it was just a mud flat, although at high tide the channel was quite deep. Uncle Lew decided to try to make a harbor out out it.*

"The Navy looked at the channel and said that they could get a destroyer in there, so one day at high tide they brought one in. Of course, in the confined space it looked like the Queen Mary. *While the Congressional committee that had come down was looking at it, the tide went out and the destroyer went over on its side. The committee went back to Washington and put a bill through Congress to make it a harbor, provided that Newport Beach would pay half of it. Uncle Lew got Orange County to call it the Orange County Harbor. They had a big barbeque, it was in all the newspapers. 'Everybody come to Newport Beach to vote for the har-*

The station of MX2B in Manchuria.

bor bonds; the government is going to pay half.' They got the boat in there again all right. The mayor dove off the yardarm, the city passed the bond issue and raised half of the money. Then they put in those two jettys. Well, once the two jettys were in place the water carried the mud out to sea. Uncle Lew was mayor of Newport Beach from about the late 20s.

"One time I was in Uncle Lew's bank and there was a whole lot of dictaphone equipment stacked up. I asked him what he was doing with it. He said, 'Aw, I don't like the damned thing.' I said,`Well, let me buy it.' 'I'll give it to you,' he replied. So he gave me the dictaphone for everybody in my office. We used it three years; it was very good for me. I found that when I read a letter, if I dictated the answer then, I didn't have to read it again. I knew what to do as soon as I had read it. The secretary would type up my response. Uncle Lew gave me the whole thing. My son Bill always used the dictaphone also. Bill is the only one I ever knew who could dictate a letter with no corrections. It didn't need to be read again, the structure was so good. Bill scored so high on the entrance exam at UCLA that he didn't have to take an English course."

One morning in the spring of 1938, Tony Gerhardt stopped in to see Don. They were old friends, as Tony had been one of Don's operators while they were in the Navy 20 years earlier. Tony was on a business trip and after a long chat about the "good ole days," Don inquired as to the purpose of Tony's trip. Tony was now the chief engineer of one of the divisions of *Press Wireless*. That was a news organization similar to *AP* or *UPI*. His assignment was to find a good receiving location in Southern California and acquire it for his company. Tony thought that since Don was familiar with the area, he would have some suggestions as to a possible site.

Tony had already tried to purchase some property on the Palos Verdes Peninsula. He had investigated a number of avenues, but so far had been unsuccessful. Tony had found it difficult even to arrange an audience with the Vanderlips, who owned virtually the entire hilltop. Compared to a family of their stature, Tony Gerhardt and *Press Wireless* were small fry.

Don recalled, "*When I was a teenager in the Navy operating the head station for the Pacific, the control station NPG, there was another operator in the station by the name of Tony Gerhardt. Tony was a crack wire operator and became a crack radio operator. In the mid-watch we didn't have too much to do, so we would send to each other. One would copy on a typewriter and the other would send on a bug (speed key). The idea was to see if we could send faster than the other fellow could get down correctly, never allowing an error. We considered it a failure if we made one error. So, we got up pretty fast, as fast as you could send on a bug, 45-48 words a minute. So there were two of us hams up there, Larry Lynde and I. Pretty soon we were both Chief of the Watch, with eight operators to supervise. Hams know how to communicate, you know. It's in our blood. So Tony Gerhardt was assigned to my watch. He was a couple of years older than I was. Tony went on to be chief engineer of* Press Wireless.

"Press Wireless *developed a printing system and they had a worldwide news reporting system. It was formed by 7 newspapers and 4 news organizations. The principal stockholder was Norman Chandler of the* Los Angeles Times. *The reason that they formed* Press Wireless *was that RCA was charging them a high rate to transmit press on their system, upwards of a dollar a word, something like that. With* Press Wireless *they figured they could do it cheaper than what RCA charged. By and by, RCA wanted the business back and eventually offered to transmit press for them for a fraction of a cent per word. So* Press Wireless *sold out after 35 years or so. In the meantime* Press Wireless *had acquired 7 or 8 transmitter sites throughout the world to send their press worldwide, and each transmitter had a receiving site located 15 or 20 miles away from the transmitters. The transmitters were so powerful that they had to locate the receivers quite a distance away. The transmitters for Southern California were down in Playa del Rey, about 15 miles north of here, by what later became Marina del Rey. My property here, in Rolling Hills, was the receiving site for Southern California. They had 5 directions of rhombics, 8 in all, because they were connected for diversity reception. One to Guam, 3 of them to Tokyo, 2 of them to Sydney, and 2 of them for Chungking."*

Diversity reception was accomplished by connecting two or three antennas that were pointed at the same overseas transmitter to one receiver. Due to the propagation path of the transmitted signal, occasionally a character of code would drop out and not be heard at the receiver. That happened periodically when an antenna with high gain and a narrow beam width were used. By connecting two or three antennas that were pointed at the same location, but separated by a short distance, the dropouts rarely occurred. The probability of a lost character became diminishingly small.

"*In the meantime,*" Don continued, "*before I bought the place, CBS wanted* Press Wireless *to receive the BBC (British Broadcasting Corporation) for their news. So* Press Wireless *put up two vee beams to receive Europe. With those antennas that gave me a total of 7 directions when I bought the place after the war. Anyway, part of the antenna system had been built up during World War II because the government badly needed the communications to those spots. In fact, the Tokyo antenna was used to receive their Manila station, and later their Corregidor station, after war broke out with Japan. MacArthur held on for quite a while, you know.*

"One day in 1938, Tony Gerhardt, the fellow I had learned to speed operate with, called me up and said, 'Don, will you have lunch with me?' I said, 'Sure.' So he came over and we had lunch, and Tony said, 'Don, I want to buy a receiving station for Press Wireless. *Where should I buy it?' I went outside and pointed to the southwest and I said, 'Up there on top of Palos Verdes Hills.' That was where I wanted to buy when I moved back to Long Beach in 1926. They wouldn't let me have any aerials up there so I bought a lot over in Long Beach on a little hill.*

"The top of the Palos Verdes Peninsula was my first choice and I tried to buy over there. After a while they finally called me up and said I could have aerials with architectural supervision. I said, 'No architect would okay what I want to put up.' That is why I wasn't up here in the first place, in 1926. So, Tony said, 'We tried to buy up there but they won't sell to us.' I said, 'Well, Tony, I know how you can buy up there. Norman Chandler (the publisher of the *Los Angeles Times*)*, a fraternity brother of mine, is a great friend of the Vanderlips. They own 16,000 acres up there. Let me ask him to tell the Vanderlips what you want.' Now, the reason I knew Norman Chandler was a great friend of the Vanderlips was because once a month there was a full-page spread of the Vanderlip's social doings in the* Times. *He didn't do that for anybody else. I knew that they must be great friends or they wouldn't get a full-page social listing. That was their life. The Vanderlips, you see, were socially-oriented. He was a big New York banker, and he brought his family out here, and they had bought all that property on the hill. All the land that was west of San Pedro.*

"So I asked Norman Chandler to tell Vanderlip about Press Wireless *and he arranged a meeting. The outcome was that Tony bought 105 acres right here in Rolling Hills, on the crest of Palos Verdes Peninsula. The Vanderlips partitioned off all of the land at the western edge of the property by the bluff so the view lots would be available. The* Press Wireless *property went up almost to the edge of the bluff but not quite to it. The property line was 100 feet back, so in future developments the Vanderlips would have the bluff property available. It didn't occur to them that property that didn't have a view would sell up here.*

"Tony, apparently in consideration of my help, gave me first crack at all the stuff that Press Wireless *had to buy for their station. I sold him all the copperweld wire, all the Johnson insulators with lead inserts at $12 a crack, the Johnson feeder-bar insulators, little stand off insulators, feeder-bar wire, everything he could buy, regardless of what he could have bought from somebody else. Business is a funny thing isn't it? Well, anyway, I did all right on the sales."*

The Vanderlips set down a number of requirements on the sale of the property to *Press Wireless*. About 200 pages of deed restrictions were imposed, not the least of which is that they would have to approve the buyer if the property was resold. By the time escrow closed, it was early fall. Tony set about building a structure to house the equipment and operators. At the same time they began installing a number of rhombic antennas. The rhombic had been recognized for some years as the dominant antenna for high-frequency operation.

By the end of 1939 *Press Wireless* had six rhombics spread across the 105 acres, pointed at their overseas transmitters .Approximately 10 miles of wire was in the air, including the antennas and the feed lines. Their site in Rolling Hills used the call sign "RH." As it turned out, war in the Asia and the Pacific was beginning to gear up, and this site would be an important link from that area to the United States.

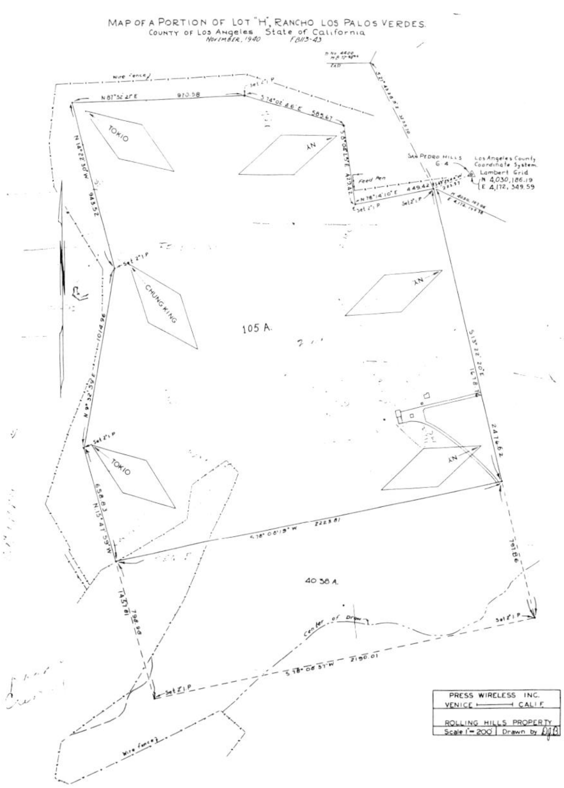

The *Press Wireless* antenna installation.

The OARA CW contest cup.

Fixed commercial and government radio installations had long realized that for high-frequency operation the rhombic antenna was superior. John Roberts, N6FXU, provided an insight to the rhombic antenna. He reflected: "I directed the first rhombic antenna at the Presideo of San Francisco to operate from WVY to WAR, Washington, D.C. in 1938. I engineered point-to-point radio stations and antennas from LF to HF in the western states and Alaska. Of all the antennae, the rhombic was always a favorite of mine, and I was responsible for the construction of over 100 of them."

An amateur in Quito, Ecuador, that with whom Don chatted from time, to time was Clarence Moore, HC1JB. Moore was the chief engineer at the pioneer missionary broadcast station, HCJB. He had recently installed a 10-kilowatt station to broadcast internationally on 25 meters. Moore put up a 5-element Yagi antenna for that purpose, but it proved to be unsuitable. Tucked in among mountains that rose to over 20,000 feet above sea level, Quito was over 10,000 in elevation. There, in the rarefied atmosphere, the tips of the Yagi emitted a large corona discharge when the full 10-kW transmitting power was applied to the antenna. The ends of the antenna glowed briefly and then pieces of the elements broke off and fell to the ground.

Clarence set about designing an antenna that didn't develop high voltage when the transmit power was applied; this is what had caused the air to ionize and a corona to form at the tips of the antenna elements. He came upon the idea of pulling open a folded dipole element, which was very similar to the bi-square antenna that Don had used some years earlier. With this approach the impedance was low along the antenna element, and therefore the high voltage was not developed. "Quad"-shaped elements were also fashioned for the reflector and director elements, the finished product being a cubical quad. The quad antenna was constructed and erected, and HCJB went on the air with full power. The antenna soon distinguished itself as a particularly high-performance design, and two decades later it would find widespread use among amateurs.

The 807 tube was introduced in late 1936. It was a beam power amplifier that had good high-frequency characteristics, would run 60 watts input and was rated at 25 watts dissipation. At $3.50 it was reasonably priced for the average amateur. By the end of 1938 it had become the standard transmitting tube used in the majority of ham stations in the United States Many amateurs pushed the tube to the limit, running the tube hot at upward of 100 watts input (a hot 807). The other standard item to be found in a ham station at that time was a cold beer, which became known as a cold 807.

In wintertime, the propagation conditions to Hawaii from the West Coast of the United States are particularly good on 20 meters. Even amateurs with a moderate station would be quite loud during the "pipeline" conditions in the evening. Don often chatted with Katashi Nose, K6CGK (KH6IJ). They had first worked four years earlier. Nose had written Don and asked him for a sked. On the QSL that Don later received, Nose replied: "U say u had been luking for me fer 1 hour? Well I've been luking fer you fer last 2 years!" Don was planning a somewhat out-of-the-ordinary trip to Hawaii in a few months and he would have an opportunity to visit this up-and-coming young DXer.

The '39 Trans-Pacific Race

The Trans-Pacific race is a sailing competition that usually starts off the coast of Palos Verdes Peninsula in Southern California and ends in Hawaii, nearly 2,500 miles to the west. The first Trans-Pac took place at the turn of the century and has since been held every two years in the month of July.

Don recalled, *"I went to Hawaii in the 1939 Trans-Pacific yacht race. Dick Loynes, who owned the* Contender, *wanted me to be his radio man. I didn't know what he had in mind, but one day I met him on the street and he said, 'Don, I wonder if I could see your amateur station?' I said 'Sure.' He came out one evening and I fired up the big transmitter. We talked to a bunch of European stations on phone. I didn't use code because Dick couldn't copy code. So we had a pleasant evening talking to various stations. A few days later Dick got hold of me and said 'I would like you to be my wireless operator on the yacht* Contender *on the Treasure Island to Honolulu Yacht Race.'" (Due to the Golden Gate International Exposition being held in San Francisco that year, the 1939 Trans-Pac was starting out from Treasure Island in the San Francisco Bay.) "I said, 'Oh, I couldn't take a month off to do that.' Dick said, 'I would buy everything that you specified, do anything you want.' I said, 'Well, that sounds pretty good.' I went home and talked it over with my daughter, Betty Jean. I told her, 'I have a chance to go to Hawaii as a wireless operator on a yacht and I would like to have your mother go over on the steamship and meet me there. When it comes up, if you can kind of say that you think that it is a good idea, it would help your mother's decision.' I took it up at dinner and my little girl, who was very bright and about 13 years old, said, 'Oh great. Wouldn't that be wonderful for you!' So Bertha thought it would be all right. Thirty-two wives went over on the* SS Lurline.

"I ordered a special transmitter for the Contender *that went all the way from 700 meters to 10 meters, with 350 watts output. Commercially-made VFOs (variable frequency oscillators) weren't in existence then, so we used 60 pairs of crystals at 60 frequencies. The transmitter was completely crystal-controlled. I could also use it to talk to other ships. I made out the license, commercial, KLRR, open for public correspondence because I had a first class commercial ticket. I thought I might as well get some experience on it in case I would ever need it. For instance, should I want to become an FCC inspector, or engineer, if I had experience on that license I would outrank the fellows who didn't.*

"So I put in the request and got a commercial license issued to the yacht. KLRR had the same license as the Lurline, *the* Matsonia, *the* Queen Mary, *and all the rest of the commercial ships. That qualified me for many years of operating commercially just because I was an operator on the* Contender. *Even later, when it was on the dock, I was in charge of KLRR. Dick signed my ticket and I think I have six years of operating as a first class operator on the* Contender. *I had done some actual shipboard operating, almost two years in the Navy during World War I. Before the war I also had quite a bit of experience as a commercial operator when I was in high school, on ships that went right by the peninsula here.*

"I put up a phased antenna on the boat for 10 meters. It was a pair of 5/8 wavelength antennas. There weren't many people around that had that good an antenna ashore, two 5/8 wavelengths long, vertically polarized. That's what many hams use today on 2 meters. That one put out by Cushcraft for 2 meters is the identical antenna I used on the yacht on 10 meters. I had a matching stub in the middle of the antenna, which is more efficient than a coil. It was tuned right to within one-eighth of an inch on the crystal frequency I used. So that antenna was a walloper and I worked all over the world with it. That was my pure fun during the trip. While we were at the dock and getting ready for the trip, all I had to do was draw sketches of the equipment and antenna installation that I wanted to have. Dick Loynes owned a garage and had his good mechanics do everything I said. They installed the radio set complete. I drew a sketch of the aerials that I wanted to put up, a total of 11 phased arrays like the one I just described. They put in six halyard lines off of the main mast and forward cross arm,and the mizzen mast and its cross arm. (The *Contender* was a yawl; the forward mast was higher than the aft mast, which is aft of the binacle.) *I could pull up any one of these phased arrays in jig time, and they would be sloped aft toward the U.S. The antennas were all broad banded except the 10*

meter array which was for that one frequency only, and I just left it set that way. Ten meters was the only amateur band that we could use mobile at that time. I couldn't use 20 meters at all.

"I wired the FCC asking for permission to talk to amateurs on yacht frequencies during the duration of the race. They wired back okay, and that gave me an idea. Every day I sent another telegram to the FCC asking for another privilege. By and by I had 20 telegrams from them, all approving unheard of privileges for a ship's station. To this day I have never heard of anybody having as many privileges as I had during the yacht race. Years later I was calling on a broadcast station in Seattle, talking to the chief engineer and selling some tubes, which was my business, and he said 'You know, I used to be with the FCC. I'm the one who gave the okay for all the wires for the yacht race that you sent in. That happened to be my department.' I said, 'For gosh sakes. I asked for everything under the sun. If I couldn't think of it, I didn't ask for it. You okayed them all.' The first one was simple, I wanted to talk to amateurs on yacht frequencies. If my operation was limited to 10 meters, I wasn't sure I could always talk to them. So he said 'I approved all the frequencies that you are licensed for.' I said, 'How on earth could you do that?' 'Well,' he said, 'I knew you, sort of. I had heard of you. I knew you a little bit, and I knew you wouldn't embarrass us.' I thought that was quite a compliment.

"D. Reginald Tibbetts, W6ITH, took frequency measurements of all my transmissions during the yacht race and mailed me the certificates. So I had those, in case there was ever any question about my transmitter being accurate. Tibbetts had precise measuring equipment of the day. He did that for a business.

"We left from Treasure Island, which was the World's Fair site of the day, in the middle of the San Francisco Bay. This was one of the features of the fair, the start of the yacht race. Ordinarily, this race leaves from the coast of the Palos Verdes Peninsula on the Fourth of July. You can go down to the bluffs and see them starting off. That year it left from Treasure Island in San Francisco Bay. The week before the race all 26 yachts were tied up side-by-side in slips at the Port of Trade Winds, by the side of the World's Fair. We had World's Fair visitors climbing all over our ships as part of the show.

"We spent about a week at Treasure Island before the race began. During that time I chased noise sources on the Contender, *which had 35 motors of various sizes aboard. Some were for pumping fresh water to the faucets, others were for the showers or pumping out the bilge. Every day I would get on a different amateur band and work the local hams in San Francisco. I would tell them about the yacht race and they would all say, 'Gee, I would like to come down and see you.' I would say fine, but bring your tools, I want to put you to work. 'Oh,' they replied, 'I would be delighted.' So I had hams clamoring all over the ship to help me chase all of the noises being generated aboard, and when we got through, I had the noise well under control.*

"Every motor on the yacht had made a racket. I was afraid the crew might be running the motors when I was on the air talking to somebody. The builder of the yacht had made all the fittings, davits, cleats, and everything else out of German silver, so they wouldn't rust. For the motors, however, nobody knew anything about cutting out interference, not then. The fellows in San Francisco brought down capacitors, coils, and resistors. I told them to bring down all the things that they could think of that chased noise. We would turn all the motors on, one at a time, to try and find the noise, and then the fellows would fix it at the source. So I had the best help in the world for the whole week. The local hams also helped me rewire things that were not serviceable. I put up two lights on the main yardarm, one on each end. I also put a Morse code key on top of

W6AM working on the antenna strung from the *Contender's* yardarm.

the cabin that I could reach from the deck, so I could send code from outside.

"While we were at Treasure Island, four or five men came down there with derby hats and black suits, and I wondered what on earth was coming off. All of them were dressed in the executive attire of the day. I thought, 'Well, I wonder if this is the FBI, or the government, or what.' In fact, I was a little apprehensive. So I tried to be nonchalant and said, 'Well, what's doing, gentlemen?' 'We are from the telephone company, and you have a license to operate high seas phone. We heard about it. There is no ship on the Pacific Ocean under the American flag equipped to test with us. We have all the operational facilities at Dixon, California ready for it, but nobody to test with. There are some foreign ships that have the equipment, but we don't want to test with them. You have the only setup on an American ship. We would like you to test with us every hour on the trip.'

"I said, 'Well, I have a lot to do. I can't do it every hour, but I'll tell you what I will do. I will call you every day and test with you.' That satisfied them and they gave me the frequencies, which I already had crystals for. So, every day after breakfast—the breakfast varied with the time of where we were—I'd go in and call their facility at Dixon on the worst frequency that I could imagine. When we were still right close to shore I called them on 10 meters. I figured that there was no use testing the right frequency, I wanted to test a bum one. So, when we were 50 or 100 miles out I called them on 10 and got them perfectly. They said, 'Well, we appreciate the test.' I said, 'I'll call you after breakfast every day.' So every day I called on the worst frequency. When I got way over near Hawaii, I called them on 80 meters, the equivalent of our 80 meter band. Always a different band every day and the worst frequency I could think of. A perfect one close in would have been 80 meters, actually 88 meters, or whatever the commercial band of the day was. When I got over to Hawaii, of course, the 12-meter band, close to the 10 meter band, would have been the best choice. I did it all backwards and got them every time. Of course, I had these fantastic aerials on the ship. The beam I had put up was for the frequency that I expected would work best for that distance on the daily broadcast.

"Pretty soon, 3 o'clock came, July Fourth, and it was time to leave. Same as they leave off the peninsula here at 3 o'clock. I hadn't had time to get a QSL card made before I left. D. Reginald Tibbetts called me. Reg was the one who built, developed and created the radio communication network for the men who built the Bay Bridge. All the barges, ships, floaters, and cranes, had one of the hundreds of these 5-meter sets that he had built. Those sets weren't commercially available in that day, so Reg was a very smart fellow. Five or six years earlier Reg and I had a contract to build and install all that equipment. Anyway, he said, 'Don, can I help you on anything?' I replied, 'Well, I didn't have time to order a QSL card'. He said, 'I'll get one for you.' So, I answered, 'Here's what the card should say' and I gave him all of the information. Yeah, he could make that. I asked him, 'How can I get them over to Hawaii so I can pass them out when I get there?' 'Oh', he said 'I will arrange that.'

"Captain Dick Loynes, who was well known in Long Beach at the time, was a great yachtsman. He's the one we all consider the father of the Long Beach Marina. That was his lifetime ambition, to put the marina through. He always had a yacht. Dick was the world champion class powerboat racer for 10 years. There was the unlimited class, and the next classification had some specifications, a handicap. That was the one he raced. Dick told me that the reason he could win every year was that during the race he never opened his engine up all the way. Dick developed engines and one time he showed me a room full of all the blueprints and castings he had for special engines. He just loved it. He said that he had never run his boat full speed. That way he could time it so he would just win over anybody else, because they didn't know how fast they had to build their boat. He would beat the other boats, but not by too much. So, one time he was racing down in Florida, and somebody offered to put up a new Chevrolet car if a world's record was broken. Well, that was duck soup. Dick just pushed it out until he broke the record. The auto dealer said, 'If someone can break that record I'll give another car.' Dick's mechanic of the day was Gus Walker, president of the Farmers and Merchants Bank in Long Beach, a neighbor of mine. The next day Dick pushed it a little bit more and broke his own record. He never had gone to tops and nobody else could outrun him. One day however, he ran into a log at high speed and was injured, and wrecked his boat. He got over it. Dick had a very colorful life.

"Dick and I talked over how to arrange the shore stations. He said, 'Wouldn't it be nice if we broadcast this race? I said, 'Well, it's okay with me. I used to work in a broadcast station in college, WLB. That is, I hung my license in there. Legally, they had to have two operators on duty after 1921. I would operate the station for a couple hours every week. I had a first class broadcast ticket.' Anyway, we went up to see the people at the Mutual network. When we mentioned our proposition they practically jumped over their desks they were so enthusiastic. But then one fellow said, 'We will have to send a professional announcer along with you.' Dick said, 'No, I have got a crew that weighs 3,200 pounds. If we have any more weight we won't come in first. I have

already got an eighth of a ton radio operator.' I weighed 245 pounds in those days. I didn't look any heavier than I do now. Dick continued, 'I have all of his gear and I can't stand the weight of another person aboard. Don will have to talk. 'I told him, 'I actually was a broadcast station operator on the first broadcast station giving regular programs in the northwest.' The Mutual fellow replied, 'I guess that will be all right.'

"I had operated the broadcast station at the University of Minnesota. They had the first broadcast station in the northwest, in 1920. The western states were so sparsely populated then that Minneapolis was considered to be the northwest. Every Wednesday night I would put on a program. We developed quite a dedicated listening audience, mainly because there wasn't anything else to listen to, except KDKA in Pittsburgh.

"Well, I told them that I had done some broadcasting, so the networks said okay, but we have to send along an engineer. Dick said, 'No. We can't do that, Don has to be our engineer.' I said, 'By the way, I know Perry, your chief engineer. I sold him his TV transmitter tubes.' He had the first TV station at the time. I said, 'Go get Perry and ask him if he has anybody who has as high a class of license as I have. The government says you have no responsibility. They say I'm okay, you don't need to worry about it. If anything goes wrong, it's the government's fault. They certified me by giving me two top licenses.' So they got Perry. I said, 'Perry, do you have any engineers who have a radio telephone first and a radio telegraph first license?' He said 'No, we don't.' Well, the network fellow says, 'You win.' So, I was the announcer, the engineer, the program arranger and the ham operator.

"For the first broadcast, I had planned to put on a program. I wrote up a typewritten story of what we were going to do for the 15-minute program. In it I included the positions of the ships. When we were tied up at the Treasure Island dock a fellow had come over in a blue denim jump suit. He wasn't dressed nearly as fancy as the other yachtsmen. He said, 'I'm a ham back in Raleigh, North Carolina. My name is Dick Reynolds. I have a yacht here, the Blitzen*. The other yachtsmen say they are not going to give their positions, they are going to be secretive about the trip. But I'm going to give you our position during the race, Don. I think it is a good idea.' I said, 'Well, that is great, we will have a good broadcast. I'll give the* Contender*'s position and I will give yours.'" The* Blitzen *was a year old and had been shipped out to San Francisco from the East Coast on a freighter. The freighter got caught in one of the longshoremen's strikes and the* Blitzen *almost didn't get off-loaded in time for the race.*

"I typed up a big sheet of paper with the broadcast times and frequencies, and distributed a hundred or more copies of it. I told the crew members of the ships to send it home to their families so they could listen to the broadcasts over the network every night at 7 o'clock PST. I thought it would be a good idea to work the yachtsmen at 4 o'clock PST.

"I kept one clock on PST, because that's when the people ashore would be listening, to not be confused with ship's time, or Greenwich Mean Time (GMT), or anything else. I had three clocks in the radio room. I had one clock set to ship's time to eat on, one set to PST to broadcast on, and one set to GMT for the ship's log, as required by the FCC.

"So, Dick Reynolds said, 'I'm going to send you my position daily anyway.' Well, Dick turned out to be the owner of the R.J. Reynolds Tobacco Company in Salem, North Carolina. He was a ham, so we got along fine, of course. Well, 4 o'clock the first day came along. On this sheet I said I was going to call all the yachts at that time and get their positions. They all had harbor phones aboard that ran from 10 to 50 watts on 2738 KC. Dick Reynolds pipes right up, 'Here is the position of the yacht Blitzen*, we just passed under the Golden Gate Bridge.' I said, 'Well, I guess you are a little ahead of us, aren't you Dick?' He answered, 'Yeah.' 'Now, is there any other yacht that wants to give a position?'*

"My neighbor from Long Beach said he would give his position because I had designed his aerial. He was on the yacht Zoe H*, that was his wife's name. His name was Ray Person. I had gone down to his yacht and designed this super aerial. He had more range than anybody with the harbor phones. Ray had a 65-foot yacht, a sloop, with about a 65-foot mast. I had told him to replace the back stay with a stainless wire that had an insulator at each end and he would have the best 2-megacycle aerial on the ocean, which he did. The back stay is usually grounded at both ends, but his was used as an aerial, and it connected to the radio in his cabin. So he had the best signal of all the yachts on 2 megacycles. Ray was a good friend, a neighbor, and he belonged to the Virginia Country Club, same as I did. He had also been the escrow agent when I bought the lot for my house in 1926. So, Ray pipes up with his position and I said, 'Boy, we will have an extra good broadcast tonight, everybody at home will know where the* Contender *is, where the* Zoe H *is, and where the* Blitzen *is. Does anybody else want to give a position?' They all came in and gave their positions. They were all listening and couldn't resist. So, for the rest of the race I had all of their daily positions. Besides the Don Lee broadcast, the position information was broadcast at 5 PM on 36-meter CW using the call sign W6XEJ. The amateurs who copied them could give the information to their local newspapers."*

Syd Rubinow, the Yachting Editor of the *San Fran-*

cisco Call Bulletin, had requested a daily story from Don. He wanted to present a running account from one of the vessels in the race and make it a daily feature. So each day Don sent them a brief story through W6USA, the amateur radio station set up at the International Exposition, or any other station in San Francisco he happened to work.

While Don was having the time of his life aboard ship, he had to find something for his boys in Long Beach to do to keep them busy and out of trouble. Since Bertha was taking the *Lurline* to Hawaii, no one would be around at home to keep an eye on them. A friend of Don's had quite a bit of property around Lake Elsinore and needed some odd jobs done for the six weeks that Don and Bertha would be away. Since the boys wanted to play football in school, Don thought that the work would be good exercise.

Bill Wallace recalled: "While the folks were gone to Hawaii, Donald and I worked for Mr. Knott out at Lake Elsinore. We irrigated walnut trees, dug potatoes, picked apricots, that sort of thing. It was over 100 degrees every day. We slept in sleeping bags in a corrugated tin building on the property. It cooled off to the 80s at night. The last day there we had to dig a 100 foot long trench that was two feet wide and three feet deep. Donald and I had a contest to see who could dig faster, and we each started at opposite ends. After eight hours we both got to the 50-foot point, a tie. I guess dad figured that it would toughen us up. Well, it did. We got $40 for the six weeks we were there. It worked out to about 16 cents per hour."

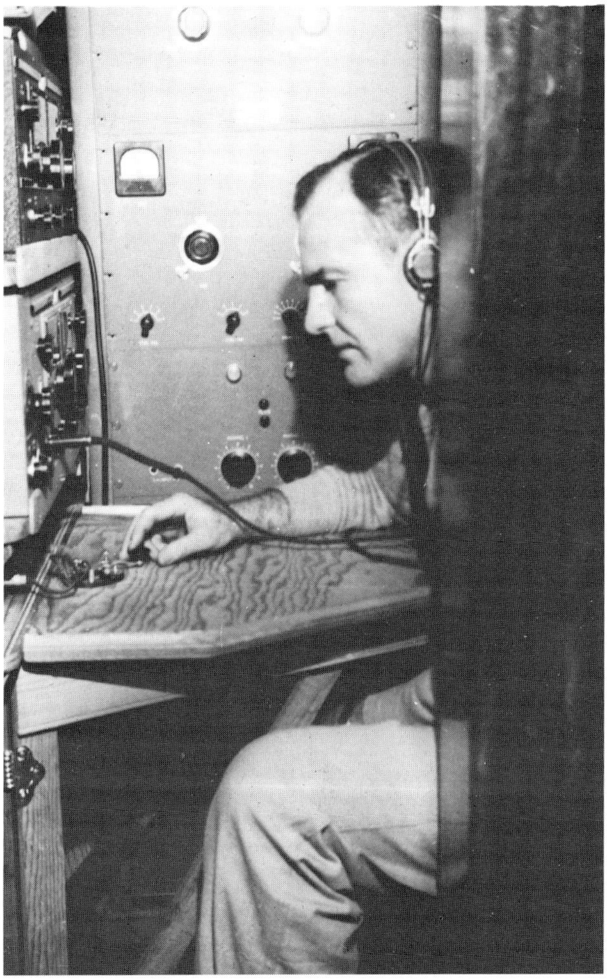

Don at the operating position on the *Contender*.

Meanwhile, Don was on the high seas. "*As we got further out to sea, the yachts spread out about 800 miles. The leaders were as much as 800 miles ahead of the slowest ones. We came in to Hawaii after about 13 days. Some of the boats came in 10 days later. The* Zoe H, *being a smaller boat, wasn't as fast as the* Contender. *They were doing very well for their classification, which was B, I suppose. We were in the A class, then there was a B class, and a C class. The* Zoe H *could race on handicaps, so they could win their division on corrected time. Well, it turned out that the* Zoe H *had the best signal and the best reception on 2 megacycles. I had Ray pick up all the reports from the boats that were back a ways. They were on 2-MC radiophone, and sometimes some of those yachts were kind of faint for me. The* Zoe H *would pick them all up and relay them to me. The antenna work had paid off. I could get the laggards' reports very neatly from Ray, because he was halfway back.*

"*Every day on the broadcast I would take some special event that had happened that day, and it would be my feature for the 15-minute broadcast. My transmissions were broadcast over the* Don Lee Mutual Broadcast System *and the* Transcontinental Mutual Broadcasting System. *The receiving stations in California were Herb Becker, W6QD, Bill Rudolph, W6OEG, and Larry Lynde, W6DEP. The first thing we said on the broadcast was, 'This is the yacht* Contender. *The entry of the Long Beach Yacht Club (LBYC) in the Trans-Pac race. Captain Dick Loynes, how far did we travel today?' In that way I got Dick in on every program. He would say how far we went and what our position was. He was entitled to it, he was the skipper of the boat. So I had him on every day. The Long Beach Chamber of Commerce, I learned later, had given Dick $1,000 for publicity purposes. So Dick wanted to be sure to mention Long Beach and the Yacht Club. Then I said 'Today we sure had an interesting trip. Some of the rigging up on top of the main mast got fouled up and Joe Petersen was up there seven hours. I thought that even though he probably was pretty tired from all that activity, you* *might like to hear his story. Joe, go ahead and tell them what you did.' 'Aw, I didn't do much,' he replied, 'I just straightened out all the lines . . .' I had to prime him to get him to talk. People ate it up, you see."*

Herb Becker, W6QD, was the relay point for the Don Lee broadcasts. He had added two more masts to the pair he already had up. From those masts he strung up seven directional antennas for the broadcast frequencies of 3492, 4797.5, 6425, 9135, 12862.5, 17310, and 23100 Kilocycles. Herb modified his 5-kilowatt transmitter (corn-fed Kw) that he built five years earlier to work on these frequencies, so his signal would have plenty of punch.

"By and by, we hit the doldrums," Don recounted. *"When you go between San Francisco and Hawaii, you hit the doldrums. When the Spaniards hit it they called it the horse latitudes, because to lighten the ship they threw their horses overboard. They got out in a rowboat in front of the big schooners and tried to row through it. The doldrums don't occur when you leave Los Angeles because you are south of them, you get the trade winds. It's kind of a neutral spot in the ocean where the winds don't blow.*

"Periodically, however, there is a rain squall, and the rain hits. In the middle of the rainstorm there is a blow. You don't have any idea which direction it might come from, it's like a little whirlwind. Well, we hit the doldrums and we were becalmed. We didn't move. So, the announcers back at the network were very concerned. 'What will we do, you're becalmed.' So I put on a broadcast: 'This is what we do in a calm. We drop little pieces of paper over the side, trying to figure out if they will move. If they move one direction an inch or so, we reset the sails to catch that one inch of wind. We also let some ropes out over the side around the ship and have a good swim.' Well, the listening audience increased during our calm. People were more interested over what happened in a calm than during the fast action, because the fine points of yachting had to come out.

"All of a sudden here came one of the rainstorms. No warning, they just hit you. The boat darn near snaps in two, the squall hits so hard. The boat isn't moving, you see, and then the wind hits and the wind can't get out of the sails. We blew out a spinnaker sail, big holes in it. It blows to shreds. We pulled out another one and had it up in 17 minutes. Pretty soon we were moving. Maybe you can move 100 miles in that darned thing before you lose it. We finally got out into the trade winds after five days in the doldrums. Some yachts go over to Hawaii in eight days when they leave from down here, but we took 13 because we had five days of the doldrums. We couldn't make our full speed the whole time.

Herb Becker's station, W6QD, transformed into a broadcast station.

Captain Loynes takes a position fix.

"Before we had left Treasure Island I visited the Federal Communications Commission. I had called on the inspector in San Francisco and told him that I was on the yacht Contender in the upcoming race. I wondered if I shouldn't give my positions to the Lurline *every day so they wouldn't run over us. 'Oh, yes,' he said, 'decidedly.' So, on the first day out, I got on 600 meters and called the* SS Lurline, *KIEK. The radio operator pushed me off, he didn't want to talk to me. I said in code, 'Now listen to me, I talked to an FCC engineer and he said it was mandatory that I send you the positions of the boats in the race every day so you won't run over us.' He grudgingly took the positions of all 26 yachts, latitude and longitude, which was a little bit of work. It took 10 or 15 minutes of his valuable time that he ordinarily spent smoking or drinking coffee, no doubt. So, he had them. The captain had these 32 yachtsmens' wives, including mine, aboard. About that time they all descended on him at once. 'We want to know where our husbands are. Where are they in the ocean?' He went down to see the radio operator and asked him, 'Where are those yachts? Can you get hold of them?' The radio operator answered, 'I have all their positions right here,' and handed the sheet to him. The captain posted it up on the bulletin board, in the main hallway on the ship. The operator couldn't do enough for me after that. I saved him embarrassment because he had the information that the captain was going to demand. He wouldn't have known what to do, what frequency to find us on, except I had already called him.*

"So, every day I contacted him and sent the information. Pretty soon I was getting messages back from the wives. They were commercial, paid messages for the yachtsmen, of all 26 boats. At the 4 o'clock schedule I said, 'If you fellows hurry up and give me all of the positions, as soon as I have all of them I will handle messages for the balance of the hour, from four to five. As long as I have time. If you can get ashore on your radiophone, go ahead. But if you can't, I have the high seas channels and I send the messages in every morning after breakfast.' Well, about 500 or 600 miles out they couldn't get back on their harborphone radios, which were on the 2-megacycle band. I had higher frequencies and I could talk to New York on five different bands. I could talk to the A.T.& T. facility at Dixon, California on five channels.

"During these tests with Dixon I'd send in the messages the other boats couldn't get through. So, the messages started coming through from the wives. I was in touch with the other yachts at four every afternoon and I gave the yachtsmen all their messages from home. Then they would give me a message to send back.

"I had taken a whole stack of tablets of yellow second sheets for scratch paper, fortunately. When I got back I had a box full of messages one foot high of traffic that I had passed. I had just scribbled them off as I went along and, after they were sent, tied them up and threw them in the box. I know that Dick Reynolds on the Blitzen *gave me a message every day for his wife, a rather long one. So I would call her on the radiophone and give it to her direct. I'd just say, 'I talked to Dick yesterday, he said that he was fine,' instead of his long message. She would say, 'Tell him hello, I love him.' I shortened everything down, no matter what they said. They just wanted to hear from each other, of course. Otherwise, I wouldn't have had time to pass all of the messages. I did that every morning after breakfast. I would clear my table of all the messages that the yachtsmen gave me that they couldn't get through on 2 megacycles. The* Contender *could broadcast pretty well on that frequency, on account of the 100-foot high aerial.*

"All of the antennas were broad-banded for three frequencies; 1/4-wavelength, 3/8-wavelength, and 5/8-wavelength. So if I had to change frequency after we tuned up but before we started the broadcast and that frequency was starting to fade out, I'd go to the next lower frequency. We always wanted to have a loud signal. We never missed a broadcast. That's unheard of on a special event. It was because I always used one frequency lower than the best one.

"A lot of traffic was handled. I had my friend, Larry Lynde, W6DEP/W6UG, of Long Beach, get up at 7:30 every morning and copy a special story for the Long Beach Press Telegram. *He and I were shipmates in the Navy and we operated NPG, the Pacific Coast Headquarters Navy station. That's where we started out in World War I. Anyway, every afternoon at 5 o'clock I sent a complete* Associated Press *message, another one to* United Press International, *another to* Press Wireless, *and another one to the* San Francisco Call Bulletin. *They were all different stories. There were enough things that happened on the yacht that I could have a different story for each publication.*

"Any time I had free I would get on 10 meters and work the hams. If 10 meters wasn't open, I'd get on the 8,000-KC yacht frequency. I had published in the ham magazines that I would transmit there and listen in the ham bands. Well, I made every ham that I worked give my transmit frequency at the end of every transmission of his, KLRR on 8,000 KC de W6BIP. I said, 'If you won't do that I can't talk to you.' A lot of them wanted to keep it secret once they found where I was transmitting. I could work a lot of them and I gave each one a story for their local paper, if they got in contact with me. It would be a different story from the one that I sent to the Associated Press. *I got this fellow, Art Sowle, W6CW, a ham in Reno, Nevada. I gave him a story and*

said, 'Give it to your newspaper, it will be a special, exclusive story for them.' Well, later on he told me that he called up the newspaper, and they said, 'So what.' But then he said a half hour later they called back and said, 'Hey, your story must be okay. It just came over the Associated Press. *We want your story.' So, he said they ran a big banner — three-inch headline the next day: 'Local Amateur Copies Yacht Race Direct.' They ran his story as the big feature of the paper, because the Associated Press story had come in. I was the same guy who sent both stories. Silly. Well, I heard yarns like that several times as we went along.*

"Wilbur Bachman, W6BIP — we call him Bip. I've known Bip since long before the yacht race. I had been up to his house, a wonderful location on one of the hills in San Francisco. His signal was tremendous. I had to tell Bip, 'When I call CQ, don't answer till the third CQ. I find that I'm sending everything through you, which is nice, but I ought to spread it around.' So, after the third CQ he would answer, and always did. He never failed me. I had a perfect contact right into San Francisco. I also told all the amateurs, they could not you cannot ask for repeats. Nothing was written down. After the first broadcast I never wrote anything down. Everything I sent was just as I thought of it. All AP, UPI, Press Wireless *and* Press Telegram *stories were also extemporaneous. When I got fellows like Sowle in Reno, I gave him another exclusive. I said, 'Don't ask for repeats. You must not ask for repeats because I can't repeat exactly what I just said. You have to copy it.' I had a bug along, but I used a straight key the entire time, sending at about 20 words per minute. It was all crystal control on the trip. It wasn't any problem to find me once the fellows knew where to look. The shore stations were the same way. They would be on a particular crystal frequency. You called CQ and combed the band for an answer.*

"Anyway, we got over to Hawaii. In the meantime we had put on 21 network broadcasts. Every one of them went through without a hitch. We found out later that some of the commercial radio services failed to get through during this time, due to a magnetic storm disrupting communications.

"Twice a week the broadcasts were coast-to-coast. Every evening at 7 PST, prime time, we were put on to the western United States. All the stations were tied together. As we got closer to the island the network wanted a noon broadcast too. The people wanted to keep up with the yacht race. It seemed like everybody was fascinated. They had never heard of a yacht race being run for two weeks with daily broadcasts. We developed a rather enormous audience. In fact, the network said it was the biggest tune-in they had ever had on a running special event in the history of radio. When we got over there it seemed like we were 25 miles off Makapuu Point and they wanted us to broadcast the entire finish, which was the whole evening. It turned out to be a three-hour broadcast. So I talked the three hours and had no help because the sailors were busy.

"We had heavy winds. The water washed over the deck. We were going so fast during a jibe after rounding Makupuu Point that the hull drove underwater and the sea came over the bow. I was drenched from head to foot. In resetting the sails three antennas and their halyard lines had to come down. In the confusion most everyone whipped out a knife and cut away part of the antennas, throwing them over the side, but I did make sure that the antenna being used on that particular broadcast was saved. I hung the loudspeaker in the rigging as high as I could reach, which was about eight feet. I hung the push -button up there too, to turn the transmitter on and off. I held the microphone up high whenever the waves came over the bow. So, I stood there and described the whole race. Nothing like Moby Dick, *but it was exciting from an amateur yachtsman's point of view, and the people loved it. We noticed automobiles all the way out to Diamond Head. It was dusk and many of them were signaling us with their headlights. We answered them with the masthead blinker system on the yacht.*

"The Contender *came in first. That was the exciting thing. There were thousands of people down at the docks, waiting for us. The* Blitzen *came in second, three hours after us and 31 seconds ahead of the third place finisher. They won the Governor of Hawaii Cup for the best corrected time. We had been behind the whole race and this heavy wind that we got, a 40-knot wind, drove the big* Contender *with its steel hull through the water lickety-split. We picked up those three hours that we were behind and came in three hours ahead to boot. The last day the heavy winds won the race for this heavy boat. The lighter boats couldn't go that fast. The* Contender *was 107 feet long and built to race. The Rohenzolern family in Germany had built the* Contender *34 years earlier. Kaiser Wilhelm used the yacht to race against King George and the six monarchs would have an annual race in the North Sea. They were friendly at that time. A special bunk was built for the crown prince, who was a very tall fellow. I think he was 6 feet 8 inches or so, something like that. There was an 8-1/2-foot bunk in there, a special room for him. They gave me that for the radio room, so I had a place to sleep. Well, all during the race I would sleep an hour here and an hour there, always by the radio.*

"The networks said that when they opened that three-hour broadcast of mine they had to cancel everything on the network. They told me that was only the second time in history that they had ever canceled

everything. The other time was when the Chancellor of Germany gave his speech.

"When we got to Hawaii my QSLs were there, exactly where I told Reg Tibbetts to deliver them, and they had come by way of an Army transport ship. He had taken them down to a wireless operator who put them on a transport, and it had beaten the yachts over to Hawaii. They were waiting for me and I had a thousand cards. Every time that I saw a ham, and there were plenty of them, I gave him a card. I air-mailed the log back to my office and everybody got a card within a few days, every contact. So that was probably one of the original DXpeditions of sorts, cause it was different than anything that had ever happened.

"I had talked to a fellow on the trip, K6OJI, and his name was Malcom Magers. The year before I had told him to put up a rhombic antenna, and he would be louder than the other hams in Hawaii. I used to talk to him once in a while and he made a special job of keeping me in touch with the Hawaiian amateurs while I was on the yacht. He had put up this big rhombic, aimed at Kansas City, where his parents lived. When I told him to put up the rhombic he said that there is nothing but jungle next door to his house. I said, 'Just whack out a hole in the jungle and run the wire through there.' Anyway, he did that and the antenna worked better than anything he had ever tried. So, Magers was down there to meet me and took me out to his place. We got on the air and I talked to some of my friends and told them that I had arrived. Then he started talking and I fell plumb asleep. I had arranged my schedules on the boat for the whole 13 or 14-day trip to have three hours of sleep a day, and it finally caught up to me.

"In preparing for the trip I went to absolute training, like I would for football or track. I drank only milk, which we had plenty of. It was frozen in big 5-gallon cans. I ate just exactly the right amount of food and gave up coffee. I also quit smoking my cigars, which I enjoyed. I went into absolute training so I could get along on three hours of sleep a night. I remember sitting next to Magers that night, and then I went to sleep. Magers wanted to drive Bertha and me around the island. That was wonderful, we had never been around Oahu.

"In the meantime, the network wanted me to put on broadcasts from Hawaii after we had arrived but I said, 'No, I'm tired. I don't want to do it any more.' So I dodged it. They sent people after me, trying to get another broadcast from me. I wanted to spend alittle time vacationing with my wife after she got there. She had gone over on the *Lurline* and met me at the dock. There were 32 wives waiting for the crew members from

The cover of *Radio Magazine* carried an aerial photo of the *Contender* taken just hours before she won the race.

the various boats. So, Magers drove us around the island. Well, it turned out that we stopped at the RCA radio station.

"I was very interested in the aerials that they had up. Here were all of these rhombics, sloping. We got hold of the chief engineer and walked around the place and I said, 'Where did you get the dope on these sloping aerials? How do they work?' He said,'They work much better than when they are flat.' He said he got the information 'Out of Terman's book.'

"After we left there, Magers kept on driving around the island. Because we had used up so much time he suggested, 'If we go through this Army base across the middle of the island we can save an hour.' That way we can get back before dark and do some more sightseeing. So he told the guard at the gate, 'Call up and tell them the Magers want to go through.' The guy called up and we went right through. They thought we were majors!

"When we drove through the base he showed us an airplane sitting on the runway, ready to go. There was a pilot in it ready to take pictures of the island when *there were no clouds overhead. Magers told us, 'About once a year there is an opening and the pilot buzzes right up and takes the pictures.'"*

While Don and Bertha were in Hawaii the July 1939 issue of *QST* arrived. "Honolulu Bound", by Don Wallace, describing the upcoming race, was the lead article in the magazine. After the excitement of the yacht race and visiting the island, the return trip on the *SS Matsonia* was rather uneventful. Don visited from time to time with the radio operator to keep up on what was happening. They arrived home in the first week of August. Bill and Don Jr. were happy to see their parents, as they had grown weary of working for a living.

Don had a QSL from Horace Greer, W6TI, waiting for him at home. He had worked W6TI at the Exposition with his station in Long Beach shortly before he went to San Francisco to join the yacht race. On the card Horace noted: "73 to the original California kilowatt!" Actually, Bartholomew Molinari, 6AWT, had a head start on Don by a couple of years with the 5-kW rock-crusher that he put on the air in 1924.

Don's secretary had put together a scrapbook of newspaper clippings that had appeared in the papers during the yacht race. To these Don added some candid shots of the crew taken before and during the race.

When Don returned there was a great deal of radio activity going on in the area. On August 22 Doc Stuart wired the FCC and asked for temporary authority to use an experimental license. He would be transmitting on the aeronautical frequencies of 6425, 8655, 12862.5, and 17310 KC with 1,000 watts of power for the purpose of experimental work in connection with the Howard Hughes trans-Atlantic flight in NX-19904. He also requested special action, as the proposed flight was due to start that week from Los Angeles.

Doc received a telegram on August 24.

> Commission granted you authority operate class two experimental aeronautical station as requested. Period until further order of the commission but in no event exceed one month. Use call W6XDV.
>
> T.J. Slowie Secretary FCC."

The purpose of Howard Hughes's flight was to demonstrate that by keeping in constant radio contact, air travel was safe for the general public. The New York-to-Paris leg of his trip set a record, 16 hours and 35 minutes, as was the time around the world, three days and 19 hours. Doc Stuart, operating as W6XDV, stayed in contact with Hughes during the entire trip. Don Wallace had provided the equipment that Hughes used during his flight, although Dave Evans had somewhat reworked it .

Just a few weeks after the Wallaces had returned home the September issue of *QST* arrived, which had Don's story, "The Yacht *Contender* Comes In First." Shortly afterward the October issue of *Radio* came out and it's cover had an aerial photo of the *Contender*. The picture was taken by Lewis Gregory, K6POR, about 80 miles off Makapuu Point, just a few hours before the race ended. There was a 7-page article that was written by Herb Becker, W6QD, describing his shore station and Don's account of the race. While everyone was trading stories of the yacht race and the round-the-world flight, a chain of events occurred in Europe that once more plunged that continent into war.

Doc Stuart, lower left, and the South Bay Amateur association.

War

On September 1, 1939, Germany invaded Poland. The British prime minister immediately asked Churchill to become First Lord of the Admiralty, roughly the equivalent of the American Secretary of War. Britain and France declared war on Germany two days later. Europe was immersed in the conflict that would become World War II.

The same month Don took his sons along on an afternoon cruise aboard the *Contender*. Dick Loynes had brought the yacht back to Long Beach after winning the Trans-Pacific yacht race in July. Bill and Don Jr., being just 17 and 15 years old, had been too young to accompany him as crew members during the race. They sailed to Catalina that weekend and anchored at the Isthmus Harbor, near the western end of the island. As they had only planned to sail for the day, no substantial provisions were brought along, just fresh pineapple that Don had brought back from Hawaii and a case of Coca Cola.

That afternoon a windstorm hit the Channel Islands. Bill Wallace recalled: "We had sailed over to Catalina, not planning to spend the night. We anchored at the Isthmus for lunch and the wind started rising. We had planned to eat at the little restaurant ashore, but it was too rough to launch the dinghy. A storm was rising so we elected to ride it out in the harbor, rather than out at sea. The wind was from the south. At times during the night it reached upward of 100 miles an hour, channeled between the bluffs. We had 300 feet of anchor chain out. That, and our engine running full bore, held us in place. The next day the eye of the storm seemed to pass through, and it was dead calm for an hour. We picked up anchor and moved a few hundred yards to the east side of Isthmus Harbor. We anchored where the bluffs afforded some protection from the wind. Then we were hit again at full force. It was two days before the storm passed. The *Contender* had a big 600-watt transmitter aboard and Dad was able to let Mother know we were okay. All we had to eat was pineapple the whole weekend. Even now, 50 years later, I don't care to eat pineapple. On Tuesday the weather cleared and the winds died down. The one restaurant ashore started cooking all the food that they had and putting it out on picnic tables. The crew of those boats who weathered the storm came ashore and were ravenous. You could eat all that you wanted. They had a big fish bowl by the door so that if you had a few dollars you tossed it in as you left. We got back to Long Beach that afternoon. We were worried about missing school, but it had been canceled that day."

If a hurricane had indeed run up the coast of Southern California and through the Channel Islands it was an extremely rare event. Although the region had been settled by the Spanish centuries earlier, adequate records on such matters have only been kept since the turn of the century. As hard as the islands were hit that weekend, the newspapers ashore did not report unusually high winds.

After the yacht race and the brief sailing cruise, it was business as usual for Don. He was out visiting companies and selling products. Charles Cook met Don then, when he worked for RCA Communications (RCAC). Fifty years later Cook recalled: "I bought an undulator (it copied CW from a continuous traveling tape that had been punched) from Don. He could sight-read the tape at a truly impressive speed as it reeled through the inker. Most of the RCAC circuits had an operator sitting by a typewriter with the tape running across a bridge track and you could read the code visually. I had put in more than a few hours doing just that myself, but Don could copy circles around me."

In the late 1930s Doc Stuart continued to be the dominant contest station in California. W6GRL was the highest-scoring station in the western United States in the 1936 ARRL DX Marathon, and in 1937 he finished fifth in the country. That year a phone section of the DX contest was added for the first time. Doc was a CW man and the phone contest did not interest him. In 1938 the CW DX contest was held when he was out of town on business, so he missed the competition. A guest operator, Dave Evans, W4DHZ (later W6SZY), used Doc's station in the phone contest and finished third in the country. Each year Doc improved his antennas and his contest score, little by little. In 1939 Doc got back into the CW Marathon again, finishing fourth in the United States. He worked an incredible 84 countries in the 10-day contest period. W4DHZ had been impressed with Doc's station the year before and returned to operate the phone section, finishing third in the country again. In the DX Marathon held in 1940, most of the European

Frank Lucas, W8CRA, achieved the first prewar DXCC in 1937.

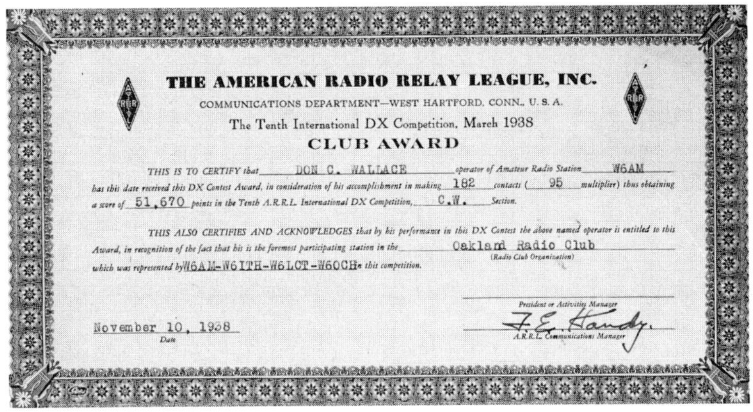

THE AMERICAN RADIO RELAY LEAGUE, INC.
COMMUNICATIONS DEPARTMENT—WEST HARTFORD, CONN., U. S. A.
The Tenth International DX Competition, March 1938
CLUB AWARD

THIS IS TO CERTIFY that DON C. WALLACE *operator of Amateur Radio Station* W6AM *has this date received this DX Contest Award, in consideration of his accomplishment in making* 182 *contacts (* 95 *multiplier) thus obtaining a score of* 51,670 *points in the Tenth A.R.R.L. International DX Competition,* C.W. *Section.*

THIS ALSO CERTIFIES AND ACKNOWLEDGES that by his performance in this DX Contest the above named operator is entitled to this Award, in recognition of the fact that his is the foremost participating station in the Oakland Radio Club *(Radio Club Organization) which was represented by* W6AM-W6ITH-W6LCT-W6OCH *in this competition.*

November 10, 1938
Date

President or Activities Manager
F. E. Handy
A.R.R.L. Communications Manager

An ARRL DX contest award.

countries and their colonies were off the air due to the escalating conflict in Europe. For the first time the West Coast of the United States was on the same level as the East Coast. That year W6GRL was the high scoring station in the United States by a large margin.

Radio magazine was to sponsor just one DX contest before the war, the 1939 DX Marathon. It was a year-long affair, with the cumulative total countries and zones worked during the 12-month period determining a station's final score. Doc Stuart had the high score in the western states. Due to events in Europe, it was the last DX contest that *Radio* sponsored.

One of the innovations that contributed to W6GRL's success was the electron-coupled oscillator (ECO). Later this would become known as the variable frequency oscillator (VFO). Instead of having just a few crystals available and calling a station until the operator found the correct transmit frequency, the ECO allowed one to adjust his frequency to where the desired station was listening when the last contact was made. Developed in the mid-1930s, many people would have stability and drift problems with their home-built VFOs for several years to come.

The first ARRL DXCC had been achieved by Frank Lucas, W8CRA (postwar W3CRA), in November 1937. Four months later Doc Stuart was issued DXCC certificate number 15. They played tag at the top spot on the DXCC for the next two years until the spreading war made it impossible to work any new countries. When Poland was invaded, Britain and France declared war on Germany. They suspended operation by their amateurs, as well as the countries that they had colonized. By September 3 nearly half of the 250 countries on the ARRL DX list (251 on the *Radio* list) were off the air (See Appendices 4 and 5).

The seat of government of the Republic of China had been driven 1,800 miles up the Yangtse River to Chungking by the attack of the Japanese, which began at Mukden on July 7, 1938. Nanking, the peacetime capital of China, fell on December 12. In 1940 the office of the Chinese Ministry of Information in the United States was searching for someone to pick up daily voice transmissions from radio station XGOY in Chungking, China. Because of the publicity Doc Stuart had received from his DXCC standing and DX contest achievements in the late 1930s, officials of the Chinese government conferred with him on this project. The Chinese Consul General in Los Angeles could not inform him of the wavelength or frequency of XGOY; Doc had to find it himself in the radio spectrum. He then set upon the task of receiving and recording in Ventura, California, the news bulletins being transmitted in the English language by the Voice of China. The recordings were then transcribed from the spoken to the written word and the copy dispatched to various offices of the Chinese government in the United States, Canada, and Mexico. After some experimental and developmental work, the first bulletin from the Voice of China was sent out by Doc on May 3, 1940.

After careful study and observation, and also to fit into the daily routine at XGOY, Doc determined that 1430 GMT was the best year-round time to receive transmission. The frequencies selected were the 49-meter band in the winter and the 31-meter band for the

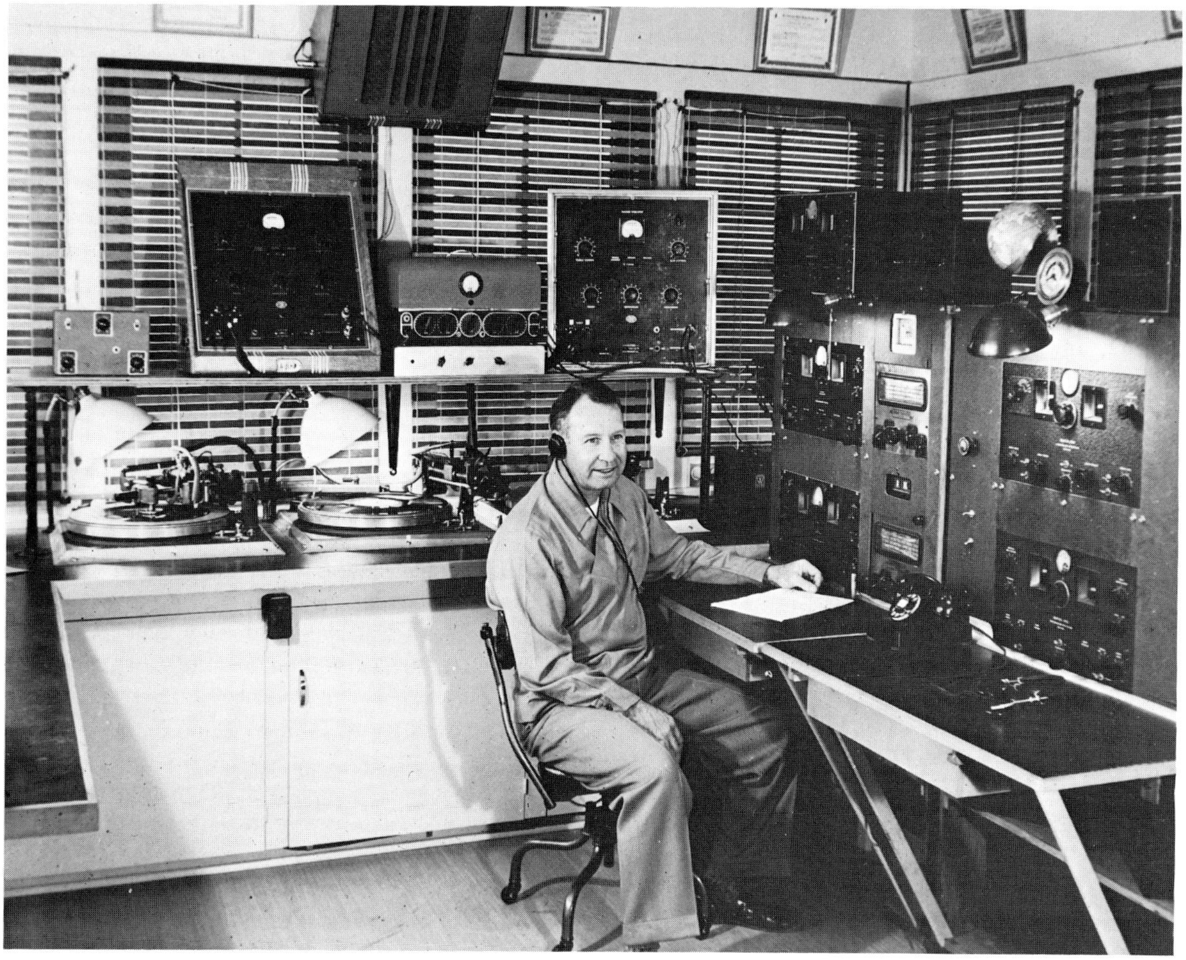

Doc Stuart at his XGOY receiving station in 1940.

rest of the year. In the winter the propagation path on the great circle route from Chungking to the United States was obscured in darkness as it passed over Manchuria, Siberia, Kamchatka, and the Aleutian Islands.

The routine adapted was to transmit by voice some 3,000 or more words, and sometimes up to 10,000 words daily, at the scheduled time from Chungking. This was recorded on standard lateral recording acetate discs revolving at 33 1/3 RPM, allowing about 10 minutes of recording on each disc. Since the discs were not being used except to transcribe their contents, and not for broadcast purposes, it was possible to cut much nearer the centerof the disc where the speech became somewhat bassy due to slow needle speed.

The recording was then placed on a play back and amplifier combination. The items of news, features, or messages were transcribed directly on a teletype; a tape was punched along with the printout. After copy was completed on the tape, the operator placed a conference call to the offices of the Chinese News Service in San Francisco, Chicago, Washington, and New York. On establishing contact, the tape was run into the teletype and copy was received simultaneously at those four locations at 60 words per minute. Feature copy was dispatched by air mail. Should the material be of immediate interest it was sent via teletype.

Receiving copy by voice over a 7,000 mile path was somewhat precarious by commercial standards. DXers, however, were more familiar with the fine points of propagation than were the theoretical scientists who consulted for the international news agencies and recommended CW. A good reader dispatched copy at the rate of 100 words per minute and accuracy was assured by spelling unfamiliar words and names. In an hour's transmission typically 6,000 words were handled. Some special programs were received over the 17,000-mile path, the long way around the world, on occasions when the shortpath propagation was out.

The longer path was more suitable for radio propagation at some times of the day. At these times the signal had an echo which was stronger than the initial signal. By eliminating the short path signal, clear reception of the longer (and stronger) path signal was possible. This was accomplished by the use of terminated rhombic antennas, which were unidirectional and reversible, so that the best path could be recognized and used.

Two double-wire rhombic antennas, each 550 feet per leg and nearly 1,100 feet in length from end to end, were used in the diversity reception. They were 80 feet high and the two antennas were separated by more than 500 feet to accomplish space diversity (the desired signal arrived at one antenna an instant later than the other antenna). The receiving unit was made up of a bank of three Hallicrafters SX-28A receivers, with their automatic voltage control (AVC) leads interconnected for diversity, and with the frequency controlled by a separate oscillator. Two Hammarlund Super Pro receivers in diversity connections were used as standbys. The audio outputs of the receivers were mixed in a standard line amplifier. From this the signal was put through a peak limiter and thence through an adjustable bandpass filter to attenuate unwanted frequencies, then to the recording amplifier and into the cutting heads of the turntable on which the records were cut. Two 87A Presto recording amplifiers and three Presto 6M turntables were used for active and standby.

The operation of Doc Stuart's station was particularly unique in that it was the first instance in which voice was used in the commercial transmission of wordage over a 7,000 mile path. It was also unique in that there was no contact or control between him and the transmitting station. It was strictly a one-way circuit; there was no facility for requesting repeats, corrections, or fills.

In its ordinary course of work, Doc's receiving station carried millions of words of copy at a cost of 1 cent per word, while common carriers were charging as much as 78 cents per word over the same route. For a period following the fall of Manila, Hong Kong, and Shanghai to the Japanese, Doc's station was the only outlet from Chungking to the outside world until commercial facilities were again established after the war. Millions of words of press and feature material were handled for news syndicates, newspapers, and magazines, in addition to the regular traffic for the Chinese government.

In 1941 a mailbag service was instituted by XGOY, through which people in China could relay personal messages to friends and relatives in the United States and Canada. A 50-word message was allowed once per month per person. In all, more than 8,500 such messages were relayed to their respective addresses and more than 1,200 letters of appreciation were received by Doc Stuart from grateful recipients. During 1942, the only year for which statistics are available, all the combined commercial telegraphic services from Chungking to the outside world totaled 900,000 words, whereas Doc handled more than 1.5 million words transmitted from XGOY. During the war years, in addition to regular copy, some 2,500 rebroadcast programs were delivered to radio stations and networks in the United States. More than 550 of these were for the *American Broadcasting Company* on its nationwide network. From records gathered it was revealed that Don's station carried on its regular reception of transmission when commercial services over the same path were out due to magnetic disturbances for periods as long as eight days in duration.

W6AM's antenna farm on Country Club Drive began shrinking in the late 1930s. As the American economy slowly improved, people began to buy lots and build homes up and down the street from Don. Once in a while this would be a lot that he had a mast installed on to support an antenna, and it would have to be taken down. Sometimes it could be relocated and sometimes not. In the late thirties Don had bought a couple of lots adjacent to his, but he could not afford to buy the whole block. By 1940 his 25 acres of antennas had diminished to just over 15 acres. At that rate his antenna farm would practically vanish in the next decade. Before neighbors moved too close, Don put up a third pole that was 95 feet high at the back of his property, on the other side of his lot from the other pole. In the long run Don knew that he would have to make do with just these three poles to support his antennas.

In June 1940 the Associated Radio Amateurs of Long Beach came down to the yacht *Contender* to operate Field Day. They brought along a battery-powered 10-meter rig that ran eight watts and they strung an antenna from the top of the main mast to a nearby oil derrick. The antenna was 100 feet high and 200 feet long. With this setup they made 38 contacts during the weekend. Using the same antenna W6AM/6 put the *Contender*'s 600-watt transmitter on the air for a brief period and made 126 contacts. A high-powered transmitter makes quite a difference in attracting attention on a crowded band. The operators were Bob Turk, W6LEE, John Koch, W6MND, Bob Kupps, W6LUA, Dave Stockham, W6NPY, Steve Stambuk, W6MEN, Dick Brock, W6LXC, Carl Louis, W6NRQ (W6SR), Ed Woolcock, W6EZL (W6FJ), Frank Ottoboni, W6NOR, Irv Emig, W6CYS (W6GC), Jim McWilliams, W6RTG, Dick Loynes and Don Wallace.

Initially the United States tried to remain neutral in the growing world conflict. In June 1940 the FCC forbade United States operators from contacting stations outside the Country. Some clandestine contacts did take place between United States and DX stations after that,

The ARALB portable rig.

The field day antenna was strung to a nearby oil derrick.

but only between operators who had known each other for many years. The penalty for anyone caught in the United States would be license revocation. A foreign operator, depending on his country, usually faced much more severe consequences. The only changes that took place in the DXCC listing after that was the occasional QSL that trickled in through the mail for an earlier contact.

Herb Becker had written the "DX and Overseas News" column for *Radio* magazine for the past several years. The state of affairs on the ham bands became so sad that commencing with the October 1940 issue he renamed the column "X-DX and Overseas News." The last ARRL DXCC listing was published in the December 1940 issue of QST. W2GT was tops with 152 countries, W6GRL had 151, W8CRA had 149, and W1FH had 138. W6AM finished at 89 countries confirmed, just short of DXCC. Don had been working the same countries over and over, rather than scratching around for new ones. Although he liked to work DX, his primary interest was still traffic handling. Actually, Don had worked 117 countries, enough to qualify for the DXCC, but he hadn't vigorously pursued the QSL cards for those contacts. With the turmoil in Europe and Asia, most of those amateurs became radio operators in their own armed forces, and the QSLs for past contacts were no longer obtainable.

In January 1941, *Radio* published the last listing of their WAZ standings (See Appendix 7). ON4AU was tops with 40 zones and 158 countries, G2ZQ 40/147, J5CC 40/130, W8CRA 39/156, W6GRL 39/151, W3AG 39/140, W6QD 39/135, and W6AM finished with 38/117. The country totals were slightly different than the ARRL because of the minor differences in the country lists between the two magazines, and, in Don's case, *Radio* would publish worked rather than confirmed country totals. In March 1941 *QST* stopped publishing the monthly "How's DX" column, and of course, there was no DX contest that year. In the July 1941 issue of *Radio* the last "X-DX and Overseas News" column appeared. It was indeed a sad state of affairs.

The last known prewar amateur radio activity in Tibet occurred at this time. Sakae Tamogami, J7CG (now JA1ATF), who had spent the past couple of years in Manchuria operating as MX3H, was on the air in mid-1941 for five months as AC4TF. His contacts were pretty much limited to Asia, as Europe and her colonies were off the air and Americans were limited to contacts within their country or with their own possessions.

When Don traveled to San Francisco he often stayed at the Whitcomb Hotel. Its location was convenient and the roof layout was ideal for stringing the antenna for his portable transmitter. The only problem was that there was already a broadcast station located on the roof and its harmonics interfered with some of the frequencies he liked to use. In June 1941 the broadcast station moved to another location and the hotel manager informed Don that he was welcome to use the 250-foot towers for his W6ZZA portable operation. Naturally, Don put his antenna at the top of the towers. It was

DX CENTURY CLUB AWARDS

These have been made to the first-listed amateurs, based on contacts with 100 or more countries, the credits all certified by examination of written evidence under the award rules.

W2GT	152	W2BHW	134	G5RV	122
W6GRL	151	W5BB	133	W8NJP	122
W8CRA	149	W2CMY	133	W9TB	122
W2GTZ	148	W8OQF	132	W6GAL	121
W2GW	146	G6RH	132	W9FS	121
G6WY	145	W4BPD	132	W2GVZ	121
W9TJ	144	W3CHE	132	W1ADM	121
W1TW	143	W2HHF	130	J5CC	120
W1BUX	141	HB9J	129	W9ADN	120
G2ZQ	141	W8LEC	128	W1JPE	120
W6KIP	140	W2UK	128	W7AMX	119
W8DFH	139	W3EPV	128	W9PST	119
ON4AU	139	W2ZA	127	G5BD	119
W8BTI	138	W9KG	126	W1IAS	119
W1FH	138	W8JMP	126	ZL1HY	118
W1SZ	137	W8DWV	125	W2BYP	118
W5VV	137	W9ARL	125	W1HX	118
W3EMM	137	W1DF	125	W8MTY	118
W1LZ	136	W3FRY	124	VK5WR	117
W8ADG	136	W4CEN	123	W8QXT	117
W6CXW	135	W5KC	123	W9EF	116
W1TS	135	D4AFF	123	W3EVW	116
W8OSL	135	W1AXA	123	W1WV	116
W2JT	135	W9GDH	122	W4CYU	116
W8DHC	134	W3EDP	122	W4DRD	116

115: W6ADP, W2CYS, W1CH, W3FQP
114: W9KA, W8BKP, W2DC, G2DH, G5BY, W3BES
113: G6CL, W2CJM, W2DSB, W2GRG
112: W6FZL, W3EVT, W3GAU, W2ARB
111: W2AAL, W1DUK, VE2AX
110: ON4UU, PAØXF, W9UM, W3AER, W8IWI, W1ICA, W1BXC W5QL, W2AV, W2IYO
109: W3DDM, W6FZY, G2MI, W9CWW, W9RBI, W8LFE
108: W6HX, W6TJ, ZS2X, HB9BG, W1BGY, W3BEN, VE3QD, HB9CE, VK3QK, W2GNQ
107: W2CBO, G5BJ, W3AG, VK2DG, VK3CX, W7DL, W2IOP, W6MVK, W3AGV, W6AHZ, W8PQQ
106: G2TR, W8EUY, W8JAH, W1IOZ, W9UQT, J2JJ, W1ZI, W1RY, W2VY, W3GEH, W9VDY
105: W2OA, G5QY, W1GDY, HB9X, W8DOD, W4TO, W4IO, W1GNE, W2BMX, W4BVD, W8LYQ, W3ZX, VK6SA, W8AAJ
104: EI5F, W1ZB, W4AJX, F8RR, W6KRI, W1GCX, W3KT, W6BAM, W8JTW, W8HGW, LU8EN
103: G6KP, W8KKG, W5CUJ, W9RCQ, W9NNZ, LY1J
102: W4CBY, W8AU, W8OXO, W1FTR, VE2EE, W2BXA
101: F8RJ, VK3KX, W6DOB, SU1WM, W1CC, SU1SG, G6MK, W4MR, W6GHU, W6KWA, W4EQK, LU7AZ, W1AB, W1AVK, W6ADT, W1CBZ, W2BJ, W4TZ
100: G6NF, VK2ADE, ZL1GX, ZL1MR, PAØQF, W8BSF, D3BMP, W9LBB, W4CCH, W8KTW, W5ASG, W8JIN, W8QDU, G6GH, W3AIU, W9LQ, W2JME, K4FCV, W6TT, W4FIJ, W2WC

Radiotelephone: **W2AZ 106; W2GW 105; W2JT 103; W6OCH 101.**

The following have submitted proof of contact with 75-or-more countries: W1KHE, W3GHD, W9AJA, W9JDP 99; G3JR, W2ALO, W3AOO, W2CTO 98; W4TP, W6MEK, W8LZK 97; G8IG, W1COI, W3FLH, W4DMB, W8BOX, W8DPS, W8IQB 96; F8LX, FB8AB, G6XL, W3EMA, W3OP 95; K4ESH, W8CJJ, W9AEH, W9BEZ 94; G6ZO, ON4GK, PAØQZ, W6FKZ, W9GBJ 93; W1KID, SP1LP 92; W1BGC, W1DOV, W8AAT, W9VKF 91; D3CSC, G6YR, ON4FE, SP1AR, W6NLZ, W8LAV, W9OVU 90; VK3HG, W2BZB, W2CUQ, W6AM, W8JFC 89; G2DZ, W3JM, W6LDJ, W8CED, W9PGS 88; PY2DN, W1APA, W6GPB 87; W2FLG, W8DAE, W8OUK, W9FLH 86; VK2TI, W4AHF, W4CFD, W6GK, W8BWC, W8GMH, W9GKS 85; SM6WL, W1BFT, W1BPN, W2AYJ, W8BWB 84; EI4J, OZ7CC, VE2GA, W2AWF, W6DTB, W6KUT, W8BFG, W9DIR 83; W1EWD, W2HTV, W3AYS, W3FUF 82; J2LL, W3EUJ, W9GY 81; G3BS, LA2X, W2BNX, W3BVN, W3EPR, W4OG, W6MHH, W8DGP, W8ITK, W9GMV, W9MRW 80; W4ZZ 79; W3DRD, W4EPV, W8FJN, W9YNB 78; W1NI, W6QAP, W9HUV 77; PAØJMW, W1EH, W3BSB, W3CRW, W3FHY, W9ERU, ZE1JI 76; HH2MC, VE3DA 75.

Radiotelephone: W4CYU 98; G5RV 92; W1ADM, W8LFE 90; W3EMM 89; W1AKY 84; W1JFG, W2IKV 80; W6IKQ, W8QXT 78; W1BLO 77; W1KJJ, W9TIZ 76; W2GRG 75.

The last pre-war DXCC list *(QST,* Dec. 1940).

unfortunate that the ban of working DX stations was in effect, but Don contacted Hawaii and the Philippines, as well as lots of stateside fellows.

Forty-five years later the author received a note from Charles Gagnon, W1LQQ. He remarked:

> I was cleaning up a batch of old QSLs and saw a couple that brought back memories. The W6AM card enclosed is from 1941, and the photo on the card was taken of Don during the 1939 Trans-Pacific yacht race.

A similar card was received from Lester Timmerman, W8TKW. He noted:

> The card was from a contact in 1940. I was licensed less than a year at the time. My equipment was a home-brew superhetrodyne receiver and a home-brew transmitter with low power output.

John Kaminski, W1NEH wrote:

> Enclosed are copies of Don's QSL card which I received in 1940 as an SWL. It was a great thrill for me to hear him on 160 meters via a Philco broadcast set and then have him verify my SWL card. This inspired me to get my ticket (license) soon after.

Don used one of the photos of himself that was taken during the yacht trip and had it printed on his QSL.

The Sixth Annual ARRL Southwestern Division Convention was held at the Hotel Huntington in Pasadena on the last weekend of August in 1941. Don organized the Oldtimer's Breakfast. By then, to qualify as an oldtimer, you had to be licensed prior to 1922. The convention always held a code speed contest and Don again took the honors, copying 35 words per minute.

Don still continued to use his home-built receiver much of the time. Commercial receivers now had a calibrated frequency dial readout that was quite accurate. However, Don often redesigned his receiver as technology progressed, and it was often more sensitive than many of the new receivers that were available. During the war the technology would advance rapidly, and his trusty receiver would be retired in 1945. It was still on the same shelf in his basement when the author discovered it 40 years later.

On May 10, 1940, Churchill was named the British

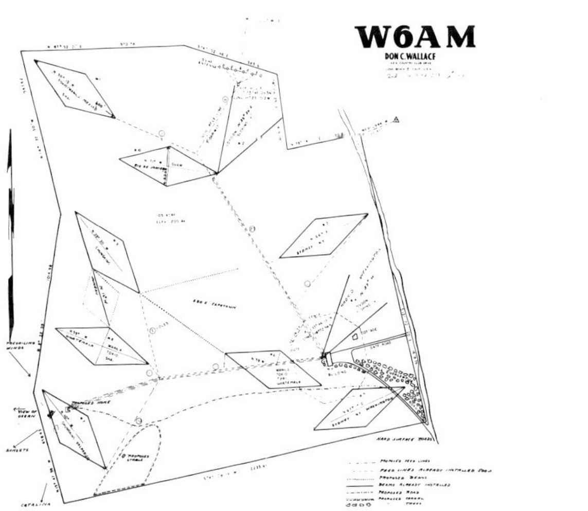

The *Press Wireless* installation in Rolling Hills had eight rhombics and 12 miles of wire in the air by 1943.

Prime Minister. The Germans marched into Paris five weeks later on June 14. Four days after that, an obscure French General, Charles De Gaulle, delivered a speech that was to rally his countrymen's resistance: "France has lost a battle, but France has not lost the war!" In the last days before the fall of France, Churchill had rallied his beleaguered people: "We shall fight in France, we shall fight on the seas and oceans . . . We shall defend our island, whatever the cost may be, we shall fight on the beaches, we shall fight on the landing grounds, we shall fight in the fields and in the streets, we shall fight in the hills; we shall never surrender!"

The Germans planned to storm England later that year, a feat that had not been accomplished since William the Conqueror in 1066. On August 13 the Luftwaffe (the German Air Force) was ordered to overcome the British Royal Air Force. That marked the beginning of the Battle of Britain, the first conflict fought entirely in the skies. The British had a newly developed system that gave them nearly 30 minutes advance warning of each attack—radar. This early warning system proved to be decisive. The blitz lasted until the next spring, when the attackers were finally beaten back.

American military men visited the British radar installations during this time. They were suitably impressed, as these were superior to the radar sets that the United States had developed thus far. The British were anxious to share their technology with the Americans because they would need the vast manufacturing resources in the United States if they were to win the war. With the drawings firmly in hand, the Americans returned stateside and began redesigning their radar sets. The Eitel-McCullough Company (Eimac), manufactured tremendous quantities of radar tubes during the war.

Don's older son, Bill, graduated from high school in 1940. He was attending college in Long Beach when the Japanese attacked Pearl Harbor on December 7, 1941. Germany and Italy declared war on the United States three days later. The next week the United States declared war on Japan, Germany, and Italy. Bill, of course, wanted to go out and enlist immediately. Don convinced him to finish that year of college and then join the Navy. *"In the Navy,"* he said, *"you always have a good meal and a warm place to sleep."* Don Jr. graduated from high school in June 1941. He also joined the Navy in June 1942, six months after Pearl Harbor. As Don Jr. had just started college he applied for admission to the Naval Academy. The Battle of Midway was fought a week earlier, and that engagement was the turning point of the war in the Pacific.

Due to the conflict in Europe in 1941, the American Armed Forces had been quietly preparing for entry into the war. Less than four months before Pearl Harbor Don received a letter from the Navy Department. It read:

> The Bureau is informed that you have had considerable experience with the operation of radio transmitting and receiving equipment carried in a suitcase. Information is requested regarding the type of equipment used and the power output. Further information is desired concerning necessary antenna sizes, distances generally covered and frequencies used. The bureau is desirous of obtaining data covering such distances as 100 miles, 500 miles, 1000 miles, and 4000 miles.

Don responded with a diagram of his 200-watt portable radio, suggestions for updating the 15 year old design, and a table of frequencies, antennas, and operating times for the required distances. After the United States entered the war, at their request he sent the portable transmitter into the Navy Department for evaluation. Its final disposition remains unknown; it is the only piece of Don's home-built equipment from the 20s that he did not retain.

Once the United States declared war, amateur radio operation was suspended. A few special exceptions were allowed for emergency work, but even these were withdrawn on January 10, 1942. The government did not require that the amateurs' radio equipment be dismantled as it had in World War I, as there were too many radios in the general population. However, transmitters that remained intact had to be registered. The Armed Forces gave out a call for equipment, as well as radio operators. The enormous logistics required for a war on two fronts, Europe and the Pacific, were staggering. Receivers, test equipment, even meters were in desperately short supply. During the first year of the war the Navy purchased a number of Don's extra RME receivers.

Don registered his transmitters with the FCC in August 1942, and received the authorizations to keep them intact. He had built another 1-kilowatt rig the year before that used a pair of the new power tubes available, 450TLs. Don still had the 5-meter set at home which ran 50 watts input to a pair of 245 tubes. Naturally, he also had the 5-meter 50-watt mobile rig using a pair of 12 tubes that he had built in 1935. It had to be removed from the car during the war. Don also had a commercially-made transmitter, an Abbott MRT-3 that ran 25 watts on 2 1/2 meters for WERS work. For 10 meters he had a pair of transmitters that were battery-operated for yachting trips under the 10-meter mobile regulation; one ran 4 watts to an F5G, the other was a Barr DB-3 that ran 2 watts to a 49 tube. Don had another 10-meter rig that ran 20 watts to a 6A6 that he used mobile for Field Day operation. Don also listed a breadboard transmitter that was the prototype to his 200-watt portable transmitter which he built in 1927. It used a single 203A that he listed at 75 watts input. He listed the water-cooled 3-phase transmitter that used the F-328A under Bertha's call, W6MA. A transmitter that ran 500 watts input to a pair of 210s was listed with Bill, his older son. Bill had become licensed as W6TCG in 1940. In all, a total of 10 transmitters were licensed at the W6AM residence. Don was ready for amateur radio operation when it would again be authorized after the war.

In 1941 President Roosevelt instructed General Donovan to form the Office of Strategic Services (OSS), which was the forerunner to the CIA. Donovan selected John Ford to head the photographic division of the OSS. The thirteenth member that Ford recruited was Sterling Barnett.

Barnett had come to Ford's attention due to the innovative camera work he had done for Lee de Forest and others in the industry. Seven years earlier Barnett had been working late one night at his lab on the sixth floor of the Bekins Building in Los Angeles and had spilled a caustic chemical. He went down to the floor below to clean up the damage, which happened to be de Forrest's lab on the fifth floor. De Forest was experimenting with TV reception from a transmitter that he had installed on Mount Lee. Barnett was a tinkerer and he developed a camera with an adjustable speed that enabled de Forest to synchronize the picture of his television.

During the war years, the amateur radio magazines were to fall on hard times. Most of their subscribers joined the Armed Forces and were stationed overseas. Consequently, their revenue was severely reduced. *QST* was well-established and, by cutting corners, managed to stay afloat. *Radio* magazine had to lean heavily toward the broadcast industry to stay in business. The smaller amateur radio publications that had somehow survived the depression simply faded into oblivion during the war. Operators could extend their station licenses, but new licenses and call signs were not issued. After the spring 1942 issue the *Radio Amateur's Call Book* ceased publication for four years.

The number of American amateurs remained relatively unchanged in the middle and late 1930s. In 1934 there were 46,000 licensed operators and at the beginning of World War II, just over 52,000. During World War I 80 percent of the amateur radio operators in the country served in the military. The average age of amateurs had increased in the past generation. Still, fully half of the amateurs in the United States answered the call to arms during World War II.

The American economy, which was stagnant during the Great Depression, began gearing up for the war effort. Only 3 percent of Americans had earned enough money to pay income tax in 1939. Due to the large infusion of capital for the war effort by the government, American industry was at full throttle by the end of 1942. There was nearly full employment again as factories began producing everything from field radios to battleships. Business was booming for Don also, as the manufacturers' lines that he represented were in great demand by the government. He had started representing Hytron Tubes in the late 1930s, and their orders increased a hundredfold in just two years.

After the attack on Pearl Harbor, *Press Wireless* was taken over by the Office of War Information (OWI) of the United States government. Overseas communications were of vital interest. Consequently, their sites were closely guarded by the military and they were under OWI direction. Japan invaded the Philippines shortly after the attack on Pearl Harbor. The *Press Wireless* sites in San Francisco and Rolling Hills were the only links to America in the coming months. After Manila fell, General Douglas MacArthur retreated into the Bataan Peninsula and directed the effort against the Japanese from his headquarters on Corregidor. The United States Army held out for another six months and probably prevented the invasion of Australia.

In the spring of 1942 MacArthur's closest allies were China's Chiang Kai-Shek's troops, located over 1,000 miles northwest of the Philippines. The Chinese were precariously supplied via the Burma Road. When it was cut off at its southern terminus at Lashio in April, they were left dependent on the even more fragile trans-Himalayan airlift, "the Hump."

There are dozens of stories of the adventures of amateur radio operators that surfaced after the war. One was of a second lieutenant who avoided capture when Bataan fell and was hiding out in the jungle. He did not know the code or have an electronic background. However, he had a tattered copy of the 1936 ARRL handbook. By dismantling an old broadcast radio and using the handbook as a guide, he used the parts to fashion a crude transmitter. He taught himself the code and would get on the air each day and transmit any information on troop or ship movements he had observed that day. He took down his antenna each night and moved his location to avoid capture. The Navy was at first skeptical of his authenticity, but as information continued to flow in eventually became convinced. Meanwhile, the Japanese mounted an intensive search for him. Fortunately, he eluded them until the islands were liberated by MacArthur three years later.

Another memorable incident occurred when a group of Navy men landed on an island in the South Pacific in 1944. They had come ashore from a submarine and their mission was to blow up the Japanese communications center on the island. One man went ahead with the dynamite and, as he slowly approached a building situated by some antennas, the door opened and a Japanese radio operator stepped outside. The American had his bolo knife ready and was prepared to quietly dispatch the operator.

Thunk! his knife hit the ground He was so startled he dropped it. The Japanese radio operator was reading an old copy of *QST*. They looked at each other for a moment and the American said "What's your call sign?" "J2—" was the response. They had worked on 20-meter CW before the war and exchanged QSL cards. Being typical hams, they chatted for a while about amateur radio and the unpleasant state of world affairs. Then they came up with an idea. The Japanese were planning to evacuate the island in a few hours. The radio shack was cleaned out and dynamite was placed both there and by the antennas. A short time later the objects were destroyed and the American returned to his comrades and the ship, mission accomplished. Both amateurs survived the war and remained friends for decades afterward.

Most people in America found the war in the Pacific confusing. Although many had a rough idea of the location of the European countries in their minds, few maps of Asia and the Pacific hung on classroom walls. Aside from an amateur radio operator or a pilot, who else had heard of Yap, Hollandia, the Bismarks or New Britain, where 16 inches of rain had fallen in a single day? The battle of the Coral Sea was actually fought in the Solomon Sea.

The ingenuity of the radio operators was remarkable. After serving on an LST to help ferry the English 8th Army from Africa to Italy in 1943, Ralph Salaway had to have surgery in Heliopolis. He recalled: "Afterward I stayed in the English barracks at Port Said for a few weeks. Their master sergeant came in at sunrise each day — the noise level sounded as though Rommel had broken through! I wanted to get back in the war, but there were a number of people waiting for lifts that were ahead of me. I was standing at the end of the barracks one day and used signal flags to talk to the signalman on the bridge of a ship passing in the canal. He told me that they had just one bunk available.

"I knew that if I was going to ship out I had to make my situation the first sergeant's problem, so I went in and asked to be paid, since I hadn't been paid in some time. He said that the paperwork generated was enormous when one army pays members of another army, which I knew. He would be filling out forms in triplicate for a month. Then I mentioned that there was a bunk available on the ship that just pulled in. Realizing my plan, he growled that I should meet the ship at sunrise and not to tell anyone about it. That's how I got out of Egypt and back into the action."

By the end of 1943 *Press Wireless* had eight rhombic antennas installed on the Rolling Hills property. Nearly 15 miles of wire was in the air. As General MacArthur pressed the Allied offensive north from New Guinea, after an area had been secured mobile press units would install rhombics pointed at the United States. They used the call sign PZ. Sixty-foot poles were assembled from kits of 10-foot long pieces clamped together and guyed at each joint. By this method the Army kept in direct contact with the Office of War Information in the United States as their command post moved forward. MacArthur had a quarter of a million men and nearly 1,000 ships under his command when he liberated Luzon. The day following the landing, *Press Wireless* had a station set up and operating.

Avron Cohen, W2SBW, worked for *Press Wireless* in Hawaii during the war. Forty years later he recalled: "We had about 18 highly directional multi-curtained rhombic antennas, each one fixed at military installations in the Pacific and on the mainland. On a recent visit to Hawaii I was able to locate the old site after much searching. The antennas and transmitters have long since disappeared and the site has been reconstructed."

Hugh Cassidy, WA6AUD, went to Europe on the troopship *George Washington* in 1944. Forty years later Cass recalled: "I had but a passing acquaintance with Don, but the first time we met I brought him to a full standstill by saying, 'You know something? You and I went over to Europe on the same ship!' He did not contradict me but there was a quizzical look which said clearly that he was trying to figure out how that could be and how old was this fellow anyhow. I finally had to tell him — same ship . . . — different war. Once, when he stopped by the house, he looked at my picture of the *George Washington* and acknowledged that it was the right ship."

The War Emergency Radio Service (WERS) was implemented in 1942. It was administered by the Civilian Defense Corps and licenses were issued to communities rather than individuals. The regulations to participate in WERS were rather strict and all operation was on 2 1/2 meters. However, it was the only radio activity vaguely resembling amateur radio in the United States for the next three years.

A small amount of amateur radio activity continued during World War II. Central and Southern Africa, as well as Central and South America, did not see appreciable fighting. Many of the operators were fairly remote, in terms of world population at that time. Little is known of the extent of that activity, as amateur radio operation was not reported in *QST* during that period.

Britain was to serve as a staging area for the largest amphibious operation in history. On D-day, June 6, 1944, the American Army and British Army landed at Normandy, France. A mobile *Press Wireless* unit came ashore on Omaha Beach with the troops. They used the call sign PX and transmitted information back to New York. In the ensuing months General Patton moved so fast across France that the unit hardly had its antenna set up when they had to tear it down and relocate. They accompanied the Army to Berlin and, by May 9, 1945, Victory in Europe Day, they had set up 24 times and strung 5 miles of wire.

An estimated 1 million casualties would have been sustained by the Allied forces if the original plan to storm the Japanese mainland had been carried out in 1945. Three scientists who were refugees from Hungary had met with Albert Einstein at Princeton University in 1939. They had urged him to inform President Roosevelt about the need for an American program to build a bomb of enormous power. Einstein complied with their request and, shortly after Pearl Harbor, the Manhattan Project was launched.

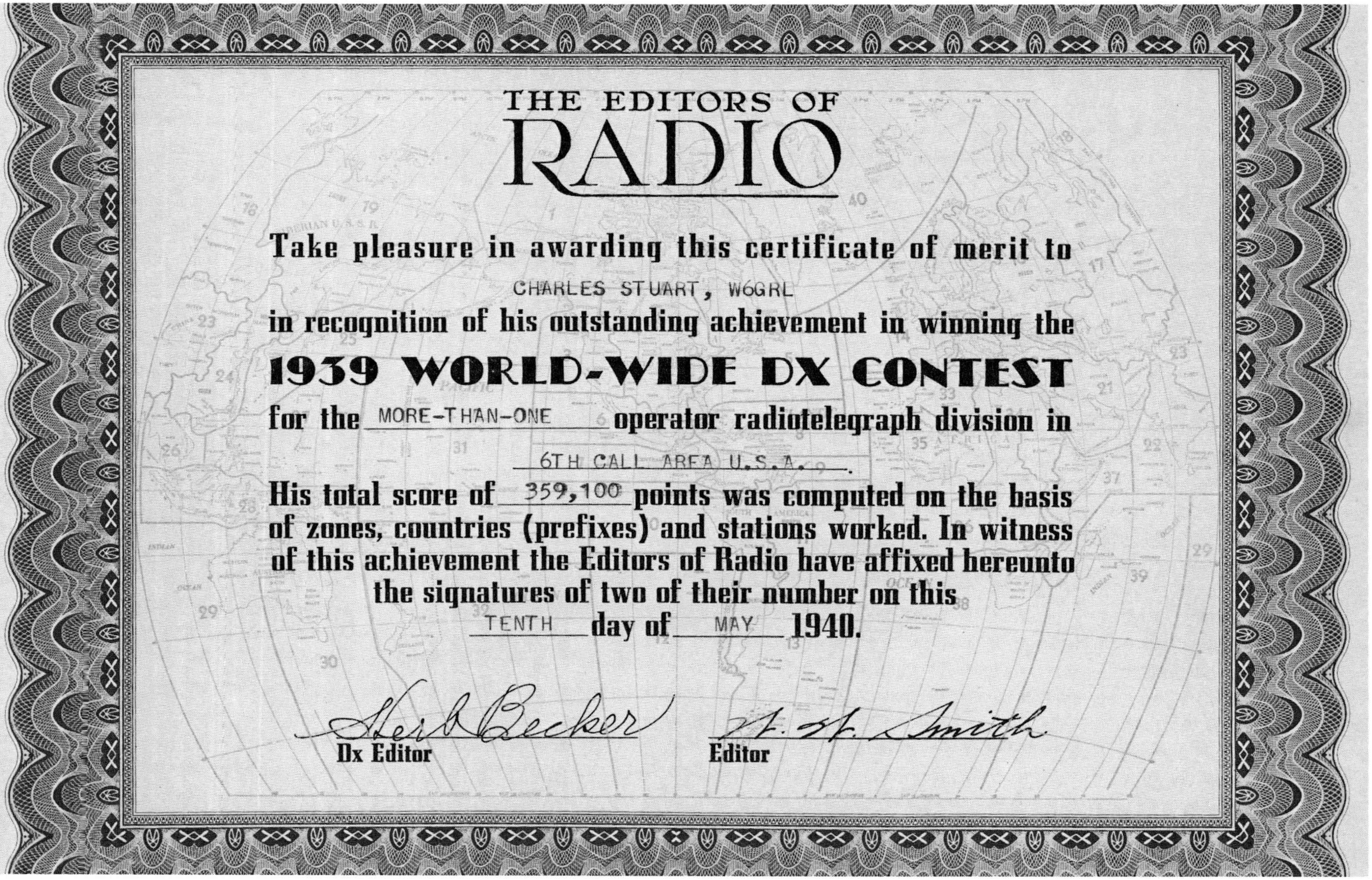
THE EDITORS OF
RADIO

Take pleasure in awarding this certificate of merit to
CHARLES STUART, W6GRL
in recognition of his outstanding achievement in winning the
1939 WORLD-WIDE DX CONTEST
for the MORE-THAN-ONE operator radiotelegraph division in
6TH CALL AREA U.S.A.
His total score of 359,100 points was computed on the basis of zones, countries (prefixes) and stations worked. In witness of this achievement the Editors of Radio have affixed hereunto the signatures of two of their number on this
TENTH day of MAY 1940.

Herb Becker
Dx Editor

W. W. Smith
Editor

A *Radio* Contest certificate.

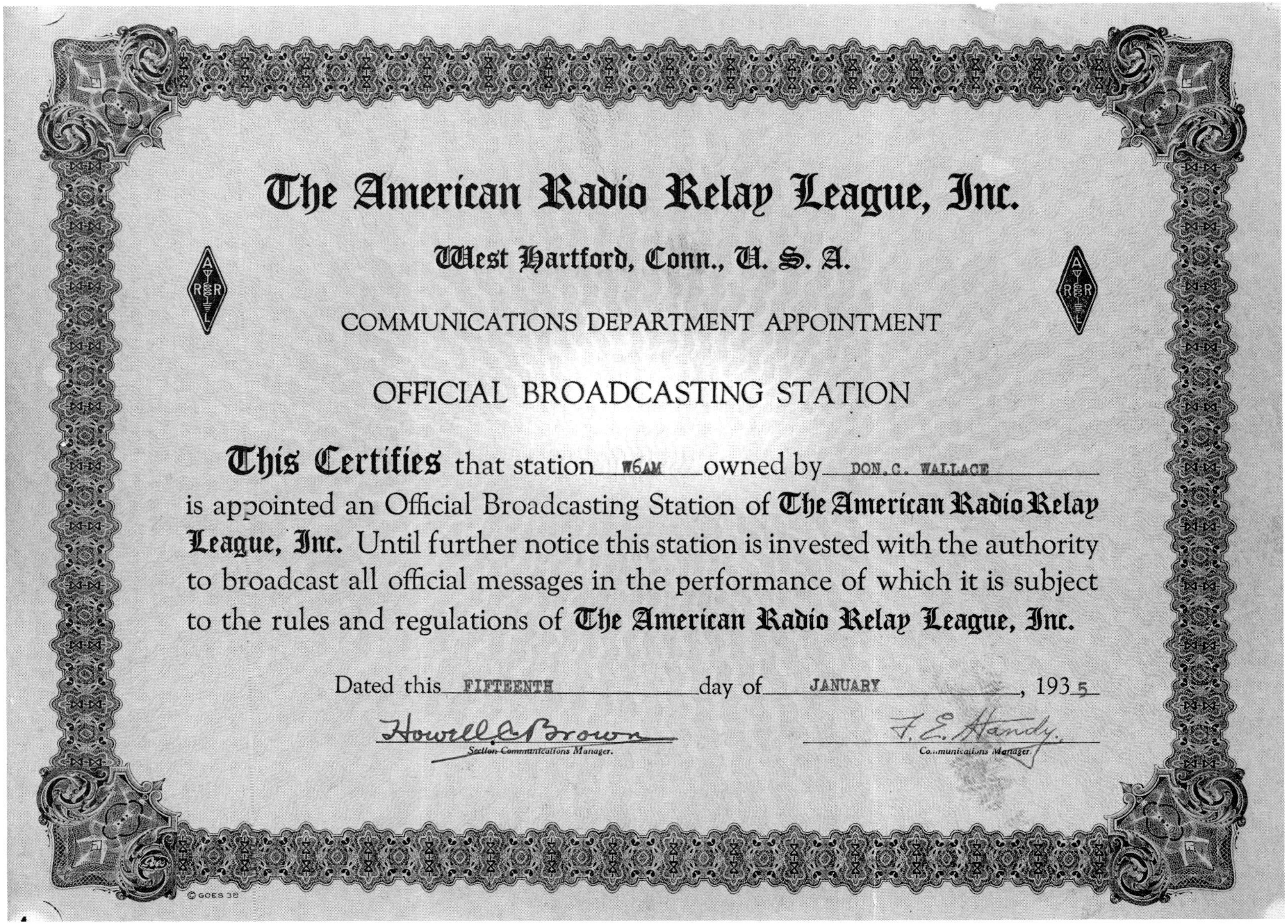

The American Radio Relay League, Inc.

West Hartford, Conn., U. S. A.

COMMUNICATIONS DEPARTMENT APPOINTMENT

OFFICIAL BROADCASTING STATION

This Certifies that station W6AM owned by DON C. WALLACE is appointed an Official Broadcasting Station of **The American Radio Relay League, Inc.** Until further notice this station is invested with the authority to broadcast all official messages in the performance of which it is subject to the rules and regulations of **The American Radio Relay League, Inc.**

Dated this FIFTEENTH day of JANUARY, 1935

Howell C. Brown
Section Communications Manager.

F. E. Handy.
Communications Manager.

© GOES 38

Don's official broadcasting station appointment.

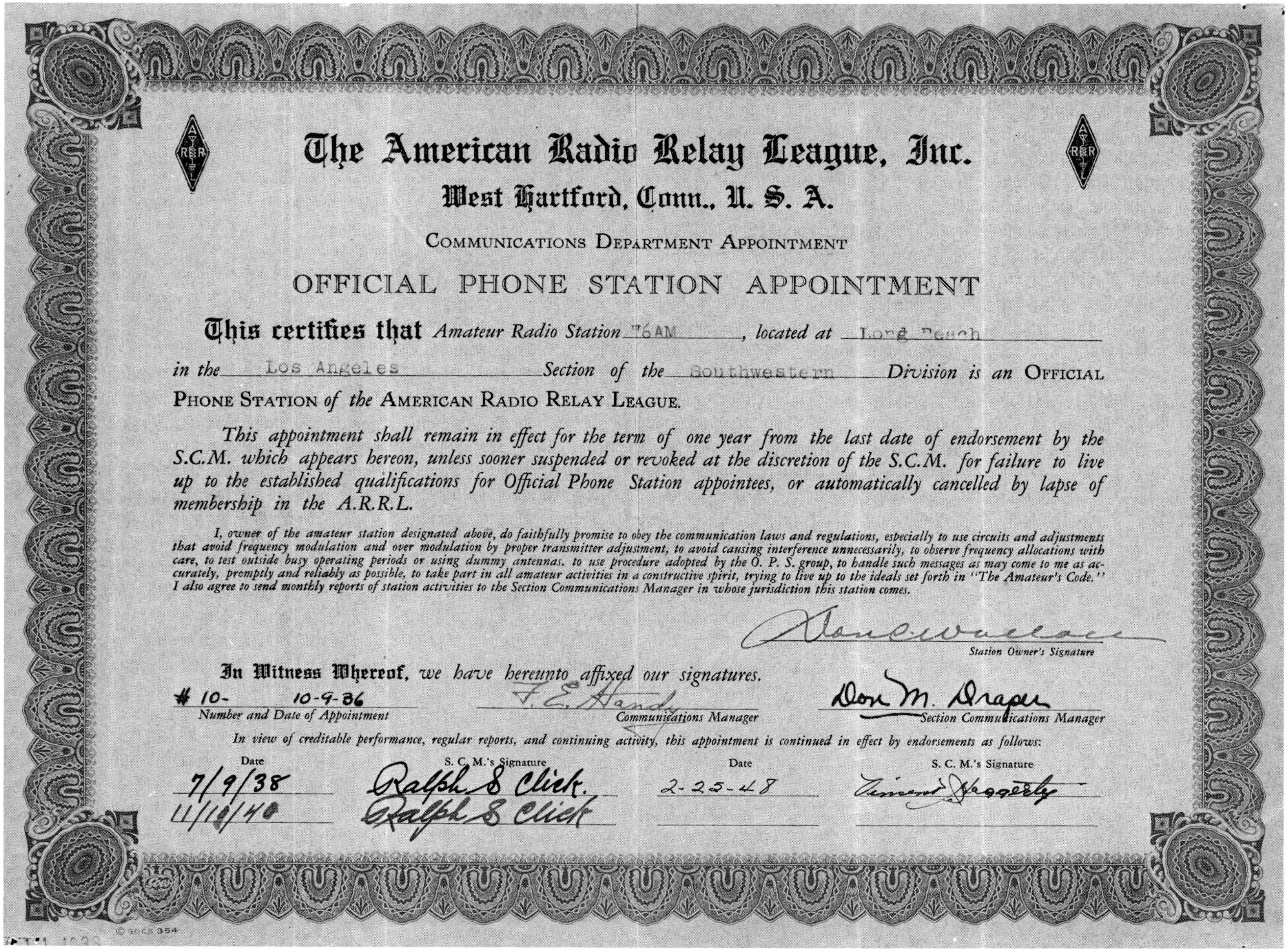
The American Radio Relay League, Inc.

West Hartford, Conn., U. S. A.

Communications Department Appointment

OFFICIAL PHONE STATION APPOINTMENT

This certifies that *Amateur Radio Station* W6AM, *located at* Long Beach *in the* Los Angeles *Section of the* Southwestern *Division is an* Official Phone Station *of the* American Radio Relay League.

This appointment shall remain in effect for the term of one year from the last date of endorsement by the S.C.M. which appears hereon, unless sooner suspended or revoked at the discretion of the S.C.M. for failure to live up to the established qualifications for Official Phone Station appointees, or automatically cancelled by lapse of membership in the A.R.R.L.

I, owner of the amateur station designated above, do faithfully promise to obey the communication laws and regulations, especially to use circuits and adjustments that avoid frequency modulation and over modulation by proper transmitter adjustment, to avoid causing interference unnecessarily, to observe frequency allocations with care, to test outside busy operating periods or using dummy antennas, to use procedure adopted by the O. P. S. group, to handle such messages as may come to me as accurately, promptly and reliably as possible, to take part in all amateur activities in a constructive spirit, trying to live up to the ideals set forth in "The Amateur's Code." I also agree to send monthly reports of station activities to the Section Communications Manager in whose jurisdiction this station comes.

Don C. Wallace
Station Owner's Signature

In Witness Whereof, *we have hereunto affixed our signatures.*

\# 10- 10-9-36
Number and Date of Appointment

F. E. Handy
Communications Manager

Don M. Draper
Section Communications Manager

In view of creditable performance, regular reports, and continuing activity, this appointment is continued in effect by endorsements as follows:

Date	S. C. M.'s Signature	Date	S. C. M.'s Signature
7/9/38	Ralph S Click	2-25-48	Vincent Haggerty
11/18/40	Ralph S Click		

Don's official phone station appointment.

Don's Rhombic Farm

Don acquired the 105-acre *Press Wireless* receiving site on top of Palos Verdes Peninsula in 1945. *"I bought the property,"* he said, *"at the end of the war. There wasn't any property selling up there at that time because no one had been allowed to build. During the war all new home construction was put on hold. Our industry was producing products for the war effort and no building materials were available. Toward the end of the war* Press Wireless *found out that they had to move to San Francisco. The FCC had told them that they were no longer going to issue any frequencies to press organizations for interior use in the United States.*

"I don't think I had ever seen their station here after they built it in 1939. Press Wireless *had a point-to-point communications link from here to San Francisco. They also had installed a point-to-point system from here to New York. Without those two circuits all the news they received here (in Rolling Hills) from overseas had to go to San Francisco via the wire service before being disseminated to the newspapers. The reason was that there were four news services:* United Press International, Associated Press, International News Association, *and* Press Wireless. *If one of them said there had been a battle and that there were a certain number of casualties and another organization reported different figures, people wouldn't believe any of the news.*

"Although the press services were competing, they had to check with each other for their own accuracy. They would write up their stories so they were fairly similar, but they were different. They did not copy each other's story, but the material would track. All these press services were in the top of the Chronicle Building in San Francisco, just down the hall from each other. Apparently, they would just walk back and forth and show each other what they had. People talk about scoops. They wouldn't dare scoop each other, they had to clear everything through their San Francisco office.

"Right away, it was going to cost Press Wireless *$35,000 a year to put the overseas messages on a wire service to San Francisco. That was an expense that they hadn't had previously. They only had one or two transmitting and receiving operators at each station at a time. As far as engineering and costs of operation were concerned, there was just one 40-kilowatt transmitter on line at a time. It didn't cost any more to send all that material to San Francisco, even with a dozen of their transmitters operating. As far as they were concerned, current was cheap, less than a cent per kilowatt hour. They would just crank the transmit power down to where the signal level was satisfactory at the receiving station. They never ran the full 40 kilowatts they were capable of at the transmitting sites unless propagation conditions were poor. They would usually run 2 or 3 kilowatts, just enough to assure solid communications. There were usually a couple of operators and about 75 receivers on these different antennas.*

"Closing this station was a desirable move for them. They had paid only $240 an acre for the land; $35,000 a year was a lot of expense compared to their initial investment. They bought the whole 105 acres up here for $25,000. The operating expense of $35,000 a year after 1945 was out of proportion. So they bought some swampland out in Napa, cheap land, and prepared to move their receiving and transmitting sites in Southern California up to the San Francisco area.

"This property going up for sale wasn't advertised, I was tipped off [by Tony Gerhardt?]. *I came up here and talked to the operators, one of whom I had worked on spark in 1914 when he lived up in Piedmont. He said they had an offer of $25,000 for the place.* Press Wireless *had told them that they wanted to recover the cost of their investment in the facility; including the buildings and aerials and that was $35,000. That gave me a hunch and I said, 'Well, I'll see what I can do.'*

"I went home and talked to my wife and said let's go for a Sunday drive. We drove up here and as we came over the hill I stopped the car. I said that I could buy that radio station up ahead. 'Humph, more radios,' she said, 'but it looks like a good investment.' So, with that authority, we finished our afternoon trip. She thought I had pulled one over on her—go for a Sunday drive and then look at a radio station.

"Anyway, I came back to the station and discussed it with the operators some more. A big shot had stood here outside the window and said, 'Hmm, I'll give $25,000 for it.' But they heard that, that's how they knew what the offer was. So, I said, 'I'll give you $35,000.' Well, the other guy froze out right away. I

could use the aerials and the buildings and he couldn't. He wanted the property for a subdivision of homes. Then I found out you couldn't buy any property up here at that time unless it was okayed by the Vanderlips.

"The Vanderlips had a corporation and the deed restriction said they had to okay every new buyer. It was primarily meant to keep out undesirables. That's now illegal, of course, but nevertheless it was effective at the time. They owned virtually the entire 16,000 acres on the peninsula. So I had to be approved by them. They didn't want a ham up there. They wanted to see those aerials that Press Wireless *had installed bulldozed. Well, I took my yacht trip scrapbook down to Kevin Vanderlip one day and said, 'You know, every community should be balanced. It ought to have a variety of people; one doctor, one lawyer, one musician, one merchant, one radio amateur, and so on. It would get awful lopsided if you keep out anybody. I think it would be a good idea to have one ham up here. I'm a good one and I'll show you why.'*

"I showed him the scrapbook and the clippings my secretary had put together during the yacht race. He said, 'Can I take that home?' I told him, 'Sure.' About four days went by and he called me. He said, 'I never knew hams did all that.' 'Well, I said,`you can see what kind of publicity you will be in for if I live up there. Norman Chandler often publishes articles about my radio activity in the Los Angeles Times. *Rolling Hills will become known the world over.'*

He replied, 'I am trying to believe you.' So from then on he helped me get the property. I hadn't heard from him for a while and I went to his office one day. There had been a big fire up in Rolling Hills the day before. It burned up about 12 or 15 houses. When I heard about it I flew right down from Seattle, where I had been visiting customers, and went in to see him. I said, 'It is a good time to buy that ranch, isn't it?' They were all down in the dumps in their real estate office because, with this fire, nobody was going to buy any property there. It just swept the hill. Now they had time to talk to me, a buyer. Vanderlip said, 'I am going to New York next week on business. If you want to pay my expenses, I'll go over to Press Wireless *and see if I can't buy it for you.' I said okay and gave him a check for $1,000. He went to New York and bought the property for himself and put it in escrow for me. That way he could put all his own deed restrictions in the transaction. He wouldn't be a wealthy man if he didn't do things in his interest. I worried about the restrictions for a while because I wanted to get the things in it that I could live with. We finally agreed on the wording in the paperwork. I had a deed that said the property is a radio station. I am the only one residing on the 16,000 acres that has that provision in their deed.*

"He said, 'How much will you put down?' I replied, `How about $20,000?' That woke him up because, at that time, nobody had any money. We had just come out of a depression. I had been working on commission and made more money the past few years than I could spend. The money I made from the Hytron Tube account in the last three years paid for the property.

"I have the only piece of property on the peninsula that doesn't have a restriction against aerials. You can't even string a clothesline. You can't have a fence unless it is a certain kind of wooden fence, painted white. The deed restrictions are about 200 pages, a whole book.

"After that, every month I called up the Times *and talked to Norman Chandler and said, 'I have a radio story for you.' He would give me a crack reporter to talk to. I told the fellow, 'I called Norman Chandler to give him a story and he always wants it on the front page.' So every month I had a story that went in the middle of the front page of the* Los Angeles Times *I would tell this reporter, 'Now, play me down but play up Rolling Hills. Buy land, Rolling Hills, California,that was the theme. You know, you can concoct a pretty good story out of anything that happens on the air. People are always interested in what goes on. So I would give him a story and when it was published I would cut it out and mail it to Kevin Vanderlip. He would see that I was doing my part when I said he would get a lot of publicity. Rolling Hills property values improved and the market went up. The place became well known and I think I had a lot to do with it. We became fast friends. I made him a lot of money and he helped me establish my status up here. We just thought a lot of each other. It was part of my personal obligation. I would have a headline on the front page and somewhere down the middle of the article would be 'Don C. Wallace of Rolling Hills.' We worked together until he died.*

"I paid $140 a month to go through the gate house down the way and that paid for my road maintenance. There wasn't any other road up there. My nearest neighbor was two miles away. When it rained I would have a hell of a time getting up there. I would stop to put on chains where the pavement stopped and I wouldn't always make it. One night I sunk into the mud up to my axle. I got on my radio and got someone down the hill to call a farmer who came out with his tractor and pulled me out. I didn't even get wet.

"All this was possible because Tony Gerhardt and I bettered our code speed together in the Navy in 1918. Tony won a big cup for code copying in the 1920s. It was three or four feet high and read' The fastest code

Don in front of the *Press Wireless* building.

copier in the United States.' He won it at a big Eastern convention. It was in all the syndicated papers. I got all the cups for the West Coast, so our practice enabled me to copy fast also. Tony and I got our speed up while we were young enough to use it.

"I had these eight rhombic antennas pointed in different directions. The Press Wireless *structure was this box of a house. It was just terrible looking. Well, I met with the Vanderlips and bought 15 acres more for $15,000. Now Bertha and I could build a house on the bluff at the western edge of our property with a protected view. I was going to re-route all the antenna feed lines down there. In those days I could work on aerials very fast. Anyway, that gave me a total of 120 acres to put antennas on.*

"The work room here in the Press Wireless *building was full of shelves and benches, with a couple of windows. There was an office and the other little room next to it was so an operator could take a nap. The large back room was for file storage. When I started removing the files from it, I ran into a rattlesnake. Mice were all over, everywhere. Wherever there were mice you would find rattlesnakes. I brought a couple of cats up, and put out a little food for them. Not a lot, just enough to keep them from going hungry if they couldn't catch any mice. The motivation was there, and they got to be pretty good mousers. After a while there were fewer and fewer rattlesnakes. They used to call this rattlesnake hill, you know."* Forty years later, Donald deNeuf, WA1SPM, the former President of *Press Wireless* recalled, "The boys at 'RH' (Rolling Hills) *Press Wireless* used to gripe about rattlesnakes up there. I always wondered if they are still around or if Don managed to chase 'em out."

F.A. (Bart) Bartlett, W6OWP, was among the hams employed at *Press Wireless*. "During the World War II years the only other buildings on those 105 acres belonged to the Carroll ranch. They farmed beans on the land under our antennas. In 1944 an Army installation was built to the southeast of our site, but its purpose was never revealed. Later I asked Don how he dealt with the rattlesnake population since they abounded in the area during the *Press Wireless* occupancy. He told me he found the answer by keeping a hunter cat to hold down the field mouse and gopher population!"

Don continued: *"I read all the books on rhombics that I could find, four of them. I had three rhombics within one degree of bearing of each other. Tokyo was one direction and Chile the reverse. In Long Beach I had found that sloped rhombics worked better in the direction that they were sloped. The third rhombic was level and I could phase with either one of the other two and improve my signal strength in either direction. The one sloped toward Tokyo was by the bluff and it was about three R-units louder in that direction than the other two antennas. The one sloped toward Chile was about three R-units louder in that direction than the one sloped toward Tokyo.*

"I made it a policy in the terms of the sale from Press Wireless *that they could rent the building and aerials from me for six months after escrow closed, at $100 a month. After that, if they wanted, they could rent it for another two months at $200 a month. During that rental period I couldn't use it as a ham station. We were not on the air anyway, except on 10 meters. I would go up here on Wednesday afternoons and find an unused receiver and they would let me use that one for a while. Boy, how thrilling. Then I would go home and work 10 meters. What I would hear up on Rolling Hills was dramatic compared to what I heard*

down in Long Beach. Well, I really wanted to get at the station and use it. After the second rental period was up they wanted to rent it some more. It was July 1946 and we had just been authorized use of 20 and 40 meters again. I said, 'Well I can't very well chase you out, but I will have to raise the rent to a figure commensurate with what I paid for the property . . .' They were out of there in five days!

"They started dismantling a lot of stuff up there, and I found them doing it. I said, 'I bought the aerials, you know, don't do that.' I didn't save all of the aerials, but the eight rhombics I did save. The two vee beams on Europe had already been taken down. They were selling all their excess hardware to the junk man rather than move it. I bought about a dozen equipment racks that were six feet tall for $5 apiece, which was more than the junk man would have given them. One of them is still up there, marked 'Shanghai.' They had marked each rack to indicate what station the receivers mounted in it were meant to receive. The rhombics would work with several receivers connected to them. There would sometimes be 25 receivers on one rhombic. They showed me how they divided the signal from the antenna to feed the receiver system. When you use resistors to divide the signal in this manner, one receiver can't pull down the signal. So I had these little ceramic knife switches to select the receiver I wanted to use. You add the resistors to divide the received signal if you have two receivers connected to one antenna, and neither one can hog the signal. The signal strength was down 3db but so was the noise. The directivity of the antenna was maintained. They had a cluster of little resistors hanging up on one aerial going to a whole bank of receivers which were all over the place. They also used an antenna matchbox to match the antenna input impedance of the receiver to the characteristic impedance of the antenna. I set up a receiver and transmitter as soon as they got some of their equipment out. Before they left they said I ought to have a matchbox, so they gave me one of theirs to use. That was the baseline for their receivers. I finally started using a Johnson KW matchbox some time later and it worked better than theirs.

"Each afternoon, for about six months of the year, a fog rolls from the ocean and engulfs the whole hilltop up there. They told me I wouldn't be able to use the antennas for about an hour in the afternoon each day, right ahead of the fog bank as it rolled in. The receivers had a tremendous fog static during that time. Gosh, it was awful. I forgot about that one day and I went out looking for the trouble because the noise was so severe. I thought somebody was sawing an aerial in two. Then I remembered that it was the fog noise. Well, I started working on that problem and I found that, if I grounded each end of all the rhombics through resistors, I hardly noticed the noise. It would get noisy just two or three times a year. I soldered the terminating resistors across the end of the antenna feed line. I grounded the center tap of the resistors and that bled off the static charge the fog put on the antennas. Then I built an impedance meter for the antennas. It was really just a battery and meter that checked the continuity of an antenna. I could switch it and measure each antenna, its feed lines, and the termination resistor before I got on the air to make sure that everything was operational.

"I always had to check the feed lines with the impedance meter. Each year, as millions of birds are migrating through, they land on the antennas and the wires. It's not unusual for a thousand birds to land on the same feed line. Since the wires are taut and just two inches apart they sometimes twist around each other several times and short together. I bolted some one-by-one pieces of wood together until I had a pole about 20 feet long, and after I located the twisted feed line I went out and tapped on it. The birds would leave and it untwisted, leaving the feed line clear and the antenna operational again.

"No one told me any of these things. When I got through reading the four rhombic books, I had about 100 questions that I had to solve myself. For instance, the first guy that wrote an antenna book said that an unterminated rhombic is bi-directional. That's the classic statement. Well, after he wrote that, others published books and they all copied it. That's wrong. If you feed the right direction of a rhombic it's about six R-units better than the other direction, even if it's unterminated. The improvement is only slight by terminating it. They make so much to-do about terminating the antenna, but that just flattens the standing waves on the feed lines a little bit. It is a little better because the impedance match is better, but that was insignificant compared to the effect of terminating the other end of the antenna and reversing its direction.

"The way I set up my tests is that I put a five-watt resistor as the rhombic termination up on the pole at the far end of the antenna. I would have previously tuned up my transmitter for about 500 watts output. I would listen to a station in one direction and then disconnect the antenna from the receiver with a knife switch and press the key to transmit momentarily. The resistor would burn open from trying to dissipate the transmitter power being applied to it, and I would then reconnect the receiver with the knife switch and observe any change in the signal strength of the station that I was monitoring. It was very enlightening.

*"*Press Wireless *told me that at the top of the pole*

there is a two-foot separation from where the antenna legs start at 80 feet up in the air, tapering down to a two-inch spacing at the lower part of the pole where the feed line from the station ends. This gradual taper over 60 feet brought the impedance from 600 ohms at the antenna feed point to 468 ohms for where the feed line stopped at the pole, 20 feet off of the ground. The balanced feed line of #12 wire was about 20 feet high for the remaining distance into the station. At the station they brought in the two wires spaced two inches apart to maintain the characteristic impedance of 468 ohms and 16 inches away from the next pair of wires. That reduced the interaction with the feed lines from other antennas to a minimum.

"Press Wireless *had performed a lot of tests over the years and they shared their results with me. They put a high-power termination resistor at the end of a half-mile long feed line of #12 wire and ran another feed line parallel to it. They transmitted on one feed line and when the spacing between the pair was moved to 16 inches, the pickup between them was negligible. The signal was unmeasurable on the feed line adjacent to the feed line being transmitted on. They then installed all of their feed lines of #12 wire with two-inch spacing for the rhombics and vee beams 16 inches apart on the feeder poles. The distance between the two was eight times the spacing of the feed line wires.*

"The greatest fallacy I found was that an unterminated rhombic antenna was bi-directional. It is still directional in the proper end where it was fed. So I started putting feed lines on both ends. I attached a feed line to each end of each rhombic and brought them into the shack. By the time I got the feed line in the antenna almost didn't need to be terminated. The thousands of feet of feed line, with a 468-ohm characteristic impedance at the end of an antenna, acted as a termination because of the long run. Some of my feed lines to the rhombics at the corner of the property were 3,000 feet long. There was very little difference in received signal strength when the long feed line was terminated, although the termination resistors will burn up if they are not big enough to handle the power. They have to dissipate about half of the transmitter power output. The impedance of the whole system flattens out a little bit with the termination. I didn't do the fancy measurements that the fellows who write the books purported to do, but I saw how wrong they were.

"I used #8 copperweld wire for most of my feed lines. Its spacing had to be two and seven-eighths inches to maintain the 468-ohm characteristic impedance, so the feed lines were installed 24 inches apart on the poles. From the redwood breakout frame outside the station to the relay system on the wall in the room, I used #12 wire spaced two inches for the feed lines and spaced 16 inches from the next pair.

"Initially, I had terminated one rhombic with resistance wire. It was connected from the top of the pole to about 30 feet down, until the wire measured a total of 600 ohms. The antenna worked very well in the one direction, but it was difficult to work out a system of reversing directions with all that wire for a termination. Just about then I found a source of non-inductive power resistors. Now I could bring the feed lines into the shack and switch in the desired termination.

"After I brought the feed line for each end of each rhombic into the radio room, I worked up a switching system to select any of the 16 directions that I had then. Eventually I could reverse any antenna — the whole works. It was all done by relays. I didn't have to do anything, just rotate a switch. I worked out a relay system for all the antennas. First, I got the best Leach relays that I could find and put those on the wall. They turned out to have a high insertion loss. The signal from the rhombic was noticeably reduced by a single relay. Then I had Leach build me some special relays

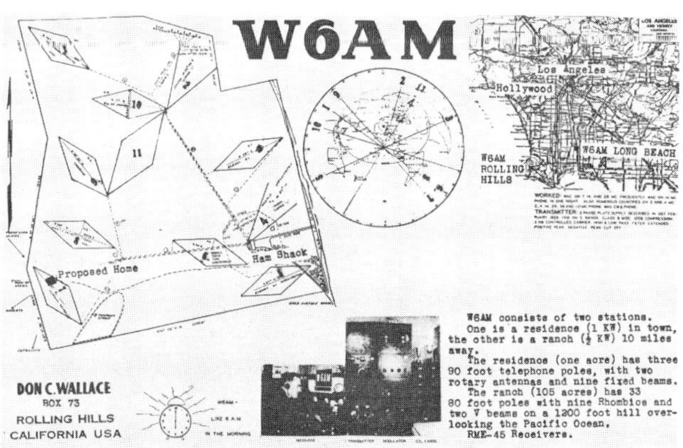

Don's rhombic farm QSL in 1946.

The rhombic feed lines were brought into a redwood frame.

and I still lost 3 dB (half the power) in the relay system between the antenna and the receiver. I finally rebuilt the relays and put all the relay contacts on insulators, and then I mounted the relays and feed lines one foot away from the wall. It took 15 hours of work per aerial. You couldn't measure the signal loss after that. I looked at the R (signal strength) meter with the relays in line, then used a ceramic knife switch to disconnect the relays and jumper two wires around them and direct from the receiver right to that aerial. Then I would switch back and run the received signal through the relay system. I couldn't measure any difference between the two signal paths.

"The insulators increased the spacing of the relay blades away from the coil and frame of the relay. Initially, when they were three-eighths of an inch away from the coil, each blade was capacitively coupled, essentially bypassing a little bit of signal to ground on all 108 relays. When I got the signal an inch away from the coil and frame in the modified relay, the coupling effect disappeared. The capacity was low enough so that the loss of the wires took over and the signal stayed on the feed line instead of coupling to ground. Then the capacity between the wires apparently dominated enough to maintain the system impedance. As I say, I had 100 questions about rhombics. No book I know of ever covered rhombic switching, and none still does. Nobody knows these results except by conversation with me.

"Terman's original book didn't take into consideration what a sloped rhombic could do. I used the slopes to my advantage. All the rhombics either had double wires or triple wires (curtains) strung on each of all four legs. They worked much better than a single wire rhombic.

"A fellow who was one of the directors of the League (ARRL), lived over in Whittier, and had an orange ranch. He had this beautiful rhombic supported with 95-foot high telephone poles, aimed at Australia. He would get on the air every night and rag chew with the Aussies, and they would all tell him he was much louder compared to any station in the United States. I had two rhombics pointed at Sydney, the reverse direction of New York, and I phased them together. I was using a BC-610 (a surplus World War II transmitter) that I had put up as soon as Press Wireless *moved out. It wasn't much power, only 600 watts on CW and 500 watts maximum on AM phone.*

"One evening when he was chatting with the Aussies, I got on the air and broke in. I said, 'When you are through I would like to talk to one of you Australians, W6AM standing by.' He said, 'Go ahead,' and I replied, 'I just wanted to check out and see if my aerials are good as the fellow's in Whittier.' Well, the Aussies said, 'You both sound just the same, but Don doesn't fade in and out. He has the louder signal.' So we kept on the air for a while, and we worked a few others, and they all said the same thing. Then he called me on the land line and said, 'Are you going to be there a while? Can I come over?' And I answered, 'Sure.'

"He wanted to see what my transmitter was, you see. He thought that I might be running more power than he was. He saw that all I had was a BC-610. He had a nice home-built transmitter of the day, it had about 800 watts power input. I ran about 500-watts power input with the BC-610. That's all the power I could run on AM. He had a little more power than I did, and he could see that. He had this marvelous aerial in Whittier, a perfect aerial. He had been so loud down in Australia compared to anybody else in the United States that they all raved about his signal strength. I just wanted to see how my two rhombics performed compared to his. Well, in just a few weeks after that he was off the air and playing gin rummy at the club. We hardly ever heard him on the air again. He was on every night prior to that, talking to hundreds of Australians. He liked being the big shot, having the loudest signal out of the country. Apparently he lost heart.

"I found that the fellows around here tried to tail end me [sign their call sign as Don was signing W6AM] after I worked a foreign station, and get a signal report. They wanted to know how loud their signal was compared to mine and once in a while one of them got a better signal report, of course. The rhombics were so sharp that if I worked a foreign station located halfway

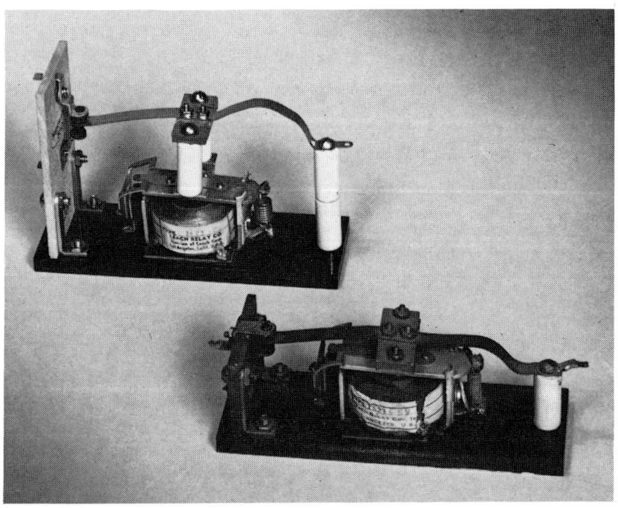

The Leach rhombic switching relays were modifieds for low loss.

Don's station at the ranch in 1946.

between the favored directions of the two closest antennas with that bearing, then I wasn't very loud at the station that I worked. That didn't bother me, but it bothered the fellow in Whittier. Eventually someone else will come along and put up a big station and be stronger than everyone else, it's inevitable. I figure that as long as the others try to get a better signal strength report than mine, then I'm still the standard of comparison. The standard of comparison is always the best. Sometimes they got a better report than me, but usually they couldn't. In those days my signal reports were usually the loudest reports on the band."

The escrow closed on the property that Don purchased from *Press Wireless* on November 7, 1945. Since Don seemed to be pretty well heeled, *Press Wireless* inquired if he was interested in purchasing their transmitting site. It was located in Playa del Rey, 15 miles north of Rolling Hills in what later became known as Marina del Rey. At the time their call sign was "VC," for Venice. But, Don wasn't interested. He had his hands full with the site on top of Palos Verdes Peninsula.

The specific chain of events that occurred in the Playa del Rey property sale is somewhat obscure. Apparently *Press Wireless* asked Don what he thought the property was worth. He said that he would think about it. The land was marshy and not suitable for building in its present condition. At some time Howard Hughes contacted Don Wallace. They had met back in 1939 when Don had supplied the communications equipment for Hughes' flight around the world, and again in 1940 when Don called on Hughes Aircraft as a manufacturer's representative. They had first met on the air in 1921 when Hughes was licensed as 5CY and Don was operating as 9DR in Minneapolis.

Hughes was interested in the property that *Press Wireless* had for sale, but he didn't want them to know it, because then the price would go up dramatically. Hughes mentioned a price he felt was reasonable for the property and asked Don to pass the information along if the opportunity arose. The next time *Press Wireless* asked Don for an estimation of the property's value, Don mentioned the price Hughes had suggested. He also said that Doc Stuart was interested in purchasing the property.

Doc Stuart's rhombic farm in Ventura had been a receiving station for Chinese broadcasts during the war. Now that the war was over, the Chinese government asked him to design a similar facility at Nanking. Doc decided that a 10-kilowatt transmitter and a rhombic antenna should be installed at their new broadcast station site. Doc also had a long association with Howard Hughes and also knew Dave Evans well. Dave started the radio (electronics) division for Hughes at the end of the war. The 10-kW transmitter was being built by the electronics division at Hughes Aircraft in Culver City. *Press Wireless* and Doc Stuart got together and decided on a price for their transmitting site in Playa del Rey, which had none of the deed restrictions of Palos Verdes Peninsula. It is thought that either Doc Stuart or Dave Evans actually purchased the land and, after escrow closed, the deed was transferred to Hughes.

A portion of the rhombic antennas and their associated telephone poles was taken down at the transmitting site. Hughes left a few antennas up for transmitter development within his company, but the rest of the antennas would not be needed. He had other plans for the property. The poles that were removed were delivered to Don in Rolling Hills. He also had big plans for his property.

Back on the Air

When World War II ended on August 15, 1945, amateur radio was authorized on 2 1/2 meters only. At that time it was a VHF band suitable only for line-of-sight communications. Everyone was excited about being back on the air, even if it was VHF. Don and a friend, Bill Lightner, W2KXW/6, went up to Mount Wilson with a 2 1/2 meter rig to see who they could work. On Labor Day, with an Abbott TR-4 running five watts, they contacted two stations located 120 miles to the southeast in San Diego. These stations were Ray Jacobs, W6OIN and H. Ascher, W6OZH.

Three months later the FCC authorized amateur operation on 10, 5, and 2 meters. The 2 1/2 meter band was deleted from amateur use and assigned for aviation. The following February 5 meters was deleted from amateur use and became television channel 2. An adjacent band, slightly lower in frequency (6 meters), was then allocated to the amateurs. Frequency modulation had been developed by Edwin Armstrong in the early 1930s, and its prewar broadcast allocation had been in the 44 to 54 MHz region. After the war it was reassigned to the 88 to 108 MHz range, even though over half a million sets had been built which were now obsolete.

New call districts were assigned by the FCC on October 15, 1945. The W6 district was now to be exclusively used by California, while Arizona, Nevada, and Utah became part of the W7 district. The W0 call district was created from most of the W9 states — North and South Dakota, Minnesota, Colorado, Nebraska, Iowa, Kansas, and Missouri. Only Illinois, Indiana, and Wisconsin remained in the ninth district. One or two states moved across boundaries in most of the other districts. The prewar call signs could be used until the individual's license came up for renewal and then a new call sign would be issued that was in line with the current call district allocations (see Appendix 2). A number of amateurs in the new districts were upset about losing their old call signs and moved across state lines so they could retain them.

Much of the map of the world had been redrawn due to the conflict in the previous six years. The prewar call of China had been XU in the late 1930s; foreign residents still used XU and now native Chinese operators signed the C prefix. Formosa had been under Japanese control prior to the war with the J3 prefix; it was now Chinese occupied and had been allocated the C3 prefix. Chosen (Korea) and Darien, on the southern tip of Manchuria, had also been occupied by the Japanese prior to the war and allocated the J8 prefix. Korea was assigned HL afterwards and Darien reverted to Manchuria, C9. The same was true of some of the European countries and many of their colonies in Africa.

W6AM and W2EXW/6 using a 2 1/2 meter rig on Mt. Wilson.

On April 1, 1946 amateur operation was again authorized on 80 meters. The 11-meter band was given to the amateurs in May and the next month the 40- and 20-meter bands were released for amateur use. It had taken the better part of a year for the FCC to sort out the new amateur and commercial frequency allocations and give the stateside amateurs authorization to transmit on the lower frequency bands. American amateurs overseas were also constrained by these regulations and, in their enthusiasm to get back on the air after the war, many operators disregarded them.

One of these operators would have quite an impact upon the postwar DXCC. When World War II started, Bill Hunton, W3AG (ex-W6ODD) joined the Merchant Marine. On VJ Day he was in Shanghai, China. When Japan surrendered, GIs across Asia and the Pacific began modifying military radios and stringing 20-meter dipoles. The long arm of the FCC didn't reach past the continental United States Prewar call signs were activated and used portable, as well as variations thereof. In 1937 Bill Hunton had built a transmitter from an article in *QST* using a pair of push-pull 807s. He had his rig with him when the war ended.

Hunton commenced operation on 20 meters from Shanghai as W3AG/XU in October 1945. He worked several GIs across China the first evening. Bill found that most of the GIs he contacted substituted XU in place of the W prefix in their call sign. XU3AG became Bill's standard call while he was in Shanghai.

Later that month Bill traveled to Penang, Malaya and operated as VS3AG. En route, he used the shipboard station and signed XW3AG, to indicate portable operation from aboard ship. The XW prefix was as yet unassigned. The country of Laos would not come into existence for another nine years, when it would be allocated the XW prefix.

In November Bill returned to the China mainland and took a boat up to Hulutao, Manchukuo (Manchuria). The next year that region would be assigned the C9 prefix. On his first night in Hulutao Bill began operation on 20 meters, using their prewar prefix, as MX3AG.

Hunton recalled, "I always enjoyed working DX rather than rag chewing locally. I tuned the bands for something interesting, rather than just calling CQ." Almost 200 stations would come to know that gut-wrenching feeling of being answered by a rare DX station. Manchuria would see operation by just half a dozen amateurs in the coming years.

Hunton remained in Hulutao for two weeks but didn't fraternize with any local people. "In Manchuria," he said, "they were sensitive to foreigners. The war was just over and their losses had been very heavy." Throughout his travels in Asia, no photographs were ever taken. He later remarked, "I never carried a camera. Many countries didn't have United States embassies and lots of local people packed pistols."

In early December 1945 Hunton opened up again from China as XU3AG. One day he worked Reg Fox, AC4YN. Reg was almost like a local, booming in from Tibet. "His signal drifted quite a bit," Bill reminisced. "Our QSO started mid-band and ended up at the far end on my Hammarlund receiver. I had tried using an HQ129X (X indicated an internal crystal filter), but it had too narrow a band pass." In the months immediately after the war, amateurs were cannibalizing whatever equipment was handy to gather the necessary parts and get a transmitter on the air. There would be some pretty ragged signals on the bands until they got their transmitters properly tuned and power supplies adequately filtered and regulated.

While he was in Shanghai, Bill often heard an Aussie operating locally. He was so loud, having a station set up just down the street, that his identity was not hard to guess. One day they happened to meet. "I'm sure that he had heard me operating also," Bill said. "We chatted for quite a while, but amateur radio was never mentioned. It was safer to keep a low profile."

Operating as XU3AG Hunton worked more than 300 stations before leaving China. He arranged stateside transportation on a boat in January 1946 and he stopped in El Salvador and Mexico on the trip home. Mostly GIs in the Pacific and a handful of Europeans were contacted, along with a few South Americans. The United States stations were still up on 10 meters. Then Bill returned to California, settling in Stinson Beach.

While Bill Hunton was crisscrossing Asia, he often worked Rod Newkirk, W9BRD, stationed in the Philippines. Rod recalls, "I was KA1KN in the P.I. after peace broke out in late 1945. Gosh, I sure had fun with an 807 home-brew and a Windom for 20 strung between tent poles. Gee, some nifties in that log: AC3SS (Sikkim), XZ3FX (Burma), HS1BK (Thailand), and VQ4ZL (Kenya). Fun! One thing I remember is the fantastically short skip on 14 MHz in that area at the time."

Shortly after VJ Day, Newkirk's tent mate, Willie Kuure, W9YNY (W8YNY), had been the first to discover some chatter on 20 meters one evening. While waiting to be shipped back stateside, they had built a home-brew 807 rig, which ran a walloping six watts input. One afternoon Willie was rag-chewing with Hal, W5EGA, who was operating as KA1ZZ aboard a ship in Manila Harbor. Into the tent came a grim-looking first lieutenant. "Sorry men, you have to shut it down. We all do want to go home, don't we?" Rod and Willie began dismantling the little rig and mentioned their contact with Hal. The officer replied "W5EGA? He's my old school chum." They fired up the rig again and the lieutenant pounded brass with Hal for the next hour. At that point the officer said, "I've still got orders to close you down, but this may take longer than I expected!" The rig was converted into a low profile operation and kept on the air until Rod and Willie returned stateside. When they left it was given to another of the deserving amateurs.

Harry Paston, W2OAA, was the first postwar amateur operating from Korea. Once authorization came through from the Supreme Commander of the Allied Forces (SCAP) he signed portable J8 and then later portable HL. Immediately after VJ Day Harry too had been on the air, using the call sign AK1MO (Asia, Korea). Harry worked Bill Hunton in Manchuria and China and on his QSL card to Bill remarked: "I learned that you got in hot water with the FCC when you arrived at the dock in the United States over operationout here. Hope you didn't get your ticket revoked." Bill didn't, but the FCC was to deliver many stern lectures as errant radio operators returned stateside. No doubt Harry was also concerned about his fate when he eventually came back to the States.

Meanwhile, there were problems continuing in the rest of the world. The civil war in China resumed almost as soon as the worldwide hostilities had ceased. French Indochina wanted their independence, but France was reticent about giving up its colonies. Russia decided to secure a perimeter of buffer countries around her borders in eastern Europe for defense against another attack, even though those countries affected were not particularly inclined to do so voluntarily. In March 1946 Winston Churchill stated: "From Stettin in the Baltic to Trieste in the Adriatic, an 'iron curtain' has descended across the continent."

Europe was left in ruins at the end of World War II. The manufacturing facilities of the large cities had been largely destroyed, along with much of the housing. The Marshall Plan (General George Marshall was appointed the United States Secretary of State after the war) was implemented, which provided American aid in helping the European countries to rebuild. Most of the early amateur radio operation in Europe was by the Allied occupation forces. As the economy improved over the next several years, native radio operators rebuilt their homes and stations and returned to the amateur bands. In the meantime amateurs in the Allied occupation forces provided much of the first postwar radio activity. FKS8AL was active from the French zone of Austria, D5AD (F9GP) in the French zone of Germany, D2GQ in the British zone of Germany, and D4ALN in the American zone of Germany.

Amateur radio activity by native operators in Japan was severely curtailed in the years after the war. The only operation was by American operators signing portable Japan (/J). Later they were issued J prefix call signs and Americans on Okinawa were assigned J9 calls. Japanese nationals would not be licensed for another seven years. Due to the limited number of stations on the air in Japan, that country was one of the more difficult places to contact in the first few years after the war. Until the mid-1950s a certificate was available for those operators who had contacted five stations in Japan, and another for those who had contacted all of the call districts in Japan.

In early 1946 amateur radio was again gearing up in the United States. New licenses were being issued and after four years of inactivity the *Radio Amateur's Call Book* had again resumed publication with the spring issue. Don recalled, *"During the war the government seized all of the call books that the Radio Amateur Call Book Company had on hand. Apparently they were concerned that the wrong people could get hold of the call books and use the information. Most of the military radio operators during the war were hams, you know. There was a delay in getting their material back from the government after the war. It was all locked up in a warehouse somewhere. Since I was their representative during the '30s, they asked me to send them whatever call books that I could so they could reconstruct their archives. Well, I sent them one of each issue I had, all the way back."* Luckily, Don had quite a few duplicates, so over half of the early issues in the 1920s and 1930s remained in his collection.

After the war Don represented a line of commercial transmitters and in an attempt to market them, he visited various offices of the Mexican government in Mexico City in late 1945. After the close of business one evening he went to the jai alai games. Forty years later Dick Rodier recalled: "Just after the end of World War II Glenn Sumerlin and I were in Mexico City for a fledging airline called Aerovias Azteca. One night Glenn and I were at the jai alai games and spotted Don Wallace several rows ahead of us, intently watching. Glenn could whistle CW perfectly, so every time there was a lull in the crowd noise he whistled W6AM very softly. At first Don thought that he was hearing things. He looked around bewildered and evidently told himself that he was imagining it. The CW continued and he stood up and tried to find the source, but we ducked down and he never saw us. The next day when he called on us he was suspicious, but we denied complicity."

Two weeks later Don received an SWL from George Hotton in Milwaukee, Wisconsin. Hotton wrote:

> You contacted W9EXT on Dec 22 and said you ran 600 watts to a pairof 450TLs to a four element beam. You said that you also have an eight element beam fixed on London. The only other eight element beam that I have heard of was used by XADK in Caserta, Italy. Your signal is definitely the best from California that I have ever heard; R9 plus on a six-tube Sky Buddy, just like a local.

A few months later Don was tuning around on 10 meters and heard W3KFB call CQ. Raymond Nether recalled: "I had just received my station license the afternoon of April 10th, and I had a home-built receiver and transmitter ready to go. I sent a shaky CQ, and lo and behold, my very first contact as a licensed ham was with W6AM. I will always remember Don's patience copying my very nervous CW and the many words of comfort he gave me."

Don spent a lot of time on 10 meters that spring as operation on the lower bands had not yet been authorized by the FCC. Donald Hyde remembers those days well. "I was sitting in a tent on Okinawa signing W6-KBH/J9. The equipment was an AR-88 receiver and a modified BC-610D transmitter. As W6AM and I were chatting, I heard the sound of his dog bark in the background. There was a small dog in my tent, an Army stray. At the sound of Don's dog, mine started barking. We agreed that it must be the first trans-Pacific dog to dog QSO!"

CQ magazine was formed near the end of the war and commenced publication in January 1945, as the successor to *Radio* magazine. *Radio* had become heavily reliant on broadcast issues and professional engineering to survive the past few years and did not return to amateur radio. Many of the editors who had contributed to *Radio* joined the staff of *CQ*, but it was in no way affiliated with the old *Radio* magazine.

The prewar editions of the *Radio Handbook* noted on the title page: "Issued annually by the editors of *Radio*," and listed the contributing editors. The ninth edition of the *Radio Handbook* in 1942 noted on the title page: "Compiled and revised from material originally prepared by" and listed contributing editors from the earlier edition. Over the next 10 years the *Radio Handbook* would expand to more than 700 pages.

Herb Becker began writing the "DX" column for *CQ* in March 1946 and, from the next month on, it was called "CQ DX." In January 1947 *CQ* announced the WAZ award. With but a few minor changes due to the redrawing of country lines after the war, the original 40 zones stood intact. Prior to the war Sakahlin Island, located north of Japan and east of Russia, was jointly occupied by both of those countries. The dividing line between zones 19 and 25 was the 50th parallel and split the island in half. After the war the entire island was occupied by Russia and the CQ zone boundary line was moved so Sakahlin Island was totally within zone 19.

The first postwar WAZ Honor Roll was published in May 1947 (See Appendix 10). Marv Gonsior, W6VFR, Gordon Marshall, W6ITA and Bill Orr, W6SAI, along with a handful of others shared the top spot with 39 zones confirmed. In September Ben Stevenson, W2BXA received postwar WAZ number 1 and a few days later W6VFR received WAZ number 2.

In July 1946 Doc Stuart worked AC4YN, becoming the first postwar contact between the United States and Tibet. Marv, W6VFR, worked Reg Fox a few weeks later. The following month By Goodman's DX column in *QST* remarked that Marv was the most recent addition to the TWA—Tibet Workers of America!

Whenever he had the opportunity Don Wallace would visit operators who were highly placed on the DXCC or WAZ lists. Occasionally he would find an aspect of their operating technique or station installation that was superior and would try to adapt it to his own. Marv Gonsior recalled: "When Don looked at my station, he was intrigued by the control that I had added to my receiver to adjust the injection level of the beat frequency oscillator (BFO). This way, when there were not any strong signals on the band, I could turn the injection level down to where the receiver was just functioning. The lower BFO level created less background noise and I could hear very weak stations. If there were strong local signals on the band, I would have to turn the level back up, of course, to prevent the receiver from overloading." Don implemented the BFO injection concept on one of his receivers and it worked very well. He sent the idea in to Ed Johnson, but they never implemented it in a commercial product.

After World War II, zone 23 still remained the most difficult of regions to work. It had remained unchanged and consisted of northwest China, Tibet, Mongolia, and a sliver of Asiatic Russia. Reg Fox, AC4YN, in Lhasa, was the only station on from that area in 1946, with a 20-watt transmitter. When Clay Murdock, W6OMC, scratched out a contact with Reg he later remarked: "AC4YN said W6s are rare there, but I didn't tell him that AC4s are rare here too!"

Marv Gonsior, W6VFR, achieved postwar WAZ #2.

The ARRL had continued to publish *QST* throughout World War II. Their financial situation was more secure that it had been in World War I. The "How's DX" column resumed publication in January 1946 by Byron Goodman, W1JPE. After much soul-searching, the headquarters staff of the ARRL had decided to wipe the DXCC slate clean The prewar country totals of the DXers did not count toward the postwar DXCC award.

It would take a full year for the ARRL to decide upon an up-to-date DXCC country list. The first postwar country list was published in the February 1947 issue of *QST* (See Appendix 8). The prime contributors to the formation of the list were Arthur Milne, G2MI, Herb Becker, W6QD, and several members of the ARRL headquarters staff. There had been 250 countries on the prewar DXCC list. Due to the fact that the world map had changed dramatically, there were a number of changes. In some cases, several countries were combined into one and, in others, some were deleted. With the additions there was a total of 257 countries on the postwar DXCC list.

The DXCC award was announced by *QST* in March 1947. The first postwar DXCC listing appeared four months later in the July issue. There were two awards: one for mixed CW and phone operation and one for phone transmission only. Charlie Mellen, W1FH, received Number 1 of both awards. The first ARRL DXCC Honor Roll listing was published in the November issue. W1FH occupied the top spot on both the Phone and CW Honor Roll.

The ARRL resumed the annual DX Marathon in the spring of 1947. It was a very similar format to their prewar contest, with one notable difference. Instead of occupying a 10-day period encompassing two weekends, the contest was held on one weekend in February and one weekend in March. The combined total score of these two weekends determined a station's final contest score. This, no doubt, relieved the family pressure on the chief operator. He would have a full month between contest weekends to placate the wife before he disappeared into the radio room for another two days.

Don entered the ARRL Marathon in 1947. Included with his logs was a note to F.E. Handy, ARRL communications manager.

> *The CW score of 66,585 points with 500 watts was enough to show that the rhombics were working well in one direction. Only one antenna is reversible, so far. My phone score was 50,540. It was surprising how many foreign stations, including Russians, spoke English. Working XU6GRL on all five bands appeared to be, according to Doc, the first time the Pacific had been spanned on all DX bands on the same day by the same pair of stations.*

Charlie Mellen, W1FH, achieved postwar DXCC # 1.

Doc Stuart traveled to China in the fall of 1946. The 10-kilowatt transmitter that had been built and tested at Hughes Aircraft was shipped to Nanking, via Shanghai. This was his first trip to Asia. Since the transfer of the capital of China from Chungking back to Nanking at the end of the war, it had been necessary to relay the transmissions of the International Department of the Ministry for Information from Nanking to Chungking by telegraph, and then the messages were sent to Doc's station at Ventura, California by XGOY. There were no facilities in Nanking at that time for such transmissions. Therefore, in late 1945 the Ministry of Information decided to install a station in Nanking to replace the Chungking circuit. Doc was given the task of designing the transmitter and antenna. The preparation of the project began in early 1946 and the equipment for the station was built in four months. Over a period of three months, five shipments of this material had arrived in Shanghai. While he was in Shanghai that summer Doc was on the air from the Broadway Hotel with a BC-610 as XU6GRL. Once all of the material arrived in China it was transported to Nanking. The 10-kW transmitter was equipped for voice modulation and carrier-shift keying (FSK) for Globe Wireless Teletype transmission at 100 words per minute.

Doc installed a large rhombic antenna in Nanking. It was 410 feet long per leg and almost 800 feet end-to-end. That gave a power gain of at least 13 dB on the lowest frequency employed, with increased gain on the higher frequencies. A 20-by-36-foot Quonset building was erected to house the transmitter. The radio teletype printers, tape puncher and robot transmitter, as well as the voice modulation studio, were installed in the Nanking office of the International Department of the Ministry of Information. From there they were connected by telephone lines to the transmitters, which were just a few miles away from downtown Nanking. A period of about three months was required for Doc Stuart to complete the installation and to make final tests, after which regularly scheduled transmissions of radio teletype were handled daily to Ventura. At that time, the present circuitous route via Chungking was completely replaced after nearly seven years of operation.

When Doc was not busy overseeing the site installation at Nanking, XU6GRL was working DX on 20 meters. After the rhombic was up he had a lot of fun using it on the ham bands. That antenna made him about five S-units louder than most other amateurs in his area of the world.

Don Wallace, Jr. had graduated from the Naval Academy in 1946. His first assignment as an ensign was on a heavy cruiser, the *USS Columbus*, off the coast of China. Since Doc had arrived in China W6AM had worked XU6GRL regularly. Don mentioned that his son was in the region and would enjoy visiting him in Nanking. Doc had some influence in the Ministry of Communications, so he stopped in one day and proceeded to pull a few strings.

Don Jr.'s ship was a part of the United States Seventh Fleet and they arrived in Shanghai in January 1947. Shortly after their arrival their skipper, Admiral Milton Miles, called Ensign Wallace into his office. The admiral had received a request from the Chinese government that Ensign Wallace be ordered to report to the Chinese Ministry of Information in Nanking for two weeks. The Admiral had agreed to the request but decided to send someone along with Donald. The civil war in the interior of China was in high gear, and infiltrators were coming into the large cities. In these troubled times, United States servicemen should travel in groups. The Admiral assigned the assistant medical officer, Lieutenant Mike Lau, to accompany Ensign Wallace.

Don Jr. and Lieutenant Lau traveled by train to Nanking and visited the Ministry of Information. The Minster, H.P. Hseng, contacted Doc when they arrived, who then picked them up and took them to his house. They stayed at Doc's for the period they were in Nanking and he took them on a tour of the area daily. Chiang Kai-Shek was out of the area during their visit, but they drove by his estate. On one of their excursions Doc introduced them to Wang Shih-Chieh, the Minister of Foreign Affairs. At an elaborate ceremony the Minister bestowed upon Doc Stuart the Special Collar of the Order of the Brilliant Star for his years of meritorious service to the Chinese people and government. It was their highest civilian award and the only other non-Chinese to receive the honor was General Claire Lee Chennault of the Flying Tigers. Doc had previously received the Victory Medal from Chiang Kai-Shek. All in all, it was a pretty good showing for a small-town dentist. After two weeks, Ensign Wallace and Lieutenant Lau returned to their ship, thoroughly impressed by their stay in Nanking.

At the end of the war, Bill Wallace had returned to college. He graduated with a bachelor of science in electrical engineering (BSEE) in 1947. Don asked Bill to join him as a partner in his company as a manufacturers' representative and then renamed the company Wallace and Wallace.

Times were as lean after the war as they were plentiful during it. All those huge orders that the military procured for spare parts were now hitting the surplus market. Bill recalled: "My commission check for my first full month on the job was $49, and I had to take expenses out of that. Business was very lean for the first four years. Then it improved and I could support a family on my salary. Margaret and I got married at that time."

Shortly after Bill had joined Wallace and Wallace he accompanied Don to a convention of manufacturers'

Doc Stuart and his wife operating XU6GRL in Shanghai.

Ensign Don Wallace, Jr., standing second from left, aboard the *USS Columbus* in China.

representatives in Los Angeles. Bill recalled, "The industry as a whole was slow. The only people doing any business were those selling test equipment. I met McMurdo Silver there. He was starting a line of test equipment. He told us that he had been relatively successful building radios in the '30s, until just before the war. One month they had a shipment packed and ready to go, and their engineer came down and told him that the radios wouldn't work. A step in the manufacturing process had been overlooked so all of the radios would have to be unpacked and modified. The financial manager overruled every one and shipped them anyway. Well, every radio came back and had to be modified. They never recovered from that financial setback."

Due to the surplus material available for moderate prices, the Yagi antenna became widely used by many city dwellers after the war. It was especially attractive because of its compactness, high gain, and ability to rotate 360 degrees, which gave worldwide coverage. The elements were usually constructed of aluminum tubing of approximately one inch in diameter and half a wavelength long, 16 1/2 feet on 10 meters and 33 feet on 20 meters. This diameter tubing was used because of the mechanical strength necessary to support the center suspended elements and also because of the high RF currents encountered at the feed point in an antenna of this type. The driven element was center fed through an impedance-matching device either by open wire or coaxial cable. Coaxial cable, developed during the war, was generally preferred, as it could wind around the supporting tower and dispensed with the necessity of the collector rings required for open wire feed lines.

Bill joined Wallace and Wallace in 1947.

The aluminum tubing elements were of the telescoping type so that they could be easily tuned for maximum forward gain and front-to-back ratio by extending or shortening them. One-tenth and fifteen hundreds wavelength spacing between the elements was used to keep the array compact and still have adequate antenna gain, 9 feet on 20 meters and 4 1/2 feet on 10 meters. The usual procedure was to use one reflector and one director in addition to the driven half-wave element. Gains from six to eight dB over a half-wave antenna were easily obtainable with the Yagi. Its greatest disadvantage was that it was essentially a single-frequency antenna. A great reduction of gain was experienced even at one percent (140 KHz on the 20-meter band) away from the frequency to which a Yagi had been tuned, so that it did not work particularly well over the entire band.

Amateurs like Doc Stuart and Don Wallace, who were fortunate enough to have a large expanse of ground, used fixed wire antennas such as rhombics, phased arrays, and curtains of various types directed to the different areas of the world. These antennas had significantly better gain characteristics than Yagis due to their large size.

Once *Press Wireless* vacated their facility in August 1946, Don set about reworking the entire station. Rather than having dozens of receivers connected to a few antennas, Don redesigned the station to operate from a single transmitter and four or five receivers connected to any one of the available antennas. He found a large 75-ohm 300-watt non-inductive resistor, available surplus, for a rhombic termination. A set of six of them mounted on insulators on a piece of wood would handle a 1-kilowatt transmitter's output. He mounted them on the wall of the station near the relays. Don brought feed lines in to the shack from both ends of each rhombic and, over the next two years, installed a relay system that would allow either end of any rhombic to be selected. Meanwhile, he reinstalled the two vee beams on Europe that *Press Wireless* had taken down. Don also installed some of the poles from the *Press Wireless* transmitting site. The clay soil on the hilltop was difficult to dig, so he purchased a case of dynamite and a quantity of blasting caps to expedite digging the holes. By the middle of 1947 Don had put up an additional rhombic. It was 100 feet high and 800 feet long, pointed at central Europe.

Of his five rhombics pointed at Japan and China, four worked well in that direction. One favored Chile, in the reverse direction. Since they were all about 600 feet long, they were quite sharp on the higher frequency bands. However, on 40 meters they were wide enough in bandwidth that the four of them could be phased together, adding to both the received and transmitted signal strength. Don put in another relay system that allowed an individual rhombic to be phased with other antennas that had the same bearing. After modifying the feed line lengths slightly, to bring each antenna's signal into the same phase as the others, each of the four antennas added to the overall strength of a desired signal from that direction as it was switched in. The phased rhombics also added to the transmitted signal power. When he worked Chinese stations on 40 meters, Don would tell them that he used four rhombics in push-pull parallel (a common power tube configuration)! Actually, they were all in the classic configuration, the nose of one antenna feeding the reverse end of the next. In any event, Don put out an impressive signal in that direction.

Gus Westerberg was the Scandinavian Airline's resident radio engineer at Douglas Aircraft after the war. He recalled: "Having the call sign SM5UM made me a VIP in the LA local ham clubs. I often visited with W6QD and W6SAI and I was invited in late 1947 to visit W6AM's radio shack on Rolling Hills. When I reached the top of the hill I saw a large, one story building surrounded by numerous rhombic antennas. Don showed me how he, by a multi-position switch, could use any of the rhombics around the building. It was an unforgettable experience to hear the relays clicking around the building and how the received signal changed accordingly. It was with some envy I left Rolling Hills. Many of us foreign hams could only install one Windom."

Some of the local hams were also envious of Don when he put the rhombic farm on the air. Don remarked: "*When I first bought the ranch, a lot of the fellows were jealous. After a few years they got over it. We have gotten along fine ever since*." Don did get along well with the great majority of local hams. All of the antenna projects he had going in the coming years would have kept even Don going around the clock, seven days a week. Groups of a dozen or more fellows would come up to the rhombic farm every couple of weeks and help him erect poles, adjust guy cables, string wire, and the thousand and one other tasks that

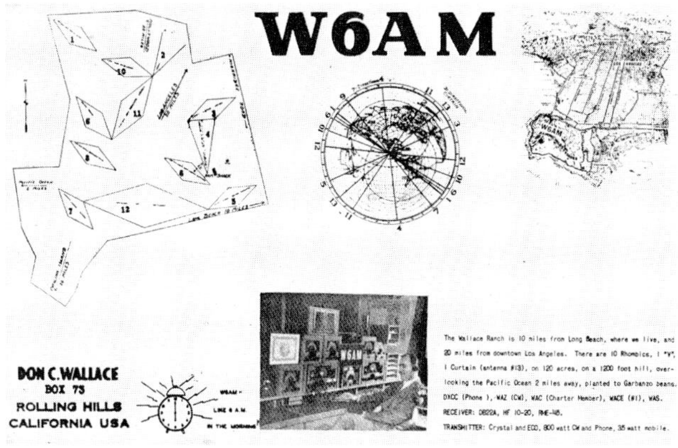

The rhombic farm QSL in 1948.

had to be done. The Associated Radio Amateurs of Long Beach put in an enormous effort at Don's in the initial years that he was building up the rhombic farm.

During the war, tremendous improvements in receiver design and construction had been made. Don's homemade receiver could no longer compete with the new receivers on the market. Although it imparted nostalgia, the home-built receiver simply took up too much valuable space at the operating table. It was placed in the basement of the Long Beach house, where it remained for the next 40 years.

Collins Radio Company was to provide both receivers and transmitters to the military during the war. Afterwards, the Collins equipment developed a reputation that was unsurpassed for the next 30 years. By the late 1940s a number of commercial receivers of high quality were available to amateurs. Tuning and adjusting a home-built regenerative receiver became a lost art.

After the war it was very unusual to see any articles written by Don in the amateur radio magazines. The era of home-built rigs outperforming commercial equipment was coming to a close. Aside from his picture occasionally appearing with contest results, most of the articles about Don in the coming years appeared in the newspapers.

Once 10 meters was again authorized at the end of the war, Don had put a mobile station in his car. In October 1947 he worked G4JZ in Gloucestershire, England. The QSL from Lindsey Coursey noted:

> Bet you were surprised when I came back, Don! Best previous mobile was a station in Massachusetts. Why all the rhombics if 6 feet on a bumper will do this?

Meanwhile, Bill Hunton, W3AG/W6ODD, located 400 miles north of Don, was in the process of fine-tuning his portable transmitter and its antenna. He traveled to

South America in the spring of 1947. He spent two days in the Virgin Islands, another two days in Brazil, and then he continued on to Trinidad. He operated from each of these locations and, after two weeks of sun and DX, returned to Stinson Beach.

Once he was back in California Hunton began to prepare for a longer trip. He made a vertical zepp antenna with an open wire feed line cut for the 20-meter band. That was a half-wave antenna (33 feet long) end-fed with a resonant quarter-wave transmission line (a multiple of 16 1/2 feet). The antenna and feed line was rolled up and stored in a suitcase. A key, ARRL minilog, HQ129X receiver, and the trusty 807 rig were added, everything barely fitting into one bag, similiar to W6ZZA's portable rig. Next year Bill had a longer trip planned.

Doc Stuart returned from China that fall. In settling back in to his rhombic station, he noticed with some consternation that several new homes were being built near his antenna masts. Although he had bought over 50 lots as they came up for sale in the '30s they were not adjoining and constituted only a small percentage of the area that his rhombic farm covered. The postwar housing boom would create the same problem for him that Don Wallace had experienced just before the war—a shrinking antenna farm.

Don often had to travel to the various military bases around Southern California to meet with some of his customers. One of these was Murdoc Field, later renamed Edwards Air Force Base. It was located in the Mojave Desert, about 75 miles north of Los Angeles. On October 14, 1947, in the sky above Murdoc Field Chuck Yeager flew an X-1 jet that had been dropped from the cargo bay of a B-29 bomber and punched a hole in the sky. Those people who were nearby on the ground heard a boom rock across the desert floor as the craft went supersonic to 1.05 the speed of sound, just as the noted physicist von Karman had theorized some years earlier. Yeager achieved the pinnacle of the aviation pilot's pyramid, similar to the position that Charlie Mellen occupied among DXers as the top dog in the DXCC Honor Roll.

When Herb Becker was putting together country totals for his CQ DX column in July 1947, he asked Don Wallace how many countries that he had worked. Don recalled: *"I got started at the request of Herb Becker, the DX editor of CQ. He got me to make a list. Herb said, 'You ought to send me a list of countries that you have worked so I can put it in CQ.' I said, 'I work all I that I want to, everything that I hear.' Herb replied, 'Take your log books and go through them, and make up a list. I will put it in CQ.' I went through my logs and all that I had was 87 countries worked. I had been talking to the same countries over and over."* Charlie Mellen, W1FH, had 130 countries confirmed and the race was on!

Don continued, *"So from then on I took the list and checked off the ones I had QSL cards from and just worked the ones I didn't have cards from, and I would keep on working them until I got a QSL. Watching the country list grow, I'll tell you, it was the most fun in radio I have ever had in my life."*

Part of the relay system for switching rhombics.

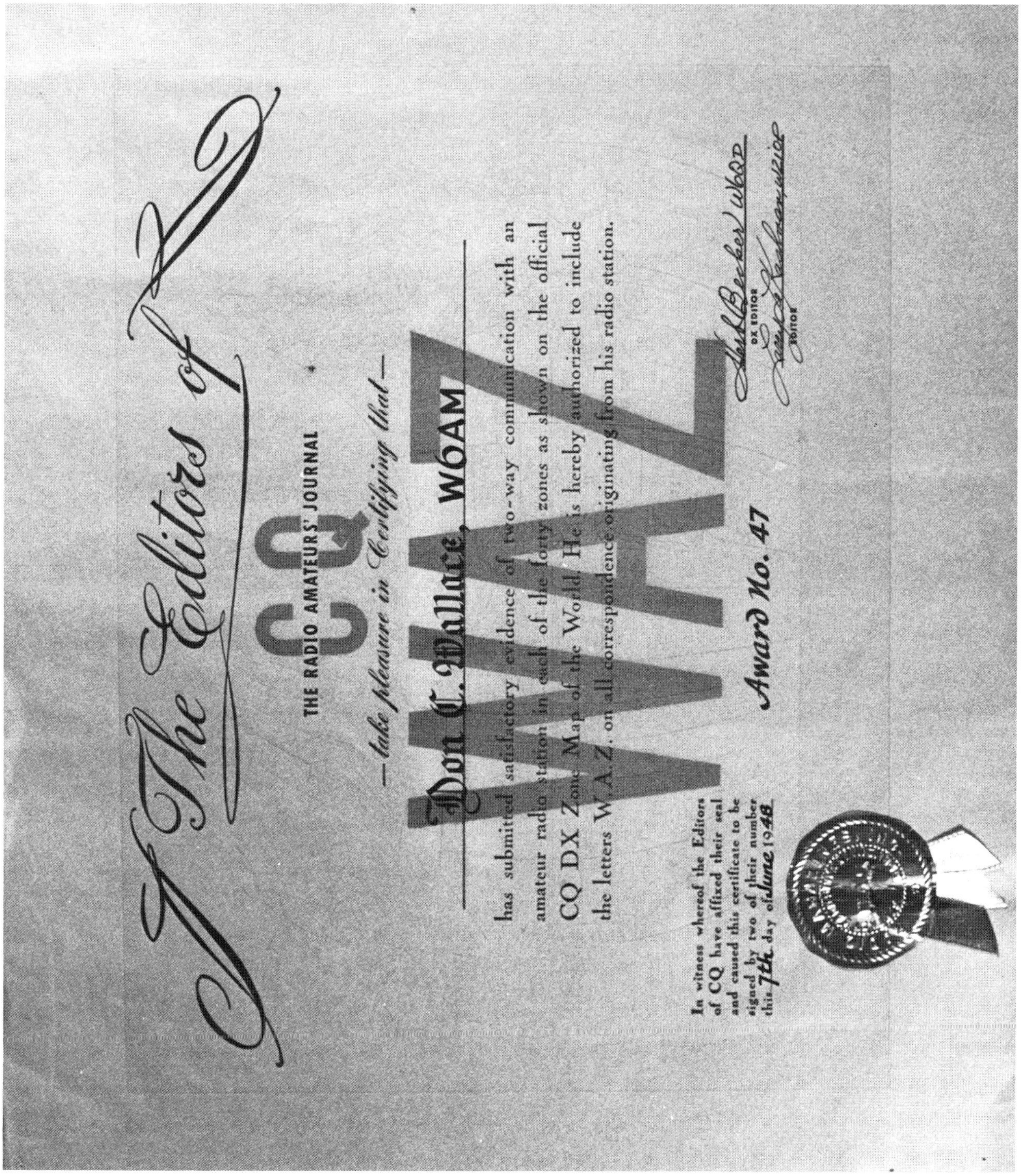

Don achieved WAZ in 1948.

The DXCC Chase

W6AM began to beat the bushes looking for rare stations. No longer did Don occupy one frequency and let DX come to him. Twenty-meter phone was practically wall-to-wall heterodynes. Lower down in the band, on CW, it was much easier to find rare stations. With all the modified military equipment coming on the air, many CW signals chirped badly and had rough notes. The art of hearing weak DX stations through the racket was acquired from decades of practice.

Working European and African stations on the short path was difficult. Even with Don's dominant antennas, the propagation advantage was with the East Coast. Their openings were longer and the signals stronger. Slowly but surely, Don picked off the French colonies in Africa: FB8BF in Madagascar, FD8AB in French Togoland, FE8AB in French Cameroons, FF8AC in French West Africa, and FQ8AC in French Equatoriale Francaise. The British also had a strong presence in Africa: ZD1PW in Sierra Leone, ZD2RGY in British Cameroons (Nigeria), ZD2KHK/NC in North Cameroons, ZD4AE in British Togoland, ZD6DH in Nyasaland, ZD9AE on Gough Island, and ZD9AA on Tristan da Cunha.

Press Wireless had no antennas installed with an east-west bearing, so Don strung a 1,400-foot-long rhombic pointed at Southern Africa. It sloped to the west and put out a potent signal. When the occasional DX station showed up on the long path, Don was so loud that he would almost knock the headphones off the operator. On the short path he also put a good signal-into Southern Africa, but the East Coast still had the propagation advantage in that direction.

One morning, shortly after Don had installed this antenna, he was listening to a station in Africa coming in on the long path. Don recalled: *"There was a fellow in Swaziland, ZS7C, running just 3 watts. I worked him six Saturday mornings in a row, then I got his QSL card, and I didn't bother him any more. I never heard anybody else work him. He was on there every Saturday morning with his 3-watt transmitter. He wasn't particularly strong, but I could copy him OK."*

Wendel Pierce, W6FSJ, and several other W6s were all on the air one Saturday morning on 20-meter CW. Don continued, *"They said, 'Well, its a dead band. We have got one station that we all can hear, a South African. Let's see who has the loudest signal.' Then they all worked him. Most of them got a 559 signal report; Wendel Pierce got 569; he had a little better antenna. I was all tuned up on their frequency. When they were through I called the South African and he said 'School's out! No report. R9 plus anything you want to say. W6AM is so loud compared to the rest of you . . .' I listened and there wasn't a sound on frequency after that . . . not a peep."*

The use of the letter R to indicate signal strength was a holdover from the 1930s when RME was one of the first manufacturers to introduce a signal strength meter on a receiver in 1935, the RME 9. Its meter was calibrated in increments of R-1 to R-9. By the 1940s most of the signal strength meters (S-meters) on receivers would be calibrated in S-units. Don, however, inevitably used the R-system the rest of his life.

"Their big delight was to prove that their rotary antennas were better than my rhombics. Once in a while they could get a better report. That was because the guys who occasionally operated my station here at the ranch would hear a good signal and they would sometimes call that station on the wrong antenna, which is worse than a dipole."

The British had several colonies in North Africa and the Middle East and Don put up a rhombic pointed in that direction that was 1,350 feet end-to-end. Then he started picking off those colonies a little easier: MD1E in Libya, MD4BPC and VQ6LC in British Somaliland, MD7DC in Cyprus, MP4KAE and VT1AF in Kuwait, MI3FG and I6ZJ in Eritrea, I5GN in Italian Somaliland, VS9GT and MP4BAB in Trucial Oman, ZB1AY in Malta, ZC1CL in Transjordan, ZC4MO on Cyprus, ZC6SM in Palestine, and ZC8PM in Arab Palestine. The majority of those stations used home-built or modified military equipment, typically a single 6L6 or 807 running 10 to 25 watts power input. The MD prefix of the call sign indicated that the holder was in the British occupation forces.

In those days the sun never set on the British empire and their military radio operators would often find themselves in remote locations. Thirty years later Harry Pain, G3ATH, recalled: "I have many memories of the 'good ole days' and I remember the contacts that I

Don at his operating position in 1949.

had with 'Rolling Hills Don.' I operated from Gibraltar (ZB2A) in 1946, Burma (XZ2HP) in 1947/48, and Ceylon (VS7PH) in 1948/49. I spent 36 years in the Royal Air Force (RAF) and in all I visited 21 countries."

Spain and Portugal had several colonies in the central region of Africa. Don phased his two rhombics that were pointed at Sydney. This gave him a commanding signal on the long path to central Africa and they also worked well in the reverse direction on the short path. In short order he worked EA9DC in Ifni, EA9AQ in Spanish Morocco, EA9DE in Rio de Oro, EA0AB in Spanish Guinea, and EK1AO in Tangier. Their antennas were usually a dipole or a random-length long wire.

In the late 40s the United States still had occupation forces in a number of European and African countries. Don worked Henry Poole, W7IOC operating from KT-1OC in Tangier, AG2AG in Trieste, and TA3FAS in Turkey. The Americans usually had a little better transmitters, sometimes even an 813 running 200 watts input.

Working stations located to the west of the rhombic farm was like shooting fish in a barrel for Don. As the hill sloped down to the ocean, two miles away, he had a dominant signal in that direction, and most operators tended to tune on to and answer the loudest station calling. CR9AG in Macao, CR10CB and CR10AA in Portuguese Timor, PK1UA in Java, PK4KS in Sumatra, PK5LK in Dutch Borneo, PK6HA in Dutch New Guinea, PK7NL and JZ0PH in Netherlands New Guinea, ZC2MC on Cocos-Keeling Island, ZC3AC on Christmas Island, and ZC5VS on British North Borneo were worked with ease.

Manchuria remained as one of the more rare countries in Asia. Although four or five stations there were licensed, only two were on the air regularly. Don worked W2WMV/C9 and C9JW which, after Bill Hunton left, were just about the only other stations on the air from there after the war. The northwest region of China was located in zone 23. In the years after the war a couple of stations became active for a brief period. C8YR and C8FP were located in Kansu province and provided the last zone for WAZ for many stations. When the Chinese civil war ended in 1949, amateur radio operation in China and Manchuria virtually ceased for the next three-and-a-half decades. Manchuria would be absorbed into China before amateur radio was again allowed.

One of the remote territories under Australian administration was known as Heard Island. It is located southwest of Australia, near the Antarctic continent. The Aussies maintained a scientific station there and occasionally one of their number was also an amateur. This was especially desirable in the remote regions, such as Heard Island, as a boat usually visits just once every 12 months. It would be the party's only contact with the outside during the ensuing year. The first known amateur radio operation from Heard was by VK3ACD in 1948. In 1949 VK1VU, VK1RA, and VK1FE were also on the air sporadically. Most of their activity was in passing traffic back to Australia and only a handful of DX stations were worked each year. VK1DY and VK1PG were also active from 1950 to 1954. Perhaps 100 amateur contacts per year occurred outside their required radio duties. Don had worked VK1VU in 1949, but Heard Island would remain rare to most other stations for another two decades.

Don would sometimes work a station that he thought should be a separate country and then make up a file of supporting data and send it to the ARRL. Don said, *"I worked the guy in the Saarland quite often, 9S4AX. He said that he was in the Saar, an autonomous region bordering Germany, and that he ought to be a separate country. He wrote me a great big long letter, two pages, about why it ought to be a country. He was the only ham I knew of there who was DXing. That was quite interesting. So I kept on working him. I said, 'I will try and get the Saar a separate country status for you.' He sent me all the dope he could, a big fat envelope of stuff. I copied it and wrote it up just like you would a college paper, in a folder, and then I made up 12 duplicate packages of data. The DX advisory committee consisted of about seven or eight* QST *staff members. They would have a meeting over lunch and decide whether or not a specific location would be a country. I found out who they were, I went to a lot of conventions, so I sent a package to each of them. Here was this book — why the Saar ought to be a country. I think I was the only one who did it. They read that information and decided it was, and pretty soon they announced 9S4 was a separate country."*

Rod Newkirk, W9BRD/1, took over *QST's* DX column from Byron Goodman in late 1947. Rod had returned stateside from his brief foray in the Philippines the year before and went to work for the ARRL. Their DX column would have Rod's guiding hand and subtle humor for the next three decades.

Among the rarest of signals found on 20 meters was that of FN8AD, Deb Shankar Seal. He lived in the city of Chandernagore, one of five French enclaves on the eastern side of the Indian subcontinent. The French colonial empire in India dated back to 1769. Shankar received his FN8AD license from the French administration in early 1949. India was pressing claims on the French Indian territory and a few weeks later Shankar was also issued VU2AX by the Indian government. As French India was the more rare and desirable location, he used the FN8AD call sign. Shankar noted: "In May 1949, FN8AD was on the 7 and 14 MHz bands, both on CW and phone, with a homebrew rig, the same old rig I am still using (40 years later)." That rig is an 807 running 15 watts input, crystal control, and modulated with a pair of 6L6s. "My receiver is still a BC-312N, converted to the AC mains," Shankar stated.

Shankar's presence on the bands caused a furor. He recalled: "Whenever I gave out CQ calls, it was very difficult to identify the stations. They came like a swarm of bees and I would wait for those who could overcome the rest. I made about 2,500 contacts in the next five years, but only received 50 percent of the QSL cards."

Although he would listen for a phone signal when asked, Shankar was reluctant to confirm one-way contacts. On W6AM's QSL card he noted: "I am very sorry to let you know that I did not contact you on phone on that date because I have no phone at present. I have seen my log but the QSO on 10 July was a CW QSO, not phone. So, please correct your log. I heard your phone signals. If you want that report I will gladly send the QSL to you." Don sent the QSL into Bob White, W1WPO, the DXCC adminstrator, and received credit for a one-way phone contact.

Shankar recalled, "There were no other licensed FN8s (after World War II) except me. There were a few [unauthorized] stations such as FN8MS, FN8DC, and FN8AB. Sometimes someone would use my portable call sign, FN8AC." According to Bob White, W1WPO (W1DX), the FN8AD card was the sole QSL ever submitted for DXCC credit after World War II. If one of the other stations was located in a French Indian city and their card submitted to the DXCC adminstrator, they too would have to be considered for FN8 credit. In those days the ARRL did not necessarily require that a station

The operating position of VU2AX/FN8AD

have a license from their place of operation to count for the DXCC.

After the war Reg Fox, AC4YN, left Tibet briefly and returned to England for a three-month leave. His relief operator was a British officer who had spent the past two years in India, Bob Ford. There were very few stations on the air yet and Bob's amateur activity at AC4YN was limited to listening. Bob was taken with the people and culture of Tibet and, when three months had passed, it was with some regret that he prepared to leave. Then came an opportunity for Bob that he couldn't pass up. A radio officer was needed at the British Residency in Gankok, the capital of Sikkim. Bob traveled from Lhasa to Gankok and spent the next two years there operating as AC3SS.

Reg Fox, AC4YN, returned to Tibet, and the amateur bands, in late 1945. His rig had improved to a 6L6, powered by a storage battery, which gave him 20 watts input. Reg was very active for the next several years. As he was usually the last station needed for CQ's WAZ award, his QSL was treasured, and AC4YN became

The operating position of AC4AX.

known as the ticket to WAZ. Reg spent a total of 14 years in Tibet.

After two years in Gankok, Bob Ford returned to England in early 1948. There he was released from the Royal Air Force and returned to Lhasa. The Tibetan government wanted to set up a radio network from the capital to its remote eastern communities and invited Bob to set up a communications link back to Reg Fox in Lhasa. Bob accepted a job with the Tibetan government and spent two months in 1949 traveling from Lhasa to Chamdo in eastern Tibet. He selected the call sign AC4RF and operated a 15-watt rig powered by a storage battery along the way. In a letter to W6VFR he remarked:

> I'm using an EL32 Xtal oscillator & 6L6 final. When I am on fone the 6L6 is screen-modulated by a 6SN7. The complete tx runs from a 6-volt vibrapak (vibrator-power supply). The antenna is a 66-foot Windom. You should hear what I hear after each contact!!! It's not all fun & games being an AC4 I assure you. The pile-ups are overwhelming at times.

Bob Ford found that he was the first Caucasian to complete the journey via the Chang-Lam (northern route) between Lhasa and Chamdo. His new home was at an elevation of 10,500 feet. Only one person in Chamdo had ever seen a white man before, a septuagenarian.

Indian independence occurred in 1947. Lord Mountbatten, British India's last Viceroy, coordinated the transfer of power from Britain to India. The British mission in Lhasa became the Indian mission. Because England no longer had an official presence in Lhasa and Reg Fox wished to remain in Tibet, he resigned from

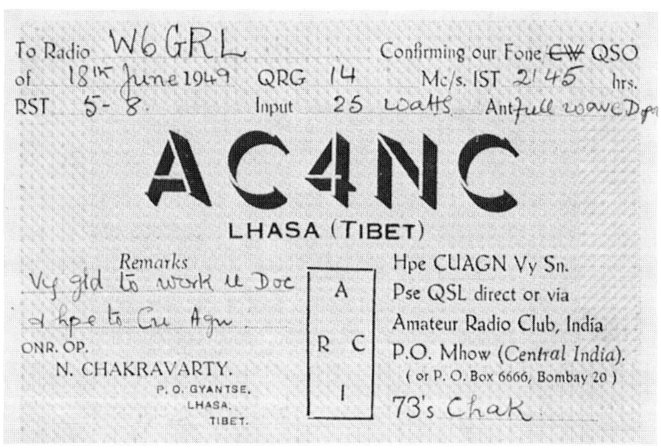

The three postwar Tibet stations before Chinese occupation.

British employment and accepted a job with the Tibetan government.

An Indian amateur, VU2NC, was assigned as the radio operator at the Indian embassy in Lhasa in 1948. He had been in Sikkim for a time, after Bob Ford had left, using the call AC3NC. N. Chakravarti (Chak) selected the call sign AC4NC to use in Tibet and was sporadically active on the amateur bands. One morning, on 20 meters, he worked W6AM and reported that his rig ran 25 watts input to a full-wave dipole antenna.

Another radio operator, AC3SQ, replaced AC3NC in Sikkim. Bob Ford remarked: "He is quite new to the ham game, so will probably need a little time to get accustomed to things, for instance 10,000 Ws calling at the same time." Actually, that was a bit of an exaggeration. Probably only a couple of hundred stations in the United States were capable of hearing AC3SQ on a good day, but all of them would probably be calling him.

Many atlases of the time showed the location of Chamdo to be in China. W6AM asked Bob Ford about this. Bob explained that the Tibetans didn't draw maps but the Chinese did. The culture, language, government, and inhabitants of Chamdo were Tibetan. Therefore, he was in Tibet.

Perry Esten, W4PN (ex-8BOX, now W6PN), worked Bob Ford one morning after his station was set up in Chamdo. Perry recalled, "AC4RF was so weak his signal sounded more like the noise level in my receiver, rising and falling slightly, rather than a CW tone. We had a good contact though, and afterwards Mitch, a W9 station, congratulated me on working Tibet." Due to the generally poor propagation conditions between America and this region of Asia, and the low-power transmitters that were used, only those stations with good antennas and quiet locations had any hope of ever hearing Tibet. Perry would never receive Bob Ford's QSL, however. That area of the world was about to experience additional turmoil.

The Chinese civil war had ended in 1949. The Communist government was installed in Peking on September 21 and the Nationalist government relocated to the Island of Taiwan, which Japan had lost at the end of World War II. On December 8, Chiang Kai-shek announced the formation of his new government in Taipei. In early 1950 Radio Peking announced that the task for the People's Liberation Army that year was to liberate Tibet.

China invaded Tibet in October 1950, and Bob Ford refused to abandon his post in Chamdo. He passed traffic back to Reg Fox as long as he could. Bob Ford was detained by the Chinese when Chamdo was occupied, and later charged with spying. Among the offenses listed were spreading separatist propaganda, claiming Chamdo was in Tibet. After five years of "rehabilitation" Bob was released. He returned to England, to the village of Rolleston-On-Dove.

During the Chinese occupation of Tibet AC4NC was relatively safe in the Indian embassy. As a precaution, however, Chak stayed off the amateur bands for two years. One morning in 1953, Chak reappeared and worked Bill Orr, W6SAI. He reported that Reg Fox was a silent key and that Bob Ford was off the air for unspecified reasons. A few days later Chak appeared on the 20-meter phone band and worked W6AM. After a few days of operation, he departed from Lhasa and returned to India. Tibet did not see amateur radio activity for another four years.

Deb Shankar Seal had worked on the embassy staff in Chandernagore, French India. He had undertaken the installation of a communications network in Sikkim in 1949. This was the same facility where Bob Ford, AC3SS, had worked the previous year. The success of this endeavor led to a similar task in Lhasa eight years later. Shankar had long wanted to visit the forbidden city of Lhasa.

In 1957 Deb Shankar Seal, VU2AX, was assigned to the Indian embassy in Lhasa. Tibet had been silent since AC4NC departed. When France ceded its colonial territory back to India in 1954, Deb's FN8AC (portable) and FN8AD (fixed) call signs were withdrawn by the French administration and he used his Indian call sign, VU2AX.

Shankar recalled: "At long last I left for Tibet in June 1957 by crossing Nathula Pass at 14,500 feet on horseback. This is northeast of the Sikkim-Tibet border. I arrived at the town of Yatung, Tibet and stayed there about a year for some specific jobs with our trade mission. (Yatung is at 9,000 feet.) The winters were supposed to have moderate snowfall, but while I was there I had to face a 14-foot snowfall in December 1957. It was a record."

The following May, Shankar was ordered to Lhasa. "I got married in Yatung and we left for Lhasa by motor vehicle." Shankar and his bride, Namita, crossed the Himalayas together. "Lhasa was the site for His Holyness, the Dalai Lama, ruler of Tibet. Both his winter palace, the Potala, and the summer palace, the Narbu-Linga are there. The Potala at 14,000 feet is the largest palace at the highest altitude in the world."

Shankar's task was to install a 500-watt wireless station and to establish the longest distance communications possible — a DXer's dream. "The reception conditions were extremely good and there was little atmospherics and static. I utilized the full power of a BC-610E with a long wire antenna. The receivers were an Eddystone and a BC-348Q." Shankar asssigned himself the call sign AC4AX. The lion's share of his operation was on a 20-meter phone. His predecessors, AC4YN, AC4RF, and AC4NC had concentrated on CW operation.

In March 1959 there was a Tibetan uprising against the Chinese occupation forces. Shankar found himself in the midst of it. "During those days of revolt we had a very crucial time. I had to maintain wireless communications under heavy overhead shelling. There was machine gun fire within 100 yards of me."

In December 1962, after five years in Lhasa, Shankar and Namita returned to India. China considered Tibet a province under rebellion and had effectively sealed her borders to all outsiders except embassy personnel since 1951. Shankar's operation was the last by an operator native to the region for the next quarter of a century. China annexed Tibet in 1974, ending their border dispute. Shankar still lives at the same location, near the Hooghly River in the West Bengal region of India and has the call sign VU2AX.

The Mongolian People's Republic was located in the northern portion of zone 23. Their first known amateur radio activity was in early 1951 when JT1AA came on the air. Ludvik Kloucek was a Czechoslovakian working in Ulan Bator, the capital city. Due to Tibet and China being off the air, he would provide most of zone 23's contacts in the early 50s.

The United Nations was about to become embroiled in a far-reaching conflict. In a speech to the National Press Club in January 1950, the United States Secretary of State remarked that the American line of defense was from the Aleutians to Japan, through the Ryukuyus, and to the Philippines. To the sharp observer this implied that Taiwan and Korea would have to look after themselves. Korea had been partitioned at the 38th parallel at the end of the war. Then, five months later, on June 25, the In Min Gun forces of North Korea invaded South Korea.

In the years after World War II a lot of military radio operators would be drawn to the amateur bands. An operator at the RAF station in the Nicobar Island group off India listened in on the amateur activity and decided to join the fun. In 1947 Jim Smith (now VK9NS) got on 20 meters using the call sign CAR, to indicate Car Nicobar, where he was stationed. Jim's first contact was with Robbie, VQ4ERR, in Kenya. Soon, the stateside amateurs discovered CAR and W6AM worked another new one. After a year Jim was posted to Singapore and there he was assigned his first authorized call sign, VS1BQ.

In the meantime, Bill Hunton, W3AG, was on the road again. In the spring of 1948 he visited the Caribbean. Bill stopped in Puerto Rico and then moved on to El Salvador. His end-fed zepp antenna had been easy to string and worked well. Bill left Central America in June and stopped briefly in Stinson Beach, heading back to Asia.

Military transports were available to those with connections and Hunton caught a plane headed west. He spent a night on Midway Island and four on Guam, then picked up a plane bound for French Indochina.

At sunset Bill Hunton put up his zepp antenna in Saigon. He had visited the chief of police earlier and had received permission to operate his amateur station. His push-pull 807s loaded up to 40 watts input. He tuned across 20 CW and found a European calling CQ. W6ODD/FI8 was on the air. Bill stayed in Saigon for three weeks, beginning July 3, 1948. At that time French Indochina consisted of what is now known as Cambodia, Laos, and Viet Nam.

Hunton recalled: "Word got around, you know. Sometimes I would turn on the receiver in the evening and find stations calling me blind. I encountered very few poor operators. Of course, most of the time I avoided pileups and looked for DX." Even though he was at one of the rarest radio locations in the world, Bill preferred to work DX stations.

In those days the ARRL forwarded QSLs to a safer address for stations operating from politically sensitive regions. Bill requested that cards for him be sent via the ARRL. In that part of the world a direct QSL drew unwelcome attention. Bill supplied the ARRL in West Hartford and the sixth district QSL manager, W6TI, with the address in Stinson Beach where his family lived. There would be a substantial stack of cards awaiting him when he returned stateside.

While he was in Saigon Bill stopped by the French embassy and met Paul Ferrand, FI8ZZ. "Paul was a generally nice sort of fellow," Bill reported, "and mentioned that due to his tenuous situation he usually didn't QSL. His rig was not apparent at the embassy." Paul was the section chief of the REF for the French embassy in Saigon. Once he was stationed in Saigon Paul made the mistake of routinely asking the minister of colonies in France for authorization to operate his amateur radio station in French Indochina, disregarding the old adage that it's easier to get forgiveness than permission.

F18ZZ was active from 1948 to 1951 but never received permission to operate from France. Bob White, W1WPO, later said that only four FI8ZZ QSLs were ever submitted to him for DXCC credit. Don Wallace was one of the lucky four with an FI8ZZ confirmation. Don recalled: *"FI8ZZ was on the air a lot and most everybody worked him. I couldn't get a QSL card from him. I kept sending him a card, every month. My average country took 17 cards sent to get a confirmation from them. I was up to my seventeenth one with him and I hadn't received a card yet. There was a ship's operator I frequently worked, Paul Fracker, W8QOH/MM, aboard the* M.V. Del Rio. *He said that the ship was going to dock at Cam Ranh Bay and that he would go into Saigon the next day. By that time I knew that FI8ZZ was in the French consulate. I asked W8QOH to go over and see the guy and get a card from him for me. In the meantime I told him to get out a sheet of paper and cut it the size of a QSL card. Then I told him what QSO information to put on it. 'Just have the fellow sign it,' I told him, 'and send it to me.' So the ship operator prepared a card with all of the data for FI8ZZ to sign. I never got the card on that try."*

Paul Fracker visited the French embassy in Saigon in late February 1949, but FI8ZZ was out of town that day. Paul operated as W8QOH/FI8 for the two days he was at Cam Ranh Bay, but it is not known if he was aboard ship or ashore.

Don continued: *"I had a list of fellows I needed cards from that I kept sending QSLs to once a month and checking off the dates that I sent them. Seventeen was the standard quantity I sent . . . I never gave up. I almost always had the card by the seventeenth try. Then I went to the Pacific Division Convention in San*

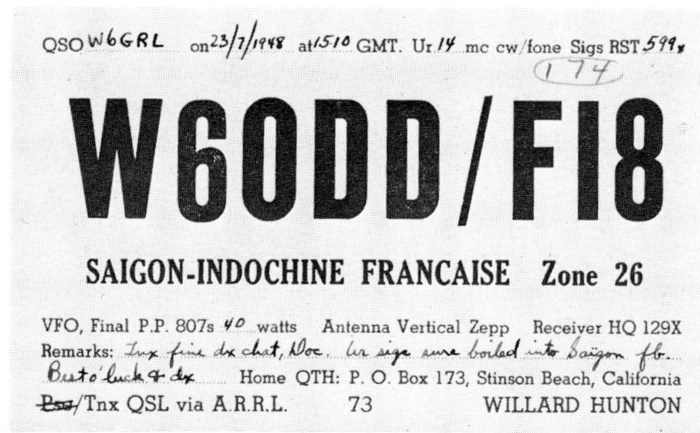

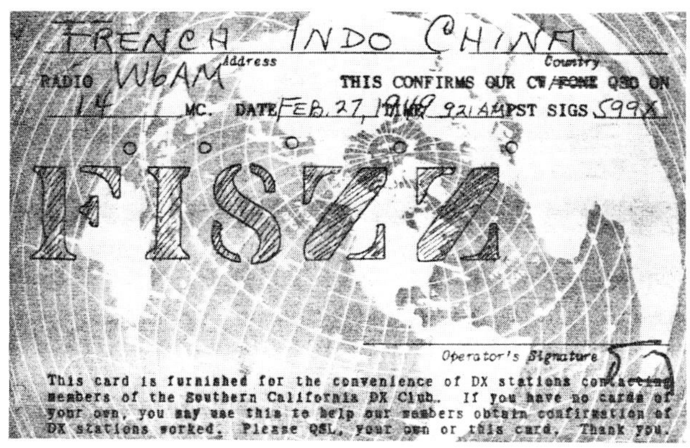

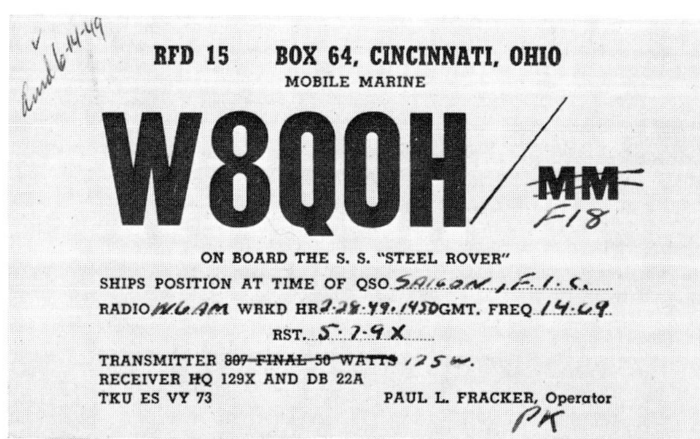

The three French Indochina QSL's that counted for DXCC.

Francisco that summer. I was sitting next to a fellow and in order to make conversation I said, 'By the way what is your business?' 'I'm with the G.E. company,' he said. I said, 'Well, I used to work for G.E.' That made us buddies. He said he now worked for the Minneapolis General Electric (formerly Peerless Electrical) *Supply Company in the radio department. I said, 'What's your job?' 'Well,' he said, 'I'm in charge of foreign business. For example, I'm sending a man over to Saigon next week.' I said, 'You know, I have been trying to get a card from FI8ZZ in the French consulate there. If I made a card out, would you tell your man to go overand ask FI8ZZ to sign it for me?' 'Oh sure, glad to do that,' he said. I said, 'Excuse me, I will go up to the room and make one out.'*

"My list of countries worked but not confirmed was always in my briefcase. I made out a club card that the SCDXC provided. I also had a stencil in there. I copied the data off my big sheet. Then I came down I made out an envelope addressed to myself, with a dollar for postage. 'This should cover the postage back,' I told him. 'Have FI8ZZ sign it and mail it.' By golly, it came back. With it was a letter explaining why he couldn't send cards. The signature on the letter under the French consulate was the same as it was on the card. So I stapled the letter to the card and sent it in to Bob White. FI8ZZ said it was a clandestine operation. He didn't want to send any cards. Bob White said it was the fourth FI8 card that he had ever seen. So two or three of my cards were pretty tough to get. When I got on top of the DXCC, those were the most important ones. They were the ones other people couldn't get."

In a letter to W6AM on embassy stationary, enclosed with Don's French Indochina confirmation, Paul Ferrand wrote: "In view of the political situation in Indochina, we are not authorized to transmit and official permission has not been given. FI8ZZ is therefore a clandestine station and he makes all his trans missions at his own risks. These requests have been made to the President of the REF and to the Minister of Defense of the French Colonies and also by myself to the local authorities, but we have not received favorable word for several months."

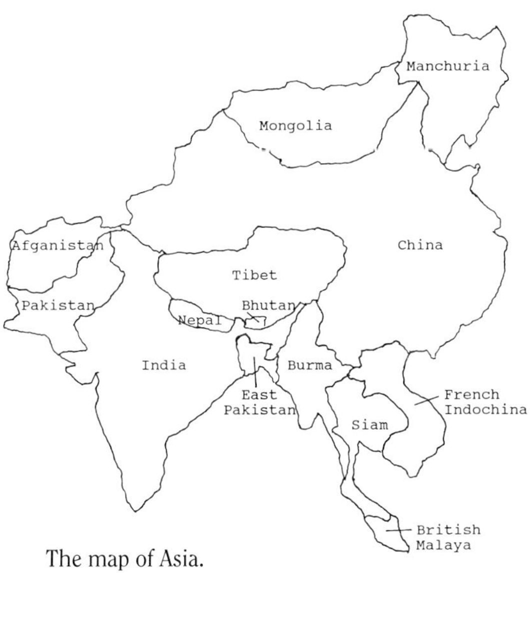

The map of Asia.

While he was in Saigon Bill Hunton worked more than 200 stations, nearly half of them W6s. One of the strongest stations he worked was W6GRL. Doc Stuart's pair of rhombics on China worked quite well into Saigon. Bill caught a plane from Saigon at the end of July and landed at Port Blair in the Andaman Islands. Permission to operate in the Andaman Islands was easy to get, according to Bill. "The RAF was there and I just contacted one of the signal officers listed in the call book." That evening Bill got on the air and Nose, KH6IJ, nabbed him on the grey line opening from Hawaii. Nose's prewar call had been K6CGK. After two days Bill picked up another plane, this one headed for Portuguese India.

Due to his frequent operations, some amateurs speculated that W6ODD was maritime mobile. Bill replied, "After 1945 I never operated from aboard a ship although some thought I did since I moved around so much."

Dawn was breaking on August 2, 1948 as Charlie Mellen, W1FH, tuned across 20 meters. He spotted a weak station working a European — W60DD/CR8! Charlie didn't occupy the top spot of the ARRL Honor Roll by lack of vigilance. He was Bill Hunton's first stateside QSO from Diu, Portuguese India. Charlie pointed out, "There were no DX bulletins then, no lists. You tuned the band daily, never knowing what you might hear. It was the golden era of DX."

W1FH was Bill's only QSO with New England from Diu, a part of Portugal's colonies on the western portion of the Indian subcontinent. As the sun rose across the United States, a handful of East Coast and midwest stations, and more than a dozen W6s also worked Bill. If only they had known . . .

After arriving at Diu, Hunton stayed four days. He worked almost 50 stations. "It was very primitive," he reported. "No English was spoken at all. I strung my vertical zepp in the trees about two miles from town. Line voltage was up and down, and the power input varied from 30 to 50 watts. Sometimes the power was out for a couple of hours at a time."

While the sun was setting in Portuguese India, it was rising in California. Don Wallace recalled, *"That morning W6KPC and a couple of the fellows were having coffee with me. We had all of the tools out and a spool of copperweld wire ready to go. I had my lineman's belt on and was preparing to go up an 80-foot pole and put up a leg of another rhombic."* At 7 AM Don got a phone call from W6CUQ, just as they were heading out the door. "Don, W60DD/CR8 is on 20 meter CW, I just worked him," Ed Hawkins (the Hawk) breathlessly reported. *"Thanks,"* Don replied, *"but we are just about to go out and put up a rhombic. I'll work him tomorrow."*

Bill was on Diu Island for two more days, but due to his operating hours and the unreliable line voltage, no more stateside QSOs were in the log. He snagged a few Europeans and South Americans, plus a couple of VK/ZLs and Africans, to bring his QSO total to over 40. "Most of my operating was after dark," Bill said, "until 2 or 3 AM. It was least conspicuous then, plus a good time for DXing. You always kept the rig in a suitcase, until you got to your room and unpacked. It wasn't wise to strike a high profile." Bill only operated CW on his trips. "Phone operation disturbed people," he explained, "and drew attention to you." Besides, Bill preferred pounding the brass.

One had to be careful when a transmitter was used in that part of the world. East and West Pakistan had recently split from the newly independent India. Nearly half a million people had perished during the turmoil. The pacifist Mahatma Ghandi had been assassinated six months earlier and tensions were high. Prime Minister Nehru was eyeing the CR8 territories, a Portuguese possession for the past four-and-a-half centuries. India viewed Portugal's presence as an illegal occupation and was determined to recover its land.

Bill left Diu and his next stop was along the Red Sea. After a brief stopover, he returned to Stinson Beach for the next 11 months. A stack of QSLs awaited him. In July 1949 Bill traveled to Europe and spent two weeks operating in the Saar as W60DD/EZ. The prefix later assigned to the area was 9S4. Europeans were locals, of course, and hundreds were worked. Bill managed to contact only a few stateside stations during his visit.

When he left the Saar, Bill returned to Stinson Beach again. He stayed there 18 months, until the spring of 1951. Then he traveled back to Saigon for a brief visit. French Indochina had recently been banned for United States stations to work. For the next two months Bill operated undercover from Saigon as FI3AG. When he prepared for departure, Bill traveled to Cap St. Jacques near the mouth of the Mekong River. On his last night in the country he worked Ed Hawkins, W6CUQ (K6ZO).

Hawkins had become an avid DXer after World War II. He had first come on the air twenty years earlier using a five-watt rig with a 210 borrowed from his father's broadcast receiver. When it was eventually discovered and reclaimed, a UV-203 was filtched from a projector in Pasadena and Ed built a 50-watt rig.

Bill stopped in Nicaragua, then he returned stateside to live permanently. He remarked: "I enjoyed operating from exotic locations. However, the world political situation was becoming very unstable. My next trip I might not make it back!"

Many of the operators Bill Hunton worked speculated on why he was traveling in some of the most volatile areas of the world. When the author visited him in Williamsport, Pennsylvania, 35 years later Bill would only remark that he liked to travel. Whatever the government security arrangements may have been that Hunton traveled under, apparently he still felt bound by them.

Bill Hunton and his wife moved to Alaska in 1957 and concentrated on raising a family; he had been "on the road" when each of their children was born. Over the next five years KL7AL was a familiar call on 20 meter CW. Equipment was difficult to procure so far north and, when the Huntons departed from Unalakleet, the 807 rig that had traveled the world with Bill was left with a local amateur.

French negotiations with Ho Chi Minh over independence had collapsed shortly after World War II, and guerrilla warfare in French Indochina had heated up. In 1950 France prohibited amateur radio activity in Indochina. The United States then placed FI8 on its banned country list. FCC Public Notice, December 21, 1950 included Austria, Indochina, Iran, Lebanon, Netherlands Antilles, Thailand, Indonesia, and Japan (Japanese nationals only). A number of radio operators in the French occupation forces in Indochina continued to operate on the amateur bands for another four years, most of them undercover. FI8AB (F3EQ) and FI8AZ (F9TX) and a few of these stations sent QSLs, but it was unusual. A scan of FN8AD's logs during this period revealed over a dozen active FI8s.

In July 1952 the ARRL communications manager, F.E. Handy, W1BDI, wrote W6AM:

> Sorry about FI8AB and any other FIs. For United States hams to work them is a 'sure way' to invite an FCC citation. The REF has refused to handle any cards to be forwarded to French Indochina since they are not licensed and work in disregard of their government.

After World War II the ARRL's policy was to accept QSLs for DXCC credit, whether or not the operator had permission to operate. Hence the term "QSL via ARRL." The League made no bones about forwarding QSLs from clandestine stations. They were happy to do so. At one point an amateur operating undercover in Saudi Arabia sent his logs to the ARRL so they could give DXCC credit to the stations he had worked. It was too dangerous for him to send direct QSLs to the individual stations.

However, the FCC left the ARRL no choice but to break with tradition at this time. No QSLs from French Indochina to American stations were accepted for DXCC credit for contacts after December 21, 1950 (See appendix 13). Prior to that date W60DD/FI8 sent out more than 200 cards, while FI8ZZ is known to have issued four (out of his thousand-plus QSOs), and W8QOH/FI8 sent out perhaps a dozen. These were the only QSLs accepted for DXCC credit of American stations.

Getting the QSL from unusual stations was always a challenge. Don recalled: *"I worked a very rare station (to remain unnamed) on CW and I got a QSL card for that contact. I worked him quite often, so one day I said, 'Would you listen for me on phone?' He couldn't transmit phone yet, but his receiver would copy AM signals okay. The ARRL counted 1-way phone QSOs for the Phone DXCC award. They said they couldn't tell the difference. If a card said 'I confirm your phone signals,' how could they tell what mode the DX station was transmitting with. So they counted 1-way phone contacts — it was published in* QST.

"Well, this fellow I worked wrote me and said he would like a National receiver, which was the big deal at the time. If I sent him a National receiver he would send me a phone QSL card. I wrote back and said, 'I couldn't very well do that, but here is a couple of dollars to put toward the receiver.' He wrote back and said I was a cheapskate. 'You will never get a phone card from me. Anyway, your phone signals sounded lousy.' So I bundled up the whole correspondence, carbons of mine which I had kept, and this fellow's letters. I sent it in and said, 'It looks like I have a confirmation for 1-way phone contact.' Bob White sent it all back and said, 'You certainly have.' That's another country that very few people have ever confirmed on phone."

Larry Brockman (N6AR) recalled another rare one on phone that Don had. "Don told me that he worked this one fellow numerous times on CW and had his QSL, and one day worked him on phone. Don sent quite a number of cards, but never got a response for his phone contact. Finally, he sent the guy a nasty letter and demanded a card. The fellow finally responded to that prodding with a letter saying, 'I didn't work you on phone on the 13th, it was the 15th. Besides, the time wasn't 1515, it 1615. And your report wasn't 57, it was only 44. And I'm still not going to send you a QSL.' With that letter Don had his confirmation of a phone QSO!"

When he was not putting up rhombics, Don was working DX at a furious pace. He finally received QSLs from over 100 countries and on June 10, 1948 he re-

The Japanese DX Radio Club in 1954. Rear: JAITD, JAIAAW, JAIEF, JA3IW. Center: JAICR, JAICJ, JAICC, JAIBK. Front: JA3AA, JAIAG, JAICO, JA5AI.

The American Radio Relay League, Inc.
COMMUNICATIONS DEPARTMENT — WEST HARTFORD, CONN., U. S. A.
Fourteenth A. R. R. L. Sweepstakes Contest, November 1947

CLUB AWARD

This is to Certify that Don C. Wallace operator of Amateur Radio Station W6AM has this date received this award, in consideration of his accomplishment in making 477 contacts with 67 A.R.R.L. Sections for a score of 63,717 points in the Fourteenth Sweepstakes Contest.

THIS ALSO CERTIFIES AND ACKNOWLEDGES that by this performances in this Sweepstakes Contest the above named operator is entitled to this award, in recognition of the fact that his is the foremost participating Phone station in the Associated Radio A. of Long Beach (Radio Club Organization) which was represented by 3 oprs in this competition.

June 1, 1948
Date

President or Activities Manager

Communications Manager, A. R. R. L.

An ARRL sweepstakes contest award.

ceived the mixed DXCC award number 210. At that time Don had credit for 115 countries and W1FH had credit for 189 (See appendix 9). It was to be a long road ahead. Phone or CW contacts could be used for that award. Stations were more difficult to work on AM phone. For every 100 stations on CW there was one operating on AM. The signal would be weaker and more susceptible to noise and interference. This was due to the wider receiver bandwidth necessary to receive AM phone signals and the poor modulation methods often used in homebuilt transmitters. On September 29, 1949 Don received phone DXCC award number 146. In the November 1950 issue of *QST* W6AM was listed with 124 countries on phone and W1FH had 195. Here was another long road ahead.

At this time Don was also climbing up *CQ* magazine's WAZ list. On June 7, 1948 he received WAZ award number 47 (See appendix 10). This was the mixed phone and CW award. The first all phone WAZ awards would not be issued for several years yet.

CQ magazine sponsored their first DX contest in 1948. It was modeled after the one contest that *Radio* magazine sponsored before the war, the 1939 DX Marathon. Entrants were permitted to work DX stations all year long and, at the end of 12 months the final total of different zones and countries worked were listed. Don didn't spend a lot of time chasing DX that year; he was putting up more antenna wire almost every weekend. He did work all 40 zones in that 12-month period, which was no mean feat.

The Southern California DX Club was formed in 1948. It was a group of Los Angeles-area DXers that had 100 countries confirmed. Bill Adams (W6ANN), was the President, K.C. Jones (W6RLN) was the Vice President, Frank Newton (W6SYG), was Secretary, Homer Biedebach (W6GFE), was Treasurer, Wendel Pierce (W6FSJ), and Don Wallace (W6AM) were the Directors, and Bill Adams also served as Editor of the club bulletin. Many of the charter SCDXC members were to become the well-known DXers in the future. The membership consisted of most of the up and coming DXers in the area. Bill Mausey (W6RT), was at the charter membership meetings also but didn't have his 100 countries confirmed yet to qualify for membership (See appendix 14).

The first joint meeting of the SCDXC and the NCDXC, the DX Convention, was held in Fresno in January 1950. The SCDXC President, Bill Orr (W6SAI), presided over the meeting, which has since remained a favorite annual event.

Somehow, in the midst of working most of the DX stations that came and went on the bands, Don found time for business and to study for technical examinations. In 1949 he passed the professional engineer's exam and became a registered P.E. This was quite an achievement for someone who had a degree in business administration!

When the Nationalist Chinese were driven from the mainland to the Island of Taiwan in 1949, their station in Nanking ceased broadcasting. Doc Stuart wound down his recording activities at his station in Ventura and sent the last of the material to the Chinese embassies. After nearly 10 years of association with the Chinese government, he was now back in the dental business full time.

Doc Stuart was working his share of DX at this time but was facing a growing problem—his diminishing rhombic farm. As more lots in his neighborhood were developed, pole after pole was coming down. The competition was getting stiffer by the day from the new DXers and his rhombic farm was down to half its original size. Doc purchased a 50-acre piece of property a couple of miles inland from his home by the ocean and started installing poles there. The location was nowhere as good, but he would not have to worry about neighbors encroaching on his antennas. Doc planned to put up a rotary Yagi and figured that he would have his rhombic farm up to its former capability in a couple of years.

Doc Stuart, W6GRL, in 1949.

The Contenders

Having acquired the site from *Press Wireless*, Don divided his time between family, business, DXing, putting up more rhombics and rebuilding the station in Rolling Hills. Five days of the week he spent with his family at the house in Long Beach. Wednesdays and Saturdays usually found him at the rhombic farm. In 1951 Don put up the sixteenth, and last of his rhombic antennas. The July 15 issue of the *Long Beach Press Telegram* carried a front-page article about Don's latest addition. At 1550 feet from end to end, it was thought to be the largest amateur radio antenna in the world. The article also noted that Don was just 10 countries away from being listed on the DXCC Honor Roll. W1FH was still top man with 238 countries.

Don now had all points of the globe covered with the 32 possible directions of his rhombics. There were 60 telephone poles between 80 and 100 feet high that had been erected across the 120 acres to support the antennas. Each rhombic was of the multi-wire curtain design and had either two or three wires in each leg. The combined total of all of the #8 copperweld wire in the air exceeded 45 miles. Another 110 poles, an average of 25 feet high, had been erected to carry the antenna feed lines from the station to the rhombics.

Almost monthly Don had an antenna party. There was no way he could have strung all of that wire and installed the poles by himself. The ARALB continued to be his mainstay. He had known many of the members for over 25 years. Don still bore the brunt of the work himself. He would be the first one up a pole in the morning and the last to call it a day. He still followed his instincts from 20 years earlier—only go up six poles a day. After that, you become so tired that it is easy to make a mistake and get hurt. Now over 50 years old, he was in remarkable shape.

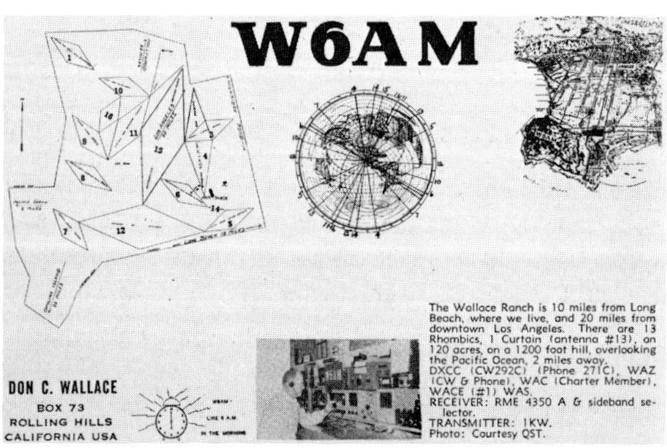

Don's rhombic farm QSL in 1951.

The feed lines from each end of each rhombic were brought to the rhombic breakout frame, located adjacent to the back of the station. The frame was constructed of redwood posts 12 inches square. By this time a second level had been added. It consisted of a matrix of two layers of crisscrossed wire mounted with about 70 insulators. The insulators were glazed ceramic, an inch in diameter and six inches long. This wiring was brought from one side of the frame through wall insulators into the ham shack. Two dozen pairs of wall feed-through insulators had been added to the 10 that *Press Wireless* had originally installed. There they connected to an overhead wire matrix in the station that was mounted on insulators and installed just a few inches below the ceiling. This wiring was connected to more wires which dropped down to the relay matrix on the back wall. An individual rhombic and its 450-ohm termination could be selected by these relays with the rotary switch at the operating position. At that time the rhombic farm was in an extremely quiet location. The power lines and telephone lines to the station were buried and Don's nearest neighbor was two miles away.

Examples of the capability of Don's station can be found among his QSLs from this era. The QSL card from VT1AC in Kuwait stated: "You are the first and only W (U.S.) station to work me so far." The Kuwait station ran just 10 watts to a long wire antenna, perhaps 15 feet high. Only a few others in the U.S had any hope of hearing him. Some months later Doc Stuart, W6GRL, was the second United States station to work VT1AC.

Tom Orr, W6EIF (W6HT), was stationed in Panama as KZ5TO for a brief time in the middle 1950s. "I didn't have much experience with DX yet. When the DX contest came along that year I got on the air and was working a pile-up of stations on my frequency. I was having trouble copying them, as I didn't know yet to work them off of my transmit frequency and spread them out. About then W6AM called. It was like an elephant stepping on a bunch of ants! What a signal. I found out later that he had phased rhombics pointed my direction." The

Canal Zone was the reverse direction of Don's antennas on Japan.

Don had brought up to the ranch some of the 1-kilowatt amplifiers from Long Beach that he had built in prior years. A pair of 450TL tubes were mounted in an old *Press Wireless* rack for operation on 80 meters. The bias supply was modified and it was rebuilt to use a pair of 450TH tubes, which were more rugged.

An acquaintance of Don's found a broadcast transmitter that was being replaced with a modern one. The old transmitter was acquired at a nominal price and Don installed it next to the BC-610. It also used a pair of 450TH tubes in class C (CW) operation. It was tuned up on 40 meters and used exclusively on that band.

About the same time Don heard of another broadcast transmitter that was being replaced with a modern one. This one was a real find. It had been built in the mid-30s and consisted of two transmitters mounted side-by-side on the same chassis. The pair of transmitters were housed in the same cabinet, which was three feet wide and six feet high. The tubes were a pair of 833s in each amplifier. The cabinet was placed next to the 40-meter amplifier and one pair of 833s was tuned up on 20 meters and the other on 10 meters, where they were left exclusively.

Don also acquired a modulator that, when it became antiquated, had been removed from a broadcast station. Although the old modulators that were built in the '30s would not meet the modern broadcast quality standards, they were quite suitable for amateur radio use. With just a few simple modifications it was adequate to modulate his single-band CW transmitters on AM phone. The use of this modulator provided Don with a kilowatt when he operated on AM phone.

The BC-610 transmitter was left in position. It would run 500 watts on all bands, phone and CW, and was kept as an emergency backup in case one of the single-band amplifiers developed a problem. On short notice a jumper wire could be installed to add the transmitter into the amplifier selection system.

Don now had the ultimate in amateur radio station design: single-band 1-kilowatt amplifiers for each of the four amateur bands. By means of a switch at the operating position he selected which amplifier was connected to the antenna system. The same switch also provided filament power to that amplifier. Another switch selected the RF drive from the exciter (transmitter) to the desired amplifier. All of the amplifiers were connected to the same high-voltage supply.

The high-voltage power supply consisted of a pole line transformer reverse connected to the 220 volt power line. The input voltage to the transformer was adjusted by a large variac which controlled the high voltage output of the power supply, up to a maximum of about 3500 volts. A surplus filter choke as big as the power transformer provided filtering, as well as a large filter capacitor.

To the casual observer it seemed that the station must be capable of running many kilowatts of power. Actually, this was not the case. A minimum of eight relays were in series with the feed line of any rhombic antenna and up to fifteen relays in the matrix were in line with the transmitted RF with some of the antennas. To try to run excessive power would be risking a breakdown. In attempting to contact a distant station it is more important to keep a signal on the air for an extended period of time than trying to get another 3 dB of signal strength by doubling the transmitter power output, which then stresses the system so much that you get knocked off the air unexpectedly.

A number of the fellows across the Los Angeles Basin believed Don was running a lot of power. They looked at the enormous power supply and remembered the big arcs that he had drawn with his water-cooled amplifier in Long Beach. When he operated from the ranch Don was always loud at stations in the area. Actually, from the top of his poles at the rhombic farm, everyone in the Los Angeles Basin was line-of-sight. It didn't occur to them that at Don's location they too were also very strong.

If the truth be known, Don wasn't running nearly the power that he formerly ran. The water-cooled tube in his transmitter in Long Beach had given up the ghost. The crack it had sustained in the 1933 earthquake had weakened it and its lack of use during the war years hadn't helped any. Its power output had diminished for the next couple of years following the war and it finally expired during a tune-up one evening. It was stored in an out-of-the-way place in the attic of the ranch, where the author discovered it forty years later.

Don rebuilt the three-phase power supply/transmitter at the Long Beach house with a pair of 450TH tubes. The F-328 water-cooled tube was no longer available commercially. Few had been produced 25 years earlier and none were to be found in the LA area. The new transmitter would barely run 2 kilowatts input, hardly enough power to carve out a spot on 20-meter phone when the band was crowded on a Sunday evening. On CW the transmitter now had a new sound, a 360-cycle tone riding on the carrier instead of 300 cycles. The line voltage in Long Beach had been changed over from 50 cycles to 60 cycles in 1947. Don's full-wave three-phase-power supply now had a slight 360-cycle ripple riding on the high voltage.

He covered his operating position at the ranch with

five receivers. Any or all of the five could be connected to the selected antenna by means of ceramic knife switches mounted above them. In this manner they could be compared as to which was receiving a particular station the best. Since he was an RME representative, Don used only RME receivers at his station. At this time Don had the largest and most powerful amateur radio station in the world.

In 1949 the FCC had authorized radio amateurs to use 160 meters. The band was to be shared with Navy LORAN signals. Amateurs were limited to 150 watts input, to minimize interference with the other service. On that band Don used a single exciter/transmitter that ran 125 watts input, and a small antenna matchbox that had to be retuned into each rhombic.

One day an amateur called from San Francisco. The ARRL 160-meter contest was coming up and he wanted W6AM to participate. Don recalled: *"Well, this fellow said that he was going to be in the contest and he wanted me to compete. He felt he had a pretty good station on 160 meters and that he could win. He heard I had the best station around so he wanted to make sure that the best competition would be on the air during the contest so that his victory would be meaningful."* Don never accepted a challenge half heartedly and he set about building a matching network that would work into all of his rhombics without retuning. Since it was broadband, to match all of the antennas, the tuning network was not as efficient as the matchbox properly tuned for a single antenna. However, since no tuning was required, the antenna switching was almost instantane. The rhombics didn't have much gain on 160 meters, perhaps 3 or 4 dB at most, but they were superior to the majority of other antennas around. Conditions were good that weekend and afterward Don sent his score off to the ARRL. The final listings of the contest revealed that W6AM had nearly twice the score of his competitor up north.

In June 1952 Don and Bertha went on a tour of Europe. Whenever possible, Don always looked up an amateur in each country that he visited. While in Scotland they visited Frank McAinsh, GM8MN, in Crieff, and then went to Tain. There they discovered in the city hall an eight-foot high painting of Don's great-grand uncle, Alexander Wallace.

Don had also been in Europe two years earlier. *"When I was over in Switzerland in 1950, I visited a number of Swiss stations, about 15. They were all using 2,500-watt transmitters except one, he was using 700 watts. One of the top DXers of the world, Jean Lips, HB9J I think it was, had 700 watts. The 2,500-watters were not as far up the DXCC as he was.*

"I asked them about their power bill. They all had these 5-kilowatt pole line transformers running each rig. 'Where do you get all these power transformers?' 'Well,' they replied, 'the power company gives them to us as soon as they are not efficient. They test them periodically, and if they are not, say, 97 percent efficient, they give them to us. We can have them for nothing.' They said their power bill was a dollar a month per house. Now there's a government designed to help the people."

Slowly but surely Don had been climbing up the DXCC listing each month. After five years of chasing DX, W6AM had finally made the Honor Roll in April 1952. *QST* listed W1FH with 246 countries, W6VFR had 238, W6AM had 231, and W6GRL had 229. Don was first listed on the Phone Honor Roll in September. Charlie Mellen was still tops on phone with 219, while Don had 171 phone countries confirmed.

The front-page stories about Don's station in the *Los Angeles Times* continued. The December 5 edition had two pictures of Don on the front page, one sitting in front of the operating bench full of RME equipment and another holding some of his DX QSL cards. The article described his 1,550-foot long rhombic and the 170 telephone poles spread across 120 acres. The article quoted: "'This was my dream all my life,' says the 54-year-old Wallace, 'but I never dreamed this good.'" The article also reported that of the 265 countries that the ARRL recognized, Don had 238 confirmed and a spot on the ARRL DXCC Honor Roll, which listed the top 10 amateurs in the world.

2 Los Angeles T

FRIDAY MORNING, DECEMBER 5, 1952

HAPPY HAM—It is a dream come true for Don Wallace of Rolling Hills, a farmer who also is one of top-ranked amateur radio operators, as he sits in front of his receiving equipment. He also has a 1550-foot radio antenna.

Times photos

The December 5, 1952 *Los Angeles Times* front page.

Not wanting to be left out when it came to recognizing the accomplishments of one of its own, the Long *Beach Press Telegram* published another article about Don in its March 20, 1953 edition. The headline read: "Rolling Hills Ham Station Center of 'Round the World' Talk Fests." The article commented on Don's forest of poles, the 45-plus miles of antenna wire, and his many radio accomplishments over the years.

Meanwhile, Doc Stuart's rhombic farm in the hills of Ventura was progressing nicely and he was working his share of DX. Then Doc met with a big setback. A decision by the State of California was to have quite an impact on the future of his station. He was notified that the 120,000-volt trunk lines from the Mandelay Generating Station to Ventura County and the surrounding area were to be routed along one edge of his new property. The noise that high-voltage lines of that nature generate around wire antennas (and Yagis) would make it nearly impossible to hear weak DX stations. That took the wind out Doc's sails. W6GRL had achieved the high score from California in the 1951 ARRL CW DX Marathon. That was his last serious contest entry. W6GRL was listed on the ARRL DXCC Honor Roll for a few months in 1952 but then dropped behind the pack as the noise level rose at his location. Doc Stuart had finally resigned himself to the fact that he would no longer be a contender in the battle for the top spot on the postwar DXCC. He would be active for a few more years, until the late '50s, and then give all of his radio equipment to a local amateur in Ventura. Doc was too motivated an individual to remain giving only a half-hearted effort. After that he immersed himself in his career field of Gnathology, which is related to the dental profession.

The W6AM rhombic farm turned out to be an enormous expense for Don. Even at the bulk rate wholesale price of 10 cents per foot, Don had $30,000 alone invested in the copperweld antenna wire up in the air. Each feeder pole had typically 30 insulators mounted on it to support the antenna feed line wire and the rhombic breakout frame was a matrix of wire and insulators. There were on the order of 5,000 high-quality insulators in the air supporting the antennas and feed lines. Even at wholesale prices the average cost per insulator was nearly $2 each. The lead-insert insulators, of which several hundred were required for supporting the legs of the rhombics, cost $12 apiece. (At two insulators per leg, plus two for each additional curtain wire, four legs per antenna, times 16 rhombics, the numbers added up fast.)

The electronic industry had become very poor in the years immediately after the war. As the spare parts and equipment from the military flooded the surplus market, new orders for these items dropped dramatically. Don had saved quite a bit of money during the war, when his income was much higher than his financial obligations. However, times had been pretty lean for the past five years. Don found that he could rent out his land and antennas on a part-time basis to help defray his enormous expenses.

The 120 acres of land on which Don had spread out his rhombic farm was suitable for dry land farming. The McCarroll family had been farming the vacant land on Palos Verdes Peninsula for years. McCarroll and Don reached an agreement in which McCarroll used the land under the antennas for farming, being careful of the poles and their guy wires when he was on the tractor. Don received a small share of the profits. The first year they raised a crop of garbonzo beans. Don's income from the crop wasn't much, but it provided a tax write-off. That was welcome, what with the big financial outlay he was investing in building up the station and antennas.

As word of Don's station spread, various electronics companies in the Redondo Beach and El Segundo areas approached him about using his antennas for assorted communications experiments. The international situation had developed into a cold war and the United States government had awarded several contracts to those companies that dealt with secure communications systems. Don's antenna system was the ideal site to perform these experiments, as the experimental station on the other end was usually located on one of the United States-administered islands in the Pacific. Due to the security arrangements required, the military installations on these assorted Pacific islands would be the ideal receiving location.

Because of business and family obligations, most of the time Don could only spend Wednesdays and Saturdays at the rhombic farm. The exception was when the occasional contest was on, but on the average that was only once every month or so.

Commencing in 1951 Don leased out his antenna farm during the week to a number of companies. His lease agreement stated that he could use the antennas on Wednesdays and from Friday night to Monday morning on weekends, and any time day or night that a new country came on the air. Most of the companies that Don dealt with only required the use of his place for just a few months.The renters included Nortronics, Aeronautical Radio Inc., Hoffman Laboratories, Decca Navigator, and Ramo-Wooldridge Company (later known as TRW). The Los Angeles County radio emergency systems (sheriff, fire, etc.) were given the use of a pole and outbuilding for their communications system for 20 years at no charge.

The steady stream of renters for the rhombic farm in Rolling Hills also helped defray the cost of maintenance for the ranch. Don knew that as the area developed he would be wise to have his ranch aesthetically acceptable. He planted eucalyptus trees along his 400-foot long driveway and oleanders for the several thousand feet that the eastern edge of his property bordered High Ridge Road. The cost of a gardener who kept the hedges pruned and trees trimmed cost $250 per month.

The first big rise in land values on the hilltop had occurred in 1948. The Vanderlip family sold 8,000 acres of land to a developer. Tracts of homes began to crop up here and there in the ensuing years and with them came rising property taxes. Were it not for the revenue and tax write-off from renting his property, Don probably couldn't have afforded to keep it.

Hughes Aircraft rented Don's antenna system for 11 years. With the kind of government contracts that they were getting, it was perfect for their requirements. Hughes Aircraft set up their equipment in the northeast corner of the building that housed Don's station and in a few months it was partitioned off into a separate room. Years later, long after they had vacated, Don still referred to it as the Hughes room.

Initially, Hughes engineers installed a 10-kilowatt transmitter in that room for some of their tests. They had a high-current power line brought in, underground, of course. In the ceiling above the transmitter in the Hughes room they mounted a high-volume fan, five feet across. When it was turned on, the rush of air being drawn into the ceiling was similar to what was experienced in a wind tunnel—one almost had to hang onto the door frame. They had also wired an antenna control switch in the room, next to the big transmitter. A small switch at Don's operating position could give the antenna control to the Hughes room. Even in this professional environment, there were to be a few exciting moments. The first time that the engineers fired up the 10-kW rig they fried the termination resistors for the rhombic antenna they were using. The remnants of the resistors resembled charcoal. After a short conference, they 'fessed up to Don what had happened before he discovered it for himself.

The Hughes engineers received a stern lecture from Don. The 300-watt, non-inductive termination resistors were very difficult to locate and, besides, his relay system could not handle that kind of power. Hughes was limited to a maximum of 2 kilowatts for their tests in the future. During the transmitter tests on that program two or three more banks of termination resistors were charred when the Hughes engineers slipped (?) with the power control. Don noticed the damage and insisted that the big rig be removed so the engineers wouldn't be tempted to turn up the power. In the future the Hughes engineers had to settle for a 2-kilowatt transmitter with which to conduct their experiments. The charred termination resistors were left on the wall as a reminder and were still there 30 years later.

Once he had the rhombic farm leased out for a few months, Don started remodeling the station. Some of the DXers in the area wanted to work the upcoming DX contests from Don's place and they decided to give it a fresh look. Don recalled: *"The fellows told me that I ought to remodel my radio room. They said to go ahead and take everything off the walls and they would do the whole radio room over. When I first bought the place I couldn't hear anything when I turned on the receiver, the echo in the room from the loudspeaker was so bad. The fellows put in an acoustical ceiling. Initially I had tacked surplus flags on the ceiling, right after the war. They were these great big Navy alphabet flags. So the ceiling was properly sound-treated by the fellows and we made the whole place look good. We also put in these shuttered windows and cleaned up the building. It didn't take too long. Fellows would be doing carpentry work at one side of the room while others painted at the far end.*

"After we were finished I never did put any of the old paperwork back up on the walls. Instead I started putting up new certificates as they arrived. So all of the old stuff is piled up behind the single-band amplifiers somewhere. All my major certificates that I won from the national contests in the early '50s are not on the wall. They are piled up back there. I just figured that the main purpose of the station was to operate it. If I did everything I could to keep all of the stuff in order, then I wouldn't have time to operate it. If I never operated it, then there wouldn't be anything to put up on the walls. I just get on the air a little bit and do what I can with the station in the meantime and that's it. You can't devote your whole life to ham radio. It's no fun then. It's the same way in Long Beach. I do a little bit of work on the station once in a while and then get on the air with what I have."

The 15-meter band was allocated for amateur use in 1953. The novice class amateur radio license had been introduced that year, with reduced code speed and technical requirements from the general class license. A portion of 15 meters was allocated for the novices to use with 75 watts of input power on CW only. The band could also be used by amateurs with general and higher class licenses for both CW and phone operation, and they could run a kilowatt, or "gallon." The term came into common use after World War II, as more and more amateurs began to build rigs that ran several hundred watts. A 500-watt transmitter was referred to as a "half-

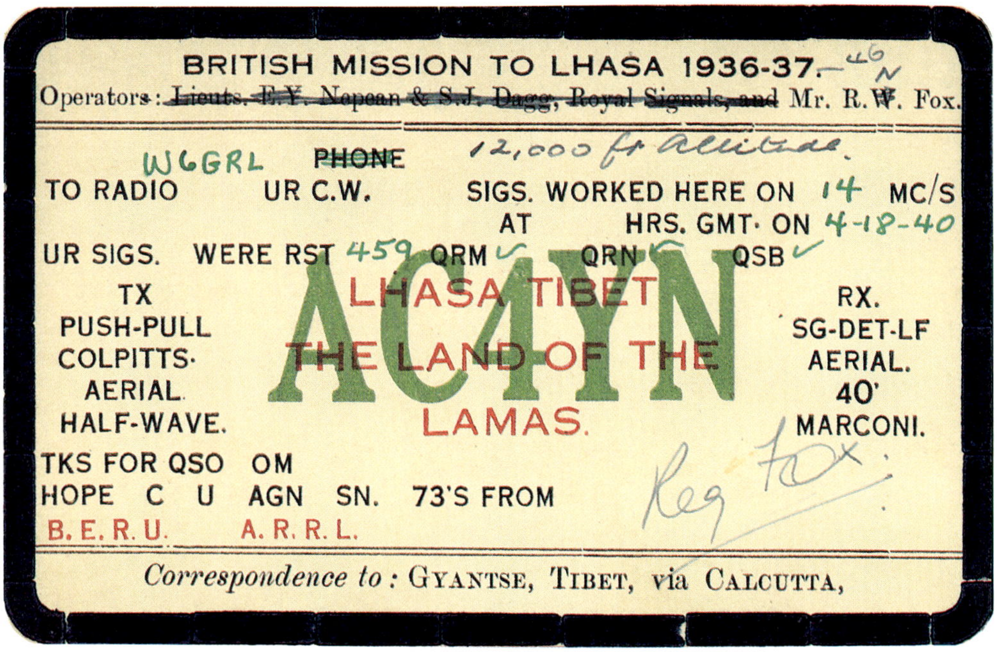

The QSL of Reg Fox, AC4YN.

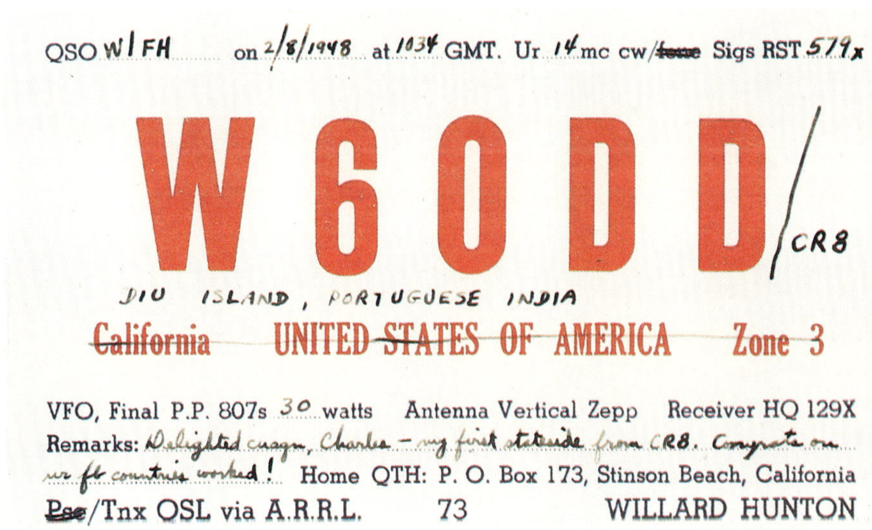

The rarest of QSLs: W60DD/D1U.

Prewar QSLs from China and Manchuria.

Postwar QSLs from China and Manchuria.

Typical DX stations of the '40's.

Typical DX stations of the '40's.

Typical DX stations of the '50's.

QSLs from Burma.

Sixty-five years of Don's QSLs.

gallon." Within a few months Don had built another amplifier for the ranch. This one also used a pair of 450TH tubes and was mounted in the rack cabinet above the 80-meter amplifier. Soon it was integrated into the amplifier selection circuitry at the operating position, along with a broadband antenna-matching network that would work into all 32 antenna directions.

It was fortunate that Don had the time to spend improving his station and getting on the air. The 1950s would see a large amount of DX activity from remote locations. Bob Roberts, G2RO, had an advisory job with the British government broadcasting service. During the decade Bob traveled to the various British territories around the world that needed technical advice on upgrading their facilities. He carried along a small crystal-controlled, 15-watt transmitter that used an 807. Wherever he stayed, Bob used the prefix of the country he was in and added the "RO" suffix. He borrowed a receiver to use in the host country and usually strung his antenna on the roof of his hotel, similar to the method W6ZZA used 20 years earlier. Bob operated from 60 countries and provided many a DXer with a new one.

Meanwhile, the map of Asia was again in a state of flux. France had ceded its French Indian (FN8) territory to India in 1954. In the coming months France would abandon Indochina after its defeat at Dien Ben Phu. India had been pressuring France for the return of their French Indian colonial territory for some years. With their situation in Indochina rapidly becoming hopeless, the French decided that it was better to give back the occupied territory and maintain good diplomatic relations with India.

The French territory in India consisted of five coastal cities. Four cities were located along the eastern coast of the Indian subcontinent: Chandernagore, Pondichery, Yanan, Karikal, and one on the western coast, Mahe. In later years it was determined that due to the size of the Indian territory separating them, each of these five locations would have counted as a separate ARRL country. The only licensed radio operation that was known to have occurred in the French Indian colonies between World War II and 1954 was Deb Shankar Seal, FN8AD, operating in Chandernagore. (FN1C in Chandernagore was very active for several years prior to the war.) Should evidence of other amateur radio activity from one or more of the other four French cities surface, a subsequent addition to the all-time ARRL country list would have to be seriously considered.

The ARRL published a criteria for accepting QSL cards for the DXCC award. Rule 7 of the DXCC prohibits counting QSL cards from countries where there are

The map of India.

normally licensed amateurs. However, the presence of, or authority for, a licensing system was not always in black and white. The political makeup of many countries was in a state of change in the years immediately after World War II. Consider the case of FN8AD. Shankar also had an Indian license for the same location while he had his FN8AD license. Had he used his VU2AX call sign from Chandernagore during 1949 to 1954, he could still count French India, as he was operating from territory that they controlled.

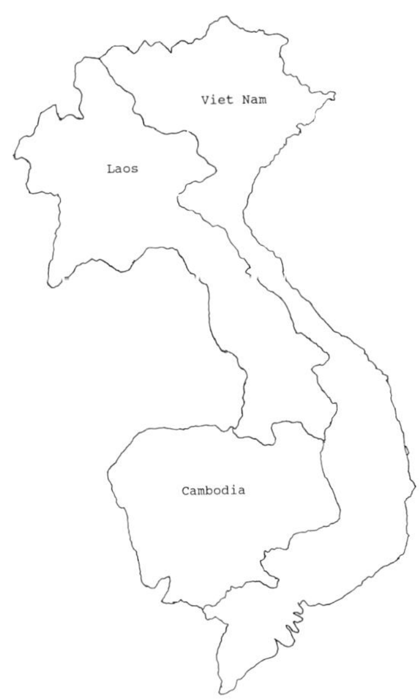

French Indochina fragmented in Laos, Cambodia, and Vietnam.

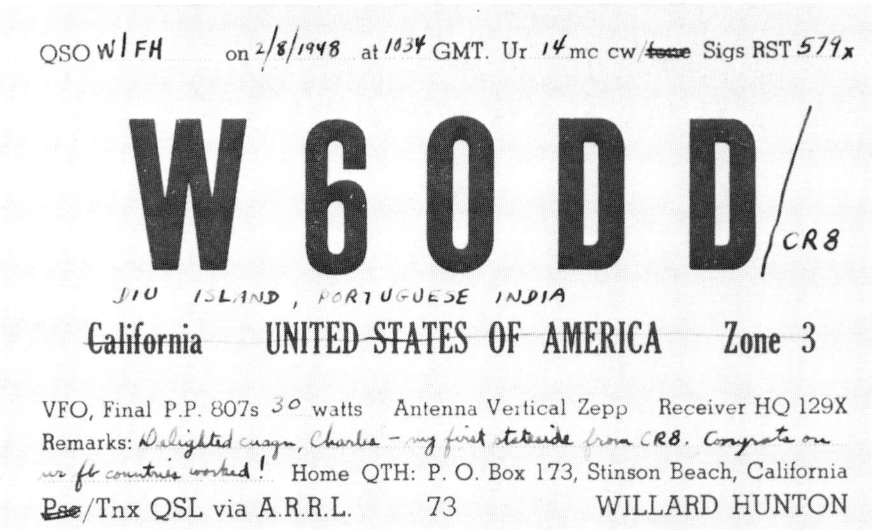
QSO W1FH on 2/8/1948 at 1034 GMT. Ur 14 mc cw/~~fone~~ Sigs RST 579x

W6ODD/CR8

DIU ISLAND, PORTUGUESE INDIA

~~California~~ UNITED ~~STATES OF AMERICA~~ ~~Zone 3~~

VFO, Final P.P. 807s 30 watts Antenna Vertical Zepp Receiver HQ 129X

Remarks: Delighted cuagn, Charles – my first stateside from CR8. Congrats on ur fb countries worked! Home QTH: P. O. Box 173, Stinson Beach, California

~~Pse~~/Tnx QSL via A.R.R.L. 73 WILLARD HUNTON

The rarest of QSLs, W6ODD/CR8, Diu.

French Indochina (FI8) had been fragmented into three countries when the French pulled out on July 19, 1955. At that time it was deleted from the ARRL country list (See appendices 11 and 12). That region was partitioned into the countries: Laos (XW8), Cambodia (3W8), and Viet Nam (FI8), which were added to the country list.

Marcel Zinck, XW8AB, was active from the French Embassy in Laos almost immediately, and Don worked him on July 30. Lucien Tchitchek (F9EW) was also active on phone from the embassy as XW8AC. Viet Nam would first see activity when Lyman (Rundy) Rutlett W3ZA (K4ZA) was in Saigon in 1958, and signed portable 3W. Cambodia would not see amateur activity for some years.

After India had assumed control of the British-Indian administration and had achieved self-determination, they were determined to recover the remainder of their territory that was occupied by foreign colonial powers. Relations had been very strained with Portugal since 1948 when India first broached the subject of the return of the Portuguese enclaves on the western coast of the subcontinent. Once France had ceded to India the five cities that made up their colonial territory, India started applying more pressure on Portugal. Indian Prime Minister Nehru entered negotiations with the Portuguese, but they were not fruitful. In 1955 he used Ghandi's approach to a resolution, trying peaceful occupation of the Portuguese enclaves on India's western coast. Unfortunately, the Portuguese did not exercise the restraint that the British had in India under Lord Mountbatten. Shots were fired by the Portuguese and India broke diplomatic relations with them. For the next six years the tension between the two countries was extremely high.

The majority of radio activity from Portugese India, outside of Bill Hunton's brief foray in 1948, was from their communications center in Goa. The Portugese communications center, for its eastern colonies of Timor, Macao, Diu, and Damao, was at its facility in Goa, on the western coast of India. The first known amateur radio operation from this location was in 1956. One of their radio operators, Joao Serpa, fired up his transmitter on 15 meters and used the call sign CR8AC. The antenna was a rhombic pointed at Portugal, so their signal strength in the United States was usually poor.

The next year another Portugese amateur radio operator was stationed in Goa. From 1957 to 1961 Raul Leitao Fernandes (ex-CR4AL from Cape Verde Island) was active as CR8AC at the Portuguese communications center. Twice, in 1958 and 1959, a Swiss amateur, Gilbert Fournier, HB9QP, visited Raul and operated his station portable from the Goa airport, seven miles southwest of Nova Goa city. His operation and Raul's would be the last from Portugese India.

In the spring of 1961 Ed Hawkins, W6CUQ, was climbing up the DXCC listings. Ed looked at a world atlas and noted quite a bit of real estate between the Portugese Indian enclave of Goa and Diu and Damao in the north.

Ed called Don Wallace and asked him if he had a QSL card from from both locations in Portugese India. Don recalled, *"Ed Hawkins asked if I had both spots covered. We were both on the Honor Roll and that was*

1954 *The Contest Committee of* 1954

The International DX Club

takes pleasure in awarding this Certificate of Merit to

W6AM

in recognition of his outstanding achievement in winning the

World Wide DX Contest

for the U.S. W6 division in

MULTIPLE OPERATOR ALL BAND PHONE

his total score of 98,100 points was computed on the basis of zones, countries (prefixes) and stations worked. In witness of this achievement, the Officers have affixed the signatures of two of their number on this 1st day of July 1955

R. C. Spenceley, KV4AA — SECRETARY

Wayne Green — PRESIDENT

TEAM W6KPC W6KSF W6YMD W6AM

A CQWW contest certificate.

Ed Hawkins, W6CUQ (K6ZO) was at the top of the honor roll in 1961.

quite courteous of him. I took out my CR8 QSLs and looked at them. I had several from CR8AC in Vasco da Gama and one from HB9QP/CR8 at the Nova Goa airport. According to my Encyclopedia Britannica Atlas the station in Vasco da Gama was located in Daman and the other in Goa. So I said, 'Yes, I have both covered.' "

The situation called for an expert opinion, Ed decided. He sent his W60DD/CR8 card off to Bob White, W1WPO, DXCC administrator, for his assessment. Bob examined the QSL card and reviewed the current DXCC criteria: 75 miles of separation by foreign territory constitutes a separate country. Thus it came to pass that the country of Diu/Damao, separate from Goa, was announced in the fall of 1961. There were now two countries in Portugese India.

Forces were in motion, however, that would change the map. In December 1961 India invaded the Portuguese enclaves on its western coast. General Manuel Antonio Vassalo e Silva was the governor of the provinces of Goa, Damao and Diu. He defied an order from the Portugese dictator Salazar to defend the colonies, which were hopelessly outnumbered, and surrendered the Portugese Indian territory to India.

On January 1, 1962 Diu and Damao (along with Goa) became a deleted country, virtually before its existence was published. Unfortunately, W6AM's atlas had been incorrect; Vasco da Gama was in Goa. The 40-odd DXers who had worked W60DD/CR8 some 13 years earlier became members of an elite club, the only ones holding a QSL from Diu/Damao.

As it turned out, Diu/Damao was the one country W6AM missed! Don had sent all of his CR8 QSLs into the ARRL, hoping that one of then might be in Diu or Damao. In his return correspondence Bob White noted: "The only Diu operation that I've heard of was W6ODD/-CR8 which was some years back." After reviewing all of the Portuguese India QSLs, HB9QP/CR8 and CR8AC became credited toward Goa and Bill Hunton's operation as W6ODD/CR8 was Diu/Damao. These were the only known QSOs from the Portuguese territories in India.

It was speculated for a time that an Indian amateur could have operated portable CR8. For a time the VU7 call sign was issued to British operators in Bahrain (VU7BR), Indian embassy personnel in Nepal (VU7AF), and Indian amateurs in Mysore State, India. It was often speculated that one of these VU7 amateurs in India could have visited Diu. VU7AH was operated by K.S.V. Rajan who was in Mysore state. Contact was established with Rajan's family and long time friends in amateur radio, but they said that he had never traveled outside of the local area. The matter was settled by T.R. Jayaramen. He wrote:

> Goa, Daman, and Diu were Portuguese occupied territories on the west coast of India. On the 19th of December 1961 the three areas were liberated. Prior to this period the relations between India and Portugal were so strained that an Indian amateur could not have operated from Diu. The strange coincidence in this case is that I became the Civil Administrator of Goa district in December 1961 when it was liberated.

DX Contesting

A team of DXers began to assemble regularly at the W6AM rhombic farm in the early '50s. They realized that here was the biggest station with the most potent signal in their part of the country, perhaps the entire world. Their thought was to modify Don's station to accommodate a multi-operator operation in the annual ARRL and CQ DX contests. Before long, most weekends found them up at the rhombic farm conferring with Don on how to bring their ideas into being. He was amenable. The task of keeping the ranch antenna system in top condition was overwhelming and, without knowledgeable manpower, the job would be hopeless.

Since there were to be a number of operators using the station simultaneously, an interlocking power system was installed for the selection of each single-band kilowatt amplifier. Once the switch for the filament power to an amplifier had been thrown, bumping a filament switch for another amplifier would have no effect. The first switch had to be returned to the off position before another amplifier could be selected. Due to Don's antenna relay matrix, only one rhombic could be used at a time. It was not practical to rewire the relay system. This sequence made certain that two amplifiers could not be connected to one antenna simultaneously, which, of course, would spell disaster for the amplifiers.

Another operating table was placed in the radio room and duplicate sets of interlocking filament power switches were installed at each of the five operating positions. With the addition of 15 meters in 1953, there was a total of six amateur bands: 160, 80, 40, 20, 15, and 10 meters. A kilowatt could be used on each band except 160 meters, then shared with the LORAN service, requiring a transmitter power limit of 150 watts input.

Don had single-band kilowatt amplifiers for each of the five bands. The output of each amplifier went to a broadband impedance matching network for that band. Due to the different height and length of each rhombic, every one had a slightly different input impedance at any particular frequency. The broadband network was tuned to the midpoint for the most representative antenna match for all of the rhombic antennas. This provided an adequate impedance match for the amplifiers and allowed the majority of the RF power generated by the amplifier to be transferred to the antenna. Retuning the broadband network when a different antenna was selected was not necessary.

John Norback, W6KFV, was one of the contest operators at Don's in the 1953 to 1957 era. "Don had some interference with the local CAA/FAA (Federal Aviation Administration) receiving station in Rolling Hills. By making some modifications to his amplifiers I managed to eliminate the interference problem. We also had to work on the transmitters a bit to clean up the key clicks."

After months of preparation the station was ready for contest operation. They entered the multi-operator single transmitter category. In actual operation several operators would be tuning a receiver for each of their respective bands, sharing the same antenna while receiving. When one operator heard a station he wanted to contact, he would press the filament switch for the amplifier of the band on which he was operating. The output of his transmitter was already connected to the input of his selected amplifier. Pressing the foot switch disconnected the receivers from the antenna and put the transmitter on the air. Once he worked the desired station he would deselect the amplifier and then another operator could select his amplifier for transmission on a different band.

Each operator often brought his own receiver and transmitter to use during the contest. That operator's station was set on a particular band and, for the remainder of the weekend, his transmitter hard-wired to Don's kilowatt amplifier on that band. It's a good thing his operating tables at the ranch were sturdy, because an immense amount of equipment was stacked on them during these contests.

This contest setup was well suited for radio conditions in that era. The number of foreign amateurs on the air was relatively small and there were no large concentrations of amateurs in any one country. Consequently the hunt-and-peck method of working one station after another across the various bands was very productive and yielded a high contest score. Native radio operators in Japan were allowed back on the air in 1952, but their numbers were relatively small for the next decade.

In 1949 Don entered the CQWW DX competition as

a single operator and finished with the high score in zone 3, which consisted of the western United States and British Columbia. It was at that time that Don decided to go as a multi-operator contest station. At the age of 51 he was getting a little old to go most of a weekend without sleep.

Frank Clement, W6KPC, joined Don for a multi-operator effort in the 1950 phone contest and they took the high score in the United States. The next year Ed Sassaman, W6UQQ, joined Don and W6KPC in the phone contest and W6AM finished second in the country. The CW section of the CQ DX contest was held the next week, in November 1951. Glen Means, W6ADP, and Conny Fleissner, W9SRB assisted Don and W6AM took the high score in the Country. In the early '50s the CQWW contest certificates continued to roll in. In the 1952 phone contest Homer Biedebach, W6GFE, Chuck Bailey, W6BXL, and W6KPC joined Don, and W6AM finished second. Actually, considering the quality of these operators and the antenna system that they had to use, they should have done a little better in the DX contests. Don's antenna farm might have put out a tremendous signal, but it was very complicated to operate. An operator would have to participate in a couple of contests up there before he really started to function at peak performance. Also, setting up equipment at the last minute left little time to iron out last-minute problems.

Don and the fellows also competed in the 1953 International DX contests. They entered the multi-operator single transmitter category. The team consisted of W6KPC, George Price, W6KSF, Bill Guimont, W6YMD, and W6AM. Don was the top United States station on CW and second on phone in 1953. In the 1954 International DX Contest the team finished third in the United States on CW and first in the country on phone.

The World Wide DX Contest was started in 1948 by the editor of *CQ,* Larry LeKashman, W2IOP, who was both an active DXer and a contest enthusiast. Perry Ferrell took over *CQ* in 1951 for three years and had little interest in either DX or contests. Before long Larry had to run the contest himself, with the help of just a few friends. He referred to himself as the International DX Club and talked Perry into at least running some of the final scores in *CQ*.

Wayne Green, W2NSD, took over *CQ* in 1955 and immediately put the contest back under the *CQ* flag. The 1954 certificates had been turned over to a friend by Larry and no amount of prodding ever succeeded in getting him to send them out. As the screams of anguish rose worldwide Wayne Green commandeered the certificates, had them filled out, and sent them to the deserving, about a year and a half overdue.

In 1955 Chuck Bailey (nicknamed Bixel), W6BXL,

A CQWW contest certificate.

W6KPC, and W6YMD joined Don in the *CQ* WW phone contest. W6AM finished third in the country. The same operators came out in the *CQ* WW CW contest the next month and took the high score in the United States. The same operators were joined by W6KFV the next year for the *CQ* WW contests. In 1956 the W6AM team scored second in the country on phone.

During these years Don often participated in the ARRL DX competition. In 1951 W6AM had the high score in California on phone. The next year he had the high score in Los Angeles in the phone section. In 1954 Don had the top entry in the Los Angeles section on CW and the following year Don had the high California score on phone in the ARRL DX marathon. In 1956 W6YMD operated Don's station, finishing fourth in the country on CW.

Don was still working DX whenever he had the time away from family and business and he was steadily climbing up the ARRL DXCC listing. By the end of 1953 W1FH was still the top dog with 253 countries, W6VFR had 246, and W6AM had 242. On phone Jayme de Campos Freiko, PY2CK in Brazil had climbed into the top spot, Charlie was now second with 224, and Don had 193. In February 1955, Marv Gonsior, W6VFR, tied W1FH at the top of the Honor Roll with 252 countries, Leslie Misch, W8HGW had 251, and W6AM 250. This brush with glory proved to be too much for Marv. He got married at that time and, as he didn't have a rig set up at his new home, soon fell out of the race for the top spot.

DX CENTURY CLUB AWARDS

HONOR ROLL

W6AM......270	W9NDA....262	ZL2GX......260
W1FH......269	W8NBK....262	W2AGW.....260
W8HGW....268	W8KIA.....261	W6SN.......260
W6ENV.....267	W6DZZ.....261	W3JTC......260
W6MX......264	W3BES.....260	W6VFR......259
W6SYG.....263	W5ASG.....260	W8BRA......259
PY2CK.....263	W6TT......260	W6MEK.....259
W3GHD....263		W7AMX.....259

Radiotelephone

PY2CK.....257	W8GZ......239	W3JNN......236
VQ4ERR....251	CN8MM....238	W1NWO.....233
W1FH......247	W8HGW....237	W6AM.......233
ZS6BW.....244	W9RBI.....237	W9NDA.....230

The Honor Roll listing from the April 1957 QST.

Don's station had become a popular spot for operators in the Long Beach area from which to come and operate contests. In the sweepstakes contest, operators across the United States and Canada were contacted during one weekend. Dawkins Espy, W6FRW, operated the phone sweepstakes from the W6AM rhombic farm in 1954 and achieved the national high score, which also set a new record.

All the years of listening to weak signals in the noise had made Don's ability to copy unusually good. Pete Kraeger, W6UYW, spent some time at the rhombic farm in the early '50s. He recalled: "One day several of us helped Don work on the antenna system and, as dinner was being fixed that evening, we were listening to a receiver. Don called over to us to listen around a certain frequency on 20 meters, as he was looking for a station that was in, for him, a new country. This station was known to operate around that spot. I was listening on the loudspeaker to what sounded like pure noise when Don came up and said, 'Oh, there he is.' He switched to another rhombic antenna and adjusted the receiver a bit, and then I could faintly hear the station that Don worked."

Dale Hoppe, W6VSS (K6UA), operated in a contest at Don's. The usual sequence when an operator wanted to transmit was to select the desired antenna direction with the big rotary switch on the table, push the filament select switch for the kilowatt amplifier on your band, and then press the foot switch to transmit. Dale recalled: "I was up at Don's in the 1956 ARRL DX phone contest. One of the fellows, W6BXL or W6YMD, had just finished transmitting. I had a DX station all staked out

The contest W6AM team.

and my transmitter adjusted to his listening frequency. I picked up the microphone and pushed the filament switch for my amplifier. Then I stepped on the foot switch to transmit and that put the one-kilowatt transmitter on the air and about 500 watts of RF to the antenna. I then rotated the rhombic antenna direction switch to the antenna position that I wanted to use. As the relays closed and opened when I rotated the switch, hot switching a half-kilowatt of carrier power, the relay blades vaporized on a half-dozen relays that were mounted on the back wall and fell to the floor. I sat there dumbfounded when I realized what I had done, while I watched the wisps of smoke from the relays curling up toward the ceiling. Don came over and checked which remaining relays and corresponding antennas were functional. Then he said to use only these certain directions for the moment and keep going in the contest. He went to the relay matrix at the back wall and began replacing the relay blades. In four or five hours we were back on the air with all 32 directions again." Even with that untimely interruption, W6AM still managed to have the high score in California.

The day before the 1957 DX contest Don had the building painted. John Norback recalled: "On Friday night it rained and we were not getting out. We called several stations that were loud, with no luck. Something was obviously wrong. I woke Don and told him of the problem. That is when Don told us the building had been painted. The insulators that carried the feed lines through the wall of the building to the rhombic breakout frame were also painted, and this was our problem. We had lots of fun, scraping those insulators in the rain in the middle of the night. After that the station returned to its usual high level of performance."

John Norback also remembers the 1958 ARRL CW DX contest well. "Saturday morning I went up high in the CW band on 20 meters, around 14.095 Mc, and called *CQ* on CW several times. After a few minutes I mentioned to Don that I heard a heterodyne on my frequency and it sounded like a station was operating on phone on my frequency. Don immediately slid in the operator's chair, adjusted my Collins 75A4 receiver for AM phone reception, and worked AC3SQ in Sikkim, a new country for him. And Don was on the top of the Honor Roll then! There were only a few countries that he needed and he didn't work a new one too often."

With all of the equipment in operation, maintenance during a contest was inevitable. However, there was a certain element of risk when working on the equipment when there were a number of operators present and some of them were tired. John continued: "During a DX phone contest Don cut power to the station (we thought) because of RF amplifier problems. Don and I were working on one of the amplifiers when Bill, W6YMD, called a DX station.

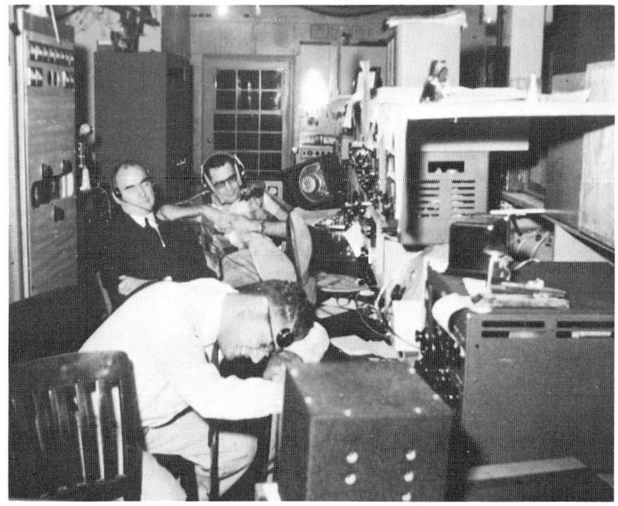

The contest team in operation (left) and, the morning after (right).

The modulator power supplies were not turned off and it put out full power when the transmitter was keyed. Don took the full modulator output of push-pull 450TLs across his body to ground. Fortunately, he was not seriously hurt."

In the late fifties the local competition in the contests got a little stiffer each year. The technology for aluminum tubing that was used in Yagi antennas had been improving all the time. There were now some very big antennas being built. Up to this time most towers consisted of a telephone pole, or a home-built one modeled after a bridge support and made of wood, or steel welded together. Commercial towers for amateur use were now becoming affordable and some of the big Yagis were mounted over 100 feet in the air. Don's rhombics still had the gain advantage over the big Yagis by 4 or 5 dB, but the owners of these installations often didn't content themselves with a single kilowatt of power. Some of the big tubes that Eimac was producing were finding their way into amateur stations. A number of 10-kilowatt amplifiers were on the air in the LA Basin, giving the operator a 10-dB signal strength advantage over a 1-kilowatt station.

The FCC had received complaints that a number of amateurs were running high power in the Los Angeles area and decided to do something about it. These stations were on the air at full power when the competition was fiercest, during the DX contests. The first weekend of the twenty-fourth ARRL DX competition in February 1958 the ax fell. Twenty stations in the Los Angeles area were visited by the FCC and examined for excessive power.

John Norback was operating 20 meters at W6AM's on Saturday night in the ARRL DX CW contest when the FCC came by. "Don had gone to bed. I was operating on one band and W6BXL was next to me operating on another. When the front door bell rang, it was very loud, like a fire alarm. Don had installed a loud bell so he could always hear it over the background racket. We remained at the operating positions and Don got up and went to the door. As he passed us, he turned the high voltage down slightly with the variac. The FCC field men came in and asked us what antenna we had been using. They had taken a field strength reading and I was to transmit in to that antenna and increase the transmitter power until that reading was duplicated, then they would measure the power input of our amplifier.

"I had worked several stations in the past few minutes, two Europeans on short path with different antennas, and one station on long path. I asked the FCC fellows if they knew the call sign of the station that I had worked when the field strength reading was taken. With his walkie-talkie he called another FCC man who was out by the edge of the property monitoring the field strength meter, but the fellow didn't know which station I had worked. I transmitted on the first two antennas, one at a time, but they were too low for a field strength reading. Then I transmitted on the third antenna, after I had thrown the switch to phase it with a second rhombic, and it pinned the meter on his field strength equipment outside. That direction we had been using just the high rhombic, rather than the two rhombics phased together. I explained the antenna system to the FCC fellows and then they insisted that we transmit on each of our 32 directions while they took measurements.

"We proceeded with their orders and spent the next three hours going through all the various directions.

Don received many awards at DX conventions.

Finally they measured our power input at where the high voltage variac was set and the power input to the amplifier was just under one kilowatt. As they were leaving, Don stepped outside and asked whether or not we were going to be disqualified or cited. If we were, he told them that we were going to call it quits, rather than continue on in the contest. They told us to go ahead and continue."

Being cited for excessive power had severe penalties. First of all, if you were in a contest, you were disqualified from submitting any score that was achieved in that contest. The more severe penalty was that you were disqualified from the ARRL DXCC and Honor Roll listing. This was the first time in nearly half a century of radio regulations that amateur stations were ever checked for excessive power with no prior warning.

Of the 20 stations in the Los Angeles area checked for excessive power, Don was the only one who didn't receive a citation. As the FCC visited other stations that night, operators turned down the power input of their transmitters and answered the door. The FCC had them turn up the power until their measured field strength reading from the antenna matched that which had been measured a few minutes previously. Each operator that was found to be running power in excess of 1-kilowatt was cited. Most stations had just one antenna per band, and not a switchable system like Don's. In all probability Don was running between one and two kilowatts of power input with his amplifier. His rhombic antenna switching system was complicated enough that it probably saved him from a citation. After that close brush with the FCC, Don's transmitter power input was kept religiously at or below 1-kilowatt.

One of the stations cited was located on top of the Hollywood Hills and was the second to be checked that night. The operator was still unhappy with the FCC for not giving any prior warning. Afterward he carefully readjusted the jets in his sprinkler system, so if the FCC came to his door again, they and their equipment were going to "accidentally" get sprayed.

In the '20s and '30s it wasn't a well kept secret that Don could run a little extra power when he chose to. Due to his strong signal from the ranch after the war, many operators assumed that Don still carried on the tradition. Jerry Tannanbaum recalled: "It was a given around Chicago in the '40s and '50s that Don was running high power. The late Ed Schmeichel, W9YFV, gave me a briefing when I was chief engineer at Allied Radio. He had returned from a tour of members of the North California DX Association and the gist of it was that Don was running a 'California Kilowatt.' In 1956 By Goodman, W1DX, personally confirmed this. From time to time W6AM would snatch a rare DX station out from underneath me, but he was never obnoxious about it. He always raised them on the first call; when he worked them he was off."

"California Kilowatt" was a phrase that had come into use back in the '30s. Most DX stations in Europe and Africa were so weak in California that, if the operator's station had the capability, he cranked up his transmit power to its maximum in an effort to get the attention of the DX station. Most of the time the extent of this power level was rather exaggerated by the operators telling the story, although there were a couple of 5 kW rigs around then. As the joke went, California stations couldn't send very fast, because the silver dollars that had been welded onto their speed keys to handle the high current of the cathode-keyed amplifiers had too much mass to vibrate very fast. . . .

Don first tied Charlie Mellen for the top spot of the ARRL DXCC Honor Roll with 267 countries in November 1956. The next spring, in April, W6AM stood alone at the top of the DXCC Honor Roll with 270 countries, while W1FH had 269. On phone Charlie still led the stateside fellows with 247, Windy, W8GZ, had 239, and W6AM had 233. It would be a see saw battle in the coming years. Jock White, ZL2GX, and W1FH were the first

Reprinted from the March 1958 *QST*.

to reach the 300 country mark in September 1960. W6AM was in hot pursuit with 299 confirmed.

Another country race was shaping up at the same time. With the advances made in frequency control and stability during the war, it was now feasible for many amateurs to use the new single sideband (SSB) type of voice communications. This method of modulation was much more efficient, but the circuitry was rather complex to adjust properly. In an era when the majority of amateur equipment was homemade, only a few stations around the world used SSB. In the mid-1950s *CQ* magazine began to sponsor an award for two-way SSB contacts. Windy Windom, W8GZ, was the first to attain 100 countries confirmed on two-way SSB. Don built an SSB rig for the ranch, but his main interest was the ARRL DXCC. Unless they were new countries for him, few stations on SSB were worked by W6AM in the late '50s.

Another article about W6AM appeared in the *Long Beach Press Telegram* in late 1956. Don was pictured in front of his bench, which was laden with RME equipment. Don had dropped one country behind Charlie Mellen when Revilla Gigedo, a small island west of Mexico, came on the air while he was in New Mexico on business. Don had hurried home, but the amateurs on the island cut short their stay due to a hurricane warning.

The rhombic farm was big news in radio clubs around the country. Bruce Kelley, W2ICE, the secretary of the Antique Wireless Association and curator of its museum in Holcomb, New York, sent Don a letter in 1956. Bruce wrote:

> Could we obtain one more slide of your antenna farm, sort of an overall shot. The "gang" gets a terrific bang out of your layout. More than ever now since your giving 1FH a run for his money.

The Radio Club of Argentina inducted Don as an honorary member when he visited them in 1954. He sent his country totals down for their member listing after that, but Don always left out a few countries. Don felt that a local member of that club should occupy the top spot. Accordingly, Alfredo Luciano, LU6DJX was tops on CW, Enrique Keen, LU6AJ on phone, and W6AM was second on both lists.

Shortly after Don attained the top listing on the ARRL Honor Roll in 1957, in a letter to RME he described how well the equipment worked at the rhombic farm. The president of RME sent Don a letter shortly afterward. He felt that since Don used RME exclusively, a picture of Don at his station would make a good spread in *QST*. The advertisement appeared inside the front cover of the December 1957 issue of *QST*: "If you can't hear them. . . you can't work them!"

A few months earlier, on September 13, Rod Newkirk, W9BRD, the DX editor of *QST*, sent a letter to the manager of the RME division. He noted:

> Don, his station, his outstanding hamming achievements over the years, and his never-flagging enthusiasm for the game long have been one of the wonders of amateur radio. I plan to use the photos and data for a pictorial vignette in the upcoming (November) *QST*. It really is high time we gave the gang a good peek at W6AM and his outfit via "How's DX."

The Western Electric Show and Convention (Wescon) was held in Los Angeles in 1958. Hugh Moore, the show director, asked Don to be the chairman of the historical exhibit. Don got together with Al Hill, W6JQB, and Howard Seefred, W6EA. They had their friends go through their junk boxes and garages, and they pulled together a display of slide tuners, spherical audions, loose couplers, and enough equipment to put one of the spark "rock crushers" on the air. The transmitter wasn't lit off, however, as there were too many television receivers in the area for the FCC to allow that.

Don had managed to save a good deal of his early equipment and paperwork, at least those he had after 1916. His mother had inadvertently thrown out his first station when he went to college that year. There was one other time when he lost some of his historical items. Pete Kraeger, W6UYW, recalled, "In the 1950s Don had an office for Wallace and Wallace in the Bendix Buildings at twelfth and Maple in Los Angeles. I visited him there in 1955. Just before that he moved to another office in the building. Don had borrowed a janitor's cart to load files on and rolled it down the hall to the new office. When he left late that evening he had partially loaded the cart with some framed certificates from the wall, some various cups he had been awarded, and so on. The next morning he discovered that the cart had been emptied and those items were never recovered."

The past few years Don had been accumulating QSLs from the various zones for *CQ*'s phone WAZ award. There were fewer stations that operated on phone. They were also much weaker and harder to work than the CW stations. In early 1957 he finally received a card from the last zone for the phone WAZ. He sent the cards off to *CQ* magazine's award manager and on the first of March Don received phone WAZ award number 5.

The annual joint meeting of the Southern California DX Club and the Northern California DX Club was also referred to as the Fresno DX convention. In 1959 the DX convention was held on January 24 and 25. At the banquet a special award, Mr. Ham Radio, was given to W6AM for his outstanding contribution to amateur radio for over 40 years (actually, it was closer to 50). The February issue of the Northern California DX Club Bulletin noted, "The ovation given Don on this presentation was the highlight of the evening."

Don was over 60 years old now and his code copying ability had dropped from over 50 WPM to the mid-40s. At the ARRL South-West Division Don won the plain pencil copy, but only at 44 WPM. That year W6AM finally relinquished the title as the reigning CW champion to an up-and-coming DXer, Frank Cuevas, W6AOA. Frank had met Don in Long Beach in 1927 and had worked for over 30 years to beat Don in a CW competition. Considering that Frank had not been a professional radio operator, and had reached this level of proficiency by sheer determination, it was a remarkable performance.

At the Fresno DX convention the year before, each member stood up and introduced himself at the evening dinner. They would also pass on any recent DX news. Roger Mace, W6RW, stood up and announced that he had worked Albania (HA5AM/ZA) that morning, which produced a good laugh from everyone. Albania was among the most rare of countries.

In the years after World War II Albania had remained silent. In 1951 a station came on 20-meter long path signing ZA1AB. W6AM worked him and eventually received a QSL card with Yugoslavian stamps, which was accepted by the ARRL for DXCC credit.

Janos Emmer, HA5AM, worked as the communications officer aboard an airplane that flew from Hungary to Albania every week. Twice in January 1958, while they were on the ground in Tirana he put the aircraft transmitter on the air and operated as HA5AM/ZA for an hour. A handful of W6s worked him on 20 meter long path each time; Roger, W6RW, worked him the morning of the DX convention. That was Janos' last operation from Albania. He had been monitored by Albania's internal communications and was no longer permitted to enter the country, even as a flight officer on official duty. Don had also worked him and both he and Roger received their HA5AM/ZA QSL cards shortly afterward.

The first station to actually be licensed in Albania was Hans Stiehm, an East German whose home call sign was DM2ACB. He spent some time in June 1957 at Durres, attached to an East German naval training craft, and he put his 30-watt home made transmitter on the air as ZA2ACB for four days. W1FH was one of the few stateside stations to make contact with Hans.

In 1959 a Czechoslovakian by the name of Jiri Hanzelka, OK7HZ, left on a year-long scientific trip. He had permission to use his transmitter portable as he traveled through different East Block European countries. He operated for two days as OK7HZ/ZA as he crossed Albania and worked a handful of stateside stations.

Albania remained among the rarest of the rare. The next operation from Albania would not occur for another 11 years. In 1960 Bob White, the ARRL DXCC administrator, asked that everyone who had a confirmation for Albania resubmit their QSL card for evaluation. Bob wanted to compare all of them and make sure that they matched others that had been submitted.

Don's ZA1AB QSL was unique when he submitted it. That wasn't unusual, as Don was often the first to work and confirm a rare country. However, no more ZA1AB QSLs are known to have surfaced, so there may not have been adequate proof that his card was from a station that was actually located in Albania. A single QSL is always suspect. If it is a valid card, more usually turn up over the years. However, Don also had a QSL from HA5AM/ZA, so his Albania confirmation withstood Bob White's watchful eye.

Clipperton Island had been on the ARRL DXCC country list since 1939. Aside from brief activity by FC8AA in 1937, no operation was known to have taken place from there in the ensuing years, as the island was uninhabited. In 1954 a group of amateurs led by Bob Denniston, W0NWX, decided to travel to the island and put it on the air. Their first landing attempt was aboard the *Sea Rider* and their sextant was lost en route, leaving them without the ability to navigate. Their second attempt was aboard the 80-foot schooner *Barco De Orco II*. However, the North Equatorial current wasn't on their charts and they couldn't overcome the four-knot current. Don contacted them while they were adrift and FO8AJ/MM reported their difficulties. They were within

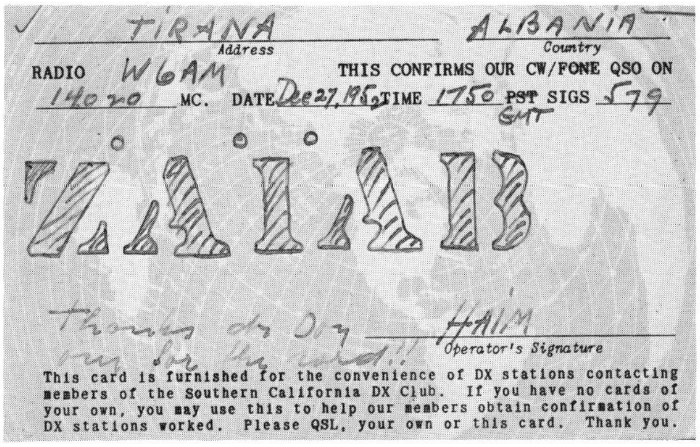

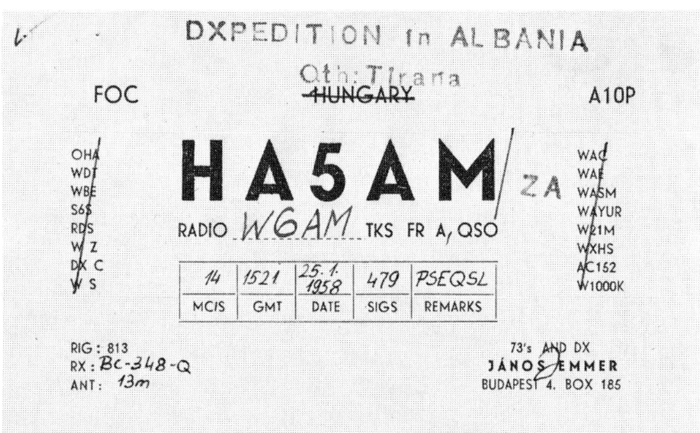

ZA QSLs from the fifties.

five miles of the island but got no closer. The story appeared in the *Los Angeles Times* the next day. Two weeks later, on April 23, they successfully landed and made a total of 1108 contacts during their brief stay. Frank Charters, W6DTE, and new novice George Sinclair, KN6ETM, visited Don thet day. Sinclair recalled, "we drove up to the ranch and knocked on Don's door. He had been tuning the frequency that the Clipperton Island station was expected to begin transmitting on. Don answered the door and while the three of us were standing there FO8AJ came on the air. Don went bach to the operation position, adjusted the transmitter frequency and gave him one call. The operator at FO8AJ answered with,'W6AM, Don, you old rock-crusher you. . . 5 and 9.' Don was their third QSO from the island rather than the first, due to his answering the door."

China had gone off the air in 1949, when their civil war ended. Shortly afterward they were assigned the BY prefix. The mainland was silent for almost 10 years and then a limited amount of radio operation was observed. BY1PK appeared on the air in 1958. They had a loud signal in the United States but would not acknowledge stateside amateurs calling them. Only stations behind the iron curtain were answered by BY1PK. Larry Brockman, WA6EPQ (now N6AR) called them by the hour but was never answered. Larry got a bit mischievous one morning and signed his call as UA6EPQ (to indicate a Russian location), hoping that BY1PK would be fooled and respond to him. Apparently his signal was too sharp and clean-sounding, for Larry never received an acknowledgment. (Most Russians either ran a BC-610 or home-built equipment, both of which had a characteristic sound.)

Lew McCoy, W1ICP, wrote many construction articles for *QST* in those days. He remarked, "One morning, when BY1PK was particularly loud and making long *CQ*s without success, I thought, `What the hell,' and gave him a long, long call and only signed my call once at the very end. He came back on, sending dit, long pause, another dit, and suddenly sent 'ICP ur RST 599 vy strong Lew de BY1PK.' I sent tnx and a report, and that was all. Sonnuvagun, I thought to myself, the guy knows me."

The operator of BY1PK in 1958 was Vesso, an eastern European. He was in Peking to get a 10-kilowatt broadcast transmitter operating for the Chinese. (The one Doc Stuart had installed 10 years earlier?) Vesso's first rig that he had built some years before was one of the 6L6/807 transmitters that Lew had described in *QST*. Some time later Lew's BY1PK QSL arrived via a diplomatic pouch from an eastern European country with another card enclosed, that one for John Knight, W6YY. These were the only known QSLs for United States stations from the original BY1PK. The next operation by an operator native to the area was not to occur for another 25 years.

Over the years Don had almost never gone to a doctor. He felt that they would always find something to treat, whether you needed treatment or not. This approach to health care had worked well for him for over 60 years and he didn't see any reason to change it. Although few people outside of his immediate family knew it, Don was about to begin another battle; this one for his life.

A Second Chance

Don was diagnosed as having cancer in 1958. In those days surgery was no sure thing and chances of a complete recovery were around 20 percent. If the disease recurred within five years, it would almost certainly be terminal. The doctors advised Don that he had no choice but to take his chances with the surgery; otherwise, he had less than a year to live.

Don went ahead with the surgery, but it was not a total success. He continued to decline, his weight dropping from 245 to 185 pounds. Another surgery was performed yet he still did not improve. His weight was down to 160 pounds when another physician, Dr. Henry Caraco, examined him. Dr. Caraco advised Don Jr. to bring in a cancer specialist, Dr. Beckstrand, and arranged a consultation. Chances of a recovery were small, but in Don's present condition and without a change in treatment, the possibility was even less.

The specialist began treating Don and slowly but surely he improved. Without the reservoir of strength from all those years of antenna work, Don probably would not have survived the second operation. He was hospitalized for three weeks and spent the better part of a year convalescing. Once he recovered he completely rearranged his day-to-day activities. There were no more all night sessions on the radio. The occasional cigar was now forbidden. A rigid, balanced diet was followed and his weight slowly rose to 185 pounds and leveled out. Although he had been in good shape, 245 pounds was too much weight to maintain after major surgery, even if he made a complete recovery. So he kept his weight to a maximum of 185 pounds. Otherwise, he went on with his usual schedule: visiting his family, the Virginia Country Club, and getting on the air daily.

The one change apparent at the rhombic farm was the addition of steps to the poles. He had worn climbing spikes for nearly half a century, but now he knew that he would have to pace himself. Once he regained his strength he would still go up and work at the top of a pole, but no energy was to be wasted by using climbing spikes to get there. Over 2,500 pole steps were installed on the nearly 3,000 feet of poles that were standing. These numbers did not include the feeder poles, which could be reached from the top of a 20-foot ladder. Don had a custom pair of boots with cork soles made, giving him even better traction.

The steps started about 15 feet from the ground to discourage local youngsters from climbing the poles. Some years later, a small flag sometimes appeared in the morning on top of one of the 140-foot poles, only to disappear the next morning. One explanation for this was that a local club made climbing the big poles an initiation rite.

The three 95-foot poles at the house in Long Beach also had steps added. By now, all of the property adjacent to the house had been developed and the homes around the Virginia Country Club were quite fashionable. Don's antennas were restricted to his property. The masts and wire antennas were gone. Commercial Yagi antenna design had come a long way and Don had one on top of each pole, single-band antennas for 10, 15, and 20 meters. Between the front pole and the back pole hung a 40-meter dipole with open-wire feeders; the only external indication that his station was one of the oldest around town.

At the time Don was in the process of winding down his contest operation. Each time the fellows had come up to the ranch and set up their equipment for a contest, another layer of wiring was added to accommodate them. When the contest ended late Sunday afternoon each weary participant picked up his equipment and left. Since the surgery, it took a lot more effort to set the station in order again. Besides, now that Don was near the top of the DXCC, he needed the station operational at all times in case a new country turned up. The contesters slowly started operating at other stations in the area and Don began to concentrate more on DXing.

In 1959 Don retired from working as a manufacturer's representative, after being in business for himself for over 35 years. He reflected, *"The boys, Bill and Don Jr., talked to me and got me to retire. They wanted me to retire, I didn't want to. I had a beautiful job, I just loved it. Anyway, when I retired I was 61 years old.*

"I never had to work, you know. In the early days, when I was a radio operator aboard ships, I was either fixing radio equipment or operating it. That was just pure fun. Then, when I was in college I was a telegraph operator, sending and receiving wire traffic, and that

was fun. When I was with the Peerless Electrical Company I would discuss various aspects of radios with my customers and then usually they would buy what I recommended. Later on, when I was in business for myself, I would go out and visit a company or a radio station most every day, usually several in one day. The fellow in charge of the place, the chief engineer or owner, was always a ham. We would talk about amateur radio for a while and I would write up an order for whatever items he needed. Then I would go on to the next company and do the same thing. I had so much fun, I don't believe that I ever got used to being paid for it. It was a wonderful job."

Even so, Don continued to keep involved with the business at a reduced level for the next couple of years. One of the new crop of DXers caught his eye. *"There was this young fellow, Ron Camp, K6EVR. He scored very well in the DX contests with a modest station at his house and I was impressed by his enthusiasm. He mentioned that he was going to transfer from the job that he currently had, so I talked it over with Bill and we offered him a job. Ron accepted and has been with Wallace and Wallace ever since."*

Once Don was up and around after the surgery, he and Bertha went to Canada and stopped in Yellowstone National Park on the way back. Their last visit to the park had been 23 years earlier, when his 171-foot sky hook had come down, and then 10 years prior to that when they had moved from Minneapolis to Long Beach. Just a few days after they departed, on August 17, 1959, the largest earthquake recorded in the Rockies in modern history struck the area. Measuring 7.5 on the Richter scale, it caused part of a mountain to collapse and form Earthquake Lake on the Madison River and also altered the timing of the Old Faithful geyser.

Now that he was retired and had sufficiently recovered from surgery, Don found time to travel more and to spend more time DXing. An article in the September 24, 1960 issue of the *Los Angeles Herald Examiner* had a story about W6AM and his rhombic farm. Don was pictured with a bench loaded with RME equipment, and an overview was given of his radio history for the previous half-century. The writer, Tom Cargo (K6UFL), noted that Don had 301 countries confirmed and had visited the Soviet Union two months earlier.

Don traveled extensively across the Soviet Union in mid-1960. At each city he visited, whenever Don spotted an amateur antenna, he would stray from the guides' watchful eye. The antennas were always wire, usually a dipole or Windom. Don visited almost 2 dozen amateur stations this way, to the consternation of his guide. Don recounted: *"When I spotted a ham antenna I would follow the feed line to see where it went, go up to the building, knock on the door, and introduce myself. The Russian amateurs were always glad to see me. It was very unusual for them to meet foreign hams. The fellow would show me his station and we would sit down to listen to the bands. Most always, they had either home made equipment or an American-made BC-610, war surplus. While we had a station tuned in, I would ask how to transmit and the operator would show me the transmit-receive switch. Then I would throw the switch and work the station that we were listening to, and maybe one or two others, always using the call of the fellow I was visiting. Their eyes would get big when I started to transmit. The operator would get a bit concerned, because that wasn't allowed in the Soviet Union. However, the guide had always remained outside and since I was transmitting on CW, no one else knew. In this fashion I made several contacts from each location I visited. When we returned from Russia I wrote a story describing my trip and put it in my files. I knew I couldn't publish it because Soviet amateurs are not allowed to let foreign hams use their stations and they would be put off of the air."* On the way back from Russia Don stopped in Finland and visited Penitte Hakala, OH2XZ. The next week his picture appeared in their radio magazine, the Finnish *Radio Amatoori*.

Prior to leaving on the trip Don typed up a set of operating instructions for the equipment at the ranch. Several locals had a key to his rhombic farm and Don figured that one of them might want to participate in a contest while he was gone. There was a vertical supporting beam behind the operating position and eight separate circuit breakers mounted on it had to be thrown to the "on" position. Don noted:

> *Receivers: RME 6900 is best all-around receiver. RME 4350A in center is best for CW, RME 4350A on right is best for SSB and AM phone.*
>
> *Transmitter: For 20 meters throw switch in table center "14 MC" down. Collins VFO is accurate. Antenna send/receive SW is to left of rotary switch. Try to get 12 ma grid drive (meter #14), dip final amplifier plate current, load antenna, adjust steering wheel powerstat to 1-kilowatt input.*
>
> *AM Phone: Toggle switch "Mod Fil" on center of table may be left on. Leave on filament power switch on BC-610. Then only the toggle switches "short key for phone" and 'modulator plate' need be thrown when going from CW to phone. Use mic on filing cabinet.*
>
> *SSB: To be used only on 20 meters. Most all DX stations operating phone on the other bands use AM.*

Additional descriptions were laboriously typed out for each band and mode. Considering the enormous amount of equipment at the ranch, this was a reasonably efficient way to put the station on the air. However, it took most operators several contests to be proficient at using Don's station.

Similarly, there was a one-page checklist on the wall by the door. The items included feeding the cats, closing the windows, bringing in the patio chairs, making sure that soldering irons, heaters, lights, and equipment were turned off, antennas grounded and, of course, that the door was locked.

Don developed a pretty efficient method of DXing. He kept a countries-needed list at hand for those countries on the ARRL DXCC list that he hadn't worked or confirmed. He also kept a second list of islands and geographic regions that he felt should count for the DXCC award. That way, when one of them was occasionally added to the DXCC list, Don would often have worked the country and have a QSL in hand.

Amateur radio remained a popular news item and the July 9, 1961 edition of the *Los Angeles Times* carried the story: "W6AM Sends Out World Wide Signal." As was his trademark, Don was pictured with a bench full of his RME equipment. In the article Don described his dream station at Rolling Hills and the DXCC chase with W1FH in Boston.

A month earlier Don had received a request for information about rhombics from Bob Layton in San Jacinto, California. Don described the basic rhombic antenna, how to add an extra curtain wire to each leg, and the construction of a tapered feed line.

> *I have found no rule that refutes the idea that the higher the better, so put them up as high as convenient, although the difference is not so great after you get above a wavelength. However, from then on the antenna is quieter. When the entire antenna is on the side of a hill, the favored direction is better. I find two or three rhombics to be better than one, but in that case I hook the second one on the end of the first and the third one on the end of the second, etc. Then I terminate the last one. I have also found no best size for a rhombic, just simply the bigger it is, the better.*

At that time an Italian exchange student and SWL, Marco Morelli, was staying with a family in Palos Verdes. He went to see the rhombic farm. Don gave him the grand tour and a QSL card as a souvenir when Marco left. "Don contributed a lot to increase my interest in radio," Morelli wrote, "and I eventually became licensed as I0MOM. Remembering him, I am always pleased when I can work a W6 station."

Don was on the air almost daily when he and Bertha were not traveling. One morning he worked a ship off the coast of southwest Africa. Don Compton, K4FS recalled: "In 1961 I was the project engineer on an Air Force system and also the radio operator on the research ship. One evening I called CQ and was startled when a powerful signal, which I thought could be another ship nearby, called me and signed W6AM. It was Don Wallace using one of his rhombic antennas, the second largest in his arsenal, aimed at Johannesburg. We had a delightful 40-minute QSO. He was interested in the fact that I had been a communications officer in the Signal Corps and had often copied press from the *Press Wireless* station and antenna farm which he later acquired. All during our rag-chew I kept marveling at his signal and I was equally impressed at how gracious an operator he was . . . obviously interested in what I was doing and not boastful about his equipment or antennas, just keenly interested in ham radio and the human contact it can provide."

The signal that Don put out from the rhombic farm would be well remembered by many that worked him. Fernando Muguerza, XE2FL, recalled: "I was on 20 meters in a contest and Don Wallace, W6AM, called me. He asked me to switch to 40 meters to give him points. I told him that I didn't have a 40-meter antenna, just a ZL special on 20 meters. He said to tune up at low power and try it; we had a QSO, S9 both ways!"

Even though Don didn't spend much time in contests, he would always work a few contacts and often open up his station or property to an interested operator for the weekend. Gene Midyett, K6VSB, recalled: "In the early '60s I went along on field day with W6ANB (W7CB), Larry Miller. W6AM had told Larry that we could set up our portable station on his property. During the contest Larry was sending like mad and I was logging when Don stopped in to see how we were doing. We welcomed the respite and soon Don was telling us stories about being a radio operator in World War I and how shy President Wilson was around a microphone. Larry and I were mesmerized by Don's stories. I was with two of the most intelligent hams one could wish for and will never forget that night when CW came alive — history and conversation at its best."

In 1956 the Southern California DX Club announced it was sponsoring an annual award for the outstanding DXer of the club. The February 1956 issue of *QST* carried all of the characteristics that were considered in selecting the recipient. The primary qualifications, besides working a volume of DX, were operating procedure and ethics. In 1961 Don was selected as the SCDXC's DXer of the Year.

The experiments by the engineers from Hughes Aircraft at Don's rhombic farm the previous decade were to have quite an impact in upcoming world events. The occupying forces in East Germany had blockaded the road between West Germany and the Allied portion of Berlin in 1961. The Berlin airlift was established, which kept up a steady stream of food and medical supplies going into the capital city. The occupying forces also attempted a serious effort at jamming Allied communications. However, a system of secure communications between Allied headquarters and the American planes involved in the airlift, by which redundant data was sent on a series of frequencies, defeated all attempts at disrupting their communications. Although it was unknown to Don at the time, Hughes Aircraft engineers had extensively tested and perfected this waveform at the W6AM antenna farm just two years earlier.

In the early 1960s Don faced a major challenge to his amateur radio hobby, the political might of the state of California. He battled the state to keep his rhombic farm and before it was all over he had to give up 75 percent of his property.

Don recounted, *"The California State College system was going to condemn my rhombic farm, all 120 acres, the whole thing. I had them all up here to visit me. The main architects of the project were also along, big names, and the whole college system staff was here. I told him this was my dream, having a rhombic farm. I had wanted this all my life and I didn't want to lose it. They said, 'That's just too bad, Don. We have to go ahead with the college.'*

"So I told my son, Don Jr., to sell 95 acres as quickly as he could. I had snooped around the real estate office and found out that I was going to be offered $12,000 an acre for the land. I told him to sell three-fourths of my property for $20,000 an acre. He, being a lawyer and my son, sold 95 acres for $22,000 an acre. So I got over $2 million for three-fourths of my property.

"That threw a monkey wrench in the plans of the state college system. Now they couldn't buy my land for $12,000 an acre because we had already established a price. You can't condemn property for less than the fair market value. No sale of property up there had ever been that high until that time. However, there was still a chance that the state college system would approve the additional funds and that they would still take the whole 120 acres by eminent domain.

"We loaned the developer who purchased the land, Ray Watt, the money to build a few houses on the property real quick, while the college system was still trying to get additional funding. This increased the value of the land even more and the state would also have to pay for the value of the houses if they took the property. So, they finally gave up. This way I still had 25 acres left for my rhombic farm.

"I talked the matter over with my neighbor, who was a county supervisor, and he said it sounded like it would be a good idea to go to the city of Dominguez to build a college. There was some vacant land over there and maybe they ought to have a college. I said, 'Ideal!' So, he maneuvered that and the Dominguez Hills College was built some time later.

"The State of California wanted to take the whole 120 acres that I owned, along with the property of my neighbor to the south, who wanted to sell. He had 100 acres that bordered mine. They were itching to sell, they needed the money. My land was the heart of the whole project. They couldn't build a college without my property. In two weeks we made the sale to Ray Watt, closed escrow, and the state couldn't meet the new price. I would have helped them with the project if they would have let me keep some of my land for antennas, but they said that they had to have the whole works, all 120 acres.

"The sale of the 95 acres gave me enough to pay the property taxes in the coming years, which soon went up to over $1,000 an acre. With the increasing property values, the taxes escalated even more. Then it would have been impossible for me to keep the whole 120 acres. I could handle the taxes on the remaining 25 acres. When I sold the 95 acres I thought, 'Well, that will give me enough money to pay the taxes for a while on what I keep."

For some time Don had been considering building a house on the bluffs at the western edge of his property. He had to decide in just a few days which 25 acres he wanted to keep. If he selected the 25-acre portion by the bluffs, then he would have to move his entire station and all of the antenna feed lines down to that location. Since Don's father had passed away when he was 62 years of age, Don wasn't sure if he would have enough time left to build his home, relocate all of the hardware, and still enjoy the station. So he elected to keep the 25-acre portion of the property where the ham shack was situated.

At first, Don was at a loss as to how to rearrange the antenna farm. At any location where he had put up antennas he always had a separate piece of real estate for each one. No matter how many sketches he tried,

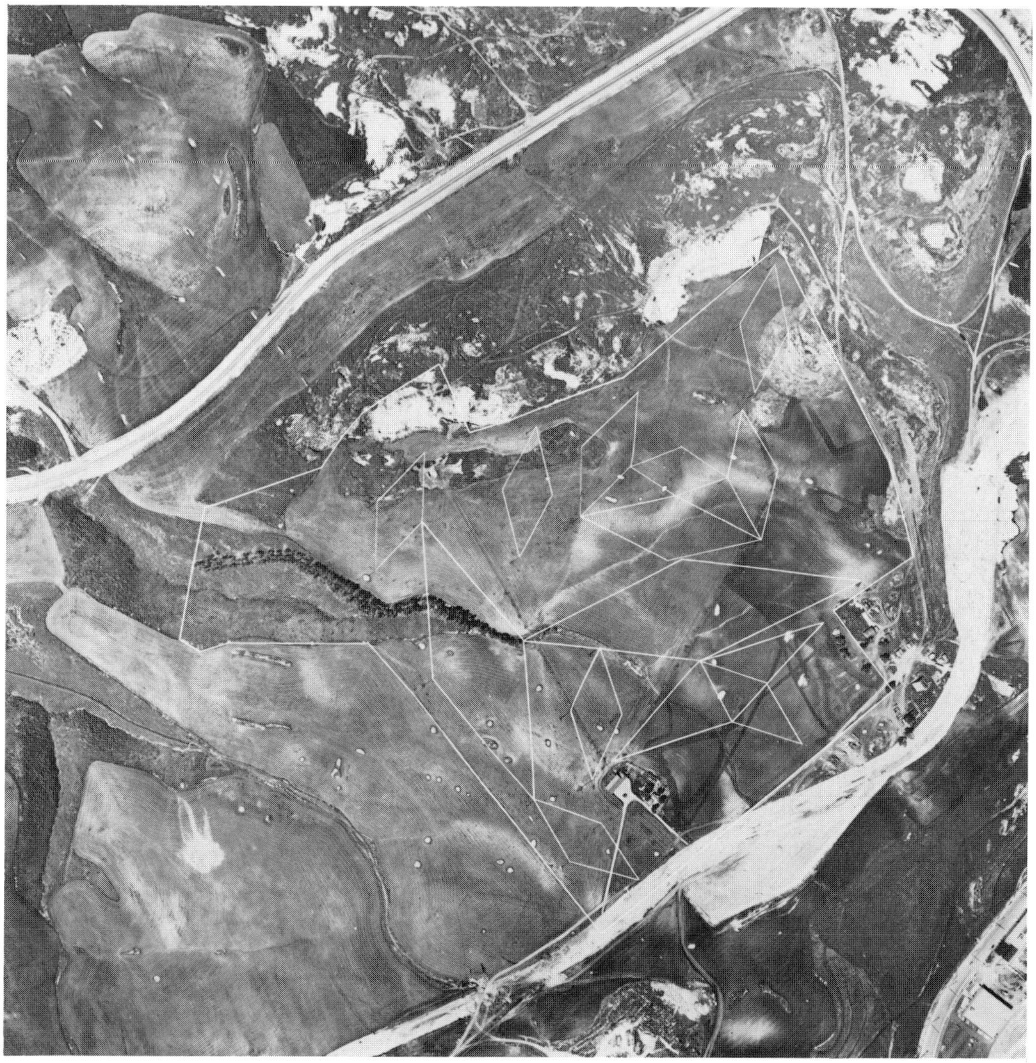

An aerial view of the 120-acre W6AM antenna farm.

there just wasn't much room for a number of large rhombic antennas. They would all have to be trimmed down a considerable amount. *"I was puzzling over the antenna problem one day and Frank Clement, W6KPC, came by. He looked at my sketches and said, 'Why don't you layer the rhombics? That way they are all still pretty good-sized and you can have most directions covered.' I thought about what he said and saw that he was right. After a bit more studying I decided to ring my remaining 25 acres with 140-foot high telephone poles. There were a total of 10 poles that I had put in. Each 140-foot pole consisted of two 75-foot poles that were spliced in the middle and had to be put up with a crane. I put up 10 of them."*

Don's 140-foot poles were really quite unique. One hundred feet was the longest pole commercially available, but that wouldn't provide enough height to layer the rhombics and still keep them high. Don's first attempt at splicing poles was to notch the bottom of a 75-foot pole and put it on top of another 75-foot pole. The splice was reinforced with heavy angle iron. Although Don left this pole up, it was not as strong as he would have liked.

His next attempt turned out to be the magic combination. Two 75-foot poles were laid end to end, with the bottom of each pole situated in the middle. Each end was less than a foot across and the middle of the pole was about 2-1/2 feet wide. Four pieces of 20-foot-long angle iron that was four inches wide and 3/8 inches thick were used to splice the poles together. Each piece of iron weighed over 300 pounds. They were drilled with 19 one-inch holes spaced a foot apart at a machine shop and then hot-dip galvanized. The one-inch through-bolts were also hot-dip galvanized.

Don had purchased the poles from the Baxter Pole Company and had about 50 poles in their lot rolled over

An aerial view of Don's remaining 25 acres. Catalina Island is visible in the background.

so he could pick out 20 of the straightest 75-footers. Longer poles were not purchased due to the fact that at that time, if a tower's height exceeded 150 feet, it would require FAA approval. The poles were then drilled for pole steps and pressure creosoted.

For guy wires Don selected a 5/16-inch cable that was rated at 13,500 pounds tensile strength and hot-dipped galvanized. The guy wires were bonded together at their joints to prevent any electrical noise from being generated. At the bottom guy anchor, a pre-formed cable termination held the guy cable to its anchor. Each pole was guyed three directions, 120 degrees between guy wires.

Don remarked: *"The 140-foot pole was completely assembled and the loose guy wires installed on the ground. I rented a 100-foot crane with two men. We put up four poles one Saturday. A month later the other six poles also went up on a Saturday."*

After he had put in the 10 poles on the perimeter of his property Don started moving the antennas from across the 120 acres and onto the 25-acre piece remaining. Rhombic number 1 was first, mounted near the top of the poles at 135 feet. Its length was extended to 1,050 feet, an almost exact fit. The next antenna moved was rhombic number 13, mounted at 130 feet, just below number 1. Its length was trimmed to 1,000 feet. In this manner the rhombics were moved one at a time from their locations across the 120 acres and layered five feet apart down the poles. Each antenna was cut with a 50-foot difference in length, in order to minimize interaction. The last antenna to be moved was number 15. It was 1,550 feet long and Don's favorite. To fit on the property it had to be trimmed to 1,100 feet and it was mounted atop of everything, at 140 feet.

Don used a numbering system that had been started by *Press Wireless*. The first antenna that they had installed was number 1, and so on. Don continued to use this system as he put up more rhombics.

The extra curtain wires that he had originally installed on the antennas had to be removed. The layering system down the poles at five-foot spacing would not allow room for the extra space required for the two and three wire curtains that he had on the antennas when they were mounted separately. Removing the curtains raised the rhombic's feed point impedance and gave them a higher Q, which resulted in a narrower bandwidth. This made the transmitters a little more difficult to tune into the antennas. Once the antennas

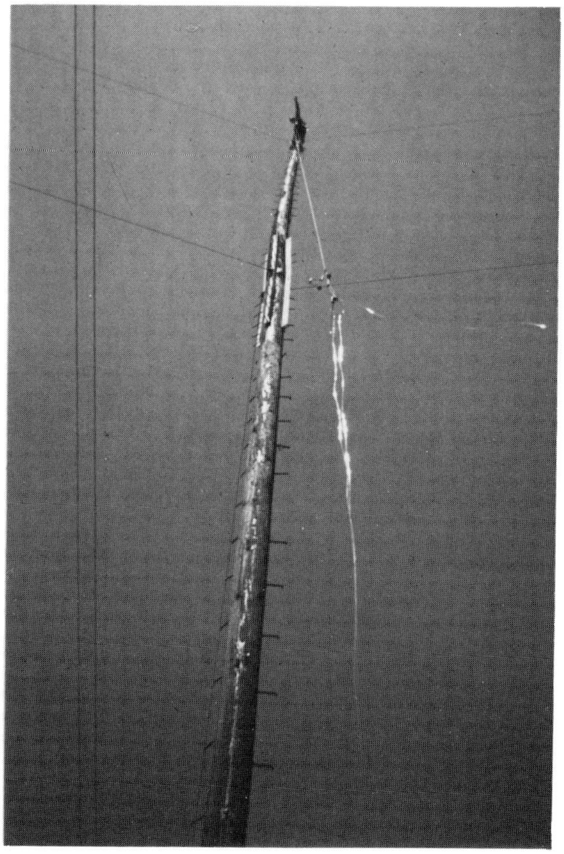

Don working on top of a 140-foot-high spliced pole.

were moved onto the 25-acre property, Don had a total of 17 miles of wire in the air, supported by 60 poles. Ten poles were 140 feet high, five were 100 feet, 12 were 80 feet, and there were another 33 lesser poles that supported the antenna feed lines which were an average of 25 feet high.

The new feed line poles that he installed were 12 x 12s, made of redwood so they wouldn't deteriorate. Some of the *Press Wireless* pine poles were falling apart. The redwood crossbars were 8-foot 4 x 4s, and about 25 feet high and spaced three feet, with two crossbars on most feeder poles. The cross arms could support up to five pairs of feed lines each. For long runs, the middle support poles were 100 feet apart and the guide insulators were mounted with the wire running through it. At each end an insulator with lead inserts tied to an eyebolt with a shackle provided the support for the run. A come-along was used to get the proper tension on the line. Every few years Don would add a few feet of wire to one end and subtract that much from the other end of the run, in order to change the wear points from the middle support insulators on the wire. The wire rubbed on the guide insulators when the wind blew.

Don was once asked about the differences between receiving and transmitting antennas. He replied: *"I never put up a transmitting antenna in my life, just receiving antennas, insulated for 1-kW. The rhombics are excellent for three degrees of beam width, usable for 15 degrees, and after that a dipole would be better."*

Don was 65 when all of this activity was in progress. He found he could no longer spend all day working up on top of the poles. All of his life he had always been able to climb six poles in the course of a day and work up on them from dawn to dark. Now he found that after four poles he was too tired to continue. That was the point to stop, he figured, rather than making a mistake 140 feet up in the air, which could result in a serious injury.

Once the antenna transition was nearly complete, the ARRL asked Don for a current description of the rhombic farm.

> *Present W6AM: All #8 copperweld wire, feeders #8, 2 7/8 inch spaced. W6AM uses 25 acres for 8 rhombics, 16 directions. The 25 acres is equipped with poles 30 feet from property lines.*
>
> *Antennas tend to be the length or width of the property and adjacent antennas are about +/- 10*

Curtain wires on a rhombic.

percent larger or smaller to reduce interaction. Number 8 copperweld wire is used for antennas and feeders. No spacers are in the feed lines. Side poles are not put up, the corners of the rhombic are pulled up with a cable to any convenient pole.

Feed lines: The rhombic impedance transformer is made up as the down lead. The spacing at the top is three-and-a-half feet at the top to two-and-seven-eighths inches at the bottom. This transforms the impedance from a nominal 750 ohms to 468 ohms. It permits all 16 feed lines to come into the station, each pair 16 inches from the next pair. That spacing eliminates reasonable interaction.

A switch at the operating position could add six feet of feed line for the cases where at a particular frequency, the rhombic didn't like that feed line length and matching network setting. At best, there was still quite a mismatch on most rhombics.

*The antenna sizes (widths, as the length was chosen by the property available) are based on the ARRL antenna handbook and computed for 14,175 KC and drawn on graph paper to determine resultant log lengths. All antennas work FB (*fine business) *on 160 to 6 meters. W6KPC computed sizes on some of the rhombics and helped put up one of them. All rhombics are installed as a single wire* (one curtain), *for then there is room for more rhombics. Lengths are based on wherever the poles are and the poles are put up as high as possible. Antenna heights are 75 feet to 140 feet. The longer rhombics are best.*

The angle between the wires of each leg of the rhombic at the feed point was 33 degrees and at the center support pole was 147 degrees.

There are now end poles up for four more rhombics, which will give a total of 24 directions. Currently there are 10 poles 140 feet high and 17 poles 75 to 100 feet high. The rhombics are as follows in tip-to-tip length: 460 feet, 580 feet, 600 feet, 600 feet, 700 feet, 750 feet, 950 feet, 1050 feet, and they cover 16 directions. The proposed rhombics to go on poles already up will be 1,350 feet, 900 feet, 1,200 feet, and 1125 feet, tip-to-tip. Then the station will be considerably better, for there will be even more choices as to direction. One direction or one slightly different angle tends to be better at any given time.

Don's guy anchors consisted of two steel plates about 1/8" x 12" x 18" and welded in an X shape. They were about 1/4-inch thick in the middle, with the last 1/2" at the ends turned up. A 1/2-inch hole in the middle allowed for a galvanized eyebolt about 8 feet long to be mounted. They were buried in the ground about two feet, at a distance from the pole where the top guy wire would form an angle of 30 degrees with the pole. The soil was of a high clay content.

Surplus 75-ohm 300-watt noninductive Globar or equivalent resistors are used as terminations, six in a series totaling 450 ohms, on each end of each rhombic in the station. The resistor center-taps are grounded on all ends all the time, except the one (antenna) being used, so lightning has never hit, due to this up-in-the-air perpetually grounded shield. About 100 special home built relays are used for fast switching by a small 24-contact rotary switch. Each relay blade is mounted at least one inch from the relay coil on one-inch standoff insulators to reduce RF bypass capacity effects to the coil, so there is almost zero loss in the relay switching system.

The rhombics appear to average better than the rotaries (Yagis) *in the vicinity, as neighboring hams frequently compare weak signals over the landline* (telephone). *Reception paths normally appear to open up 20 to 30 minutes ahead of 100-foot high rotaries in the vicinity and stays in longer. W6AM has been on top of the DXCC a part of each of five consecutive years and has the all-time top AM SS score, was in three DX contests last year, and won first place for the area in the classification used in each contest.*

Scarcely a week goes by that several hams and/or engineers have not asked W6AM to write up the W6AM rhombics. There is already a heavy correspondence on rhombics reaching W6AM, as apparently my use viewpoint is easier to comprehend than the usual description that many rhombic textbooks put forth.

Don Jr. had a long talk with Don at this time. Don Jr. recalled, "I told Dad that the ARRL had always been interested in the rhombic farm. We knew Herbert Hoover Jr. pretty well, who had just become the president of the ARRL. I said that we should approach the ARRL through him about their having a long-term involvement in the rhombic farm. We could arrange a financial agreement with favorable terms for them and this would take the strain of the taxes and maintenance off Dad. He would remain on the property until his estate was eventually settled in the future and then the ARRL would have the rhombic farm for one of their stations."

Herbert Hoover Jr. presented Don Jr.'s plan to the

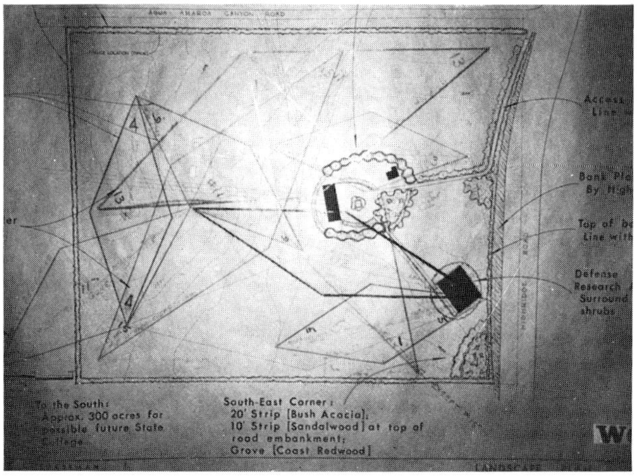

Chart of interlaced rhombics on 25 acres.

ARRL board of directors and they studied it. Considering the property values by then, the purchase price would have been in the neighborhood of $1 million, with a low down payment and interest rate. However, at this time there were opposing political forces at work within the League. Hoover was the first to rise to the ARRL presidency by virtue of his national stature, rather than working his way up through the ranks within the organization. There was opposition, perhaps from board members who felt they had been passed over, and the League declined to follow through on this generous offer to acquire the W6AM rhombic farm.

Don put up one more rhombic antenna, bringing the total to nine. This gave him 18 possible directions. Cleyon Yowell, AD6P, attended some of Don's rhombic farm antenna parties and has many recollections from those days. "There was no nonsense when we were outside, it was all work. We had been at it all afternoon one day and Don had a big barbecue for all of the guys at the end of the day. Everyone was pretty tired and we all had a cold beer. I had finished my glass when a another fellow walked up with a pitcher of beer. I thought it was for refills, but he was drinking out of it. I guess he figured that way he had to make fewer trips back to the keg."

Although Don planned to install several more rhombics, a unique problem had cropped up within his antenna system and adding more antennas and wire would only make the problem worse.

When each antenna had occupied its own piece of real estate, the only problems Don had with them was when a feed line twisted or a relay didn't close properly. Rarely, a wire in one of the legs of the antenna would break and fall down. Each of these problems was easily located and fixed. The meter that checked each antenna would indicate an open or short at that antenna position and Don would either look at that antenna's relays, which were mounted on the back wall, then walk along its feed lines, or go out to the antenna and take a look at it.

With the rhombics now layered up the poles, there were seven distinct levels of interlaced wire. Ideally, the antenna at each level was spaced five feet from the antenna that was above or below it. However, each time it rained a particular pole might settle perhaps 1/32 of an inch in the mud. Up 100 feet that pole could lean half an inch in a given direction. With the reduced tension the wire in the middle of a 500-foot long leg could then drop two feet. Once this process had repeated itself a few times one rhombic would be touching another in the middle of their legs, 130 feet up in the air and 250 feet from the nearest pole.

When his antenna meter indicated a short in a particular antenna, and it wasn't a twisted feed line, Don tried a number of ways to isolate the problem. He used field glasses from the ground level, but the angle was wrong to spot touching wires. He even climbed a pole for an additional vantage point, but the wires that crossed were too far out to determine if they were touching or just close. Don's final solution was to mark the installed height of each antenna on his rhombic layout chart. Between the chart and the antennas that read shorted, and disconnecting the termination resistors in the station, he could usually determine which legs of the two (or more) antennas were touching.

Once the antenna legs had been identified, Don loaded his car with a come-along, bolt-cutters, and assorted tools that added up to around 75 pounds. He would go over to a corner pole with the higher antenna on it and cinch up its top guy wires a bit. Then he would go back to the operating position and see if the short had cleared. Sometimes it would clear, but the higher antenna was now touching the one above it. Then Don would go back out to the pole, return the guy wire to its original position, go over to a pole that had the lower antenna installed on it, and proceed to let out its guy wire a bit. This lowered the antenna. He would proceed until he found a combination that left all the antennas clear. The guy wires on the poles were adjusted so that the six layers of rhombics were clear of each other, with little respect to the angle of the pole. The tension on the top antenna was so high that one of the 140-footers was bent like a drawn bow and remained so for many years to come. The poles, being made of wood, tended to give during high winds. The layered rhombics were to require a great deal more attention.

In his station Don had originally soldered the wires to the termination resistors from their associated relays and also the center-tap wire to ground. That system had worked fine when the antennas were separate. To isolate the wires that were touching in an antenna, Don often had to disconnect several of the terminations or one-half of a termination. It wasn't feasible to continue cutting and soldering wires, so he installed copper clips on them. This way they could be easily disconnected and reconnected, which expedited the isolation of antenna problems.

Don recalled, *"I have everything on clips. I had soldered them originally. When I moved the antennas onto 25 acres I found that I had to take all the joints apart and put the clips on. Now I can measure everything on the meter when I check the antennas. When I get a short I can pull off clips until I can find the two particular antenna wires. You can't see it from the ground, no way. I've tried and tried with field glasses, climbed poles, and everything else. You just can't see it. You can't tell where they cross and touch. This way you can disconnect wires until you know which two are hitting and then you can go pull a guy wire and clear one of them. Usually you just have to pull or loosen one guy wire."*

Although the better amateur stations in the country were putting up high Yagis and building substantial amplifiers, Don remained one of the dominant signals on the bands. At that time Jim McCook, K6GLC (W6YA), visited the United Nations station in Geneva. He recalled: "When I was visiting 4U1ITU I got on the air for a brief time with their station. Afterward I looked back at the recent contacts in their station log. One day there had been an opening on 80 meters and they had worked a number of California stations. Most of them received a 559 signal report, there were a few with 459 or 569, and one station had received a 599—W6AM. I was very impressed!"

Amateur radio was first sanctioned in Saudi Arabia in 1964. Ned Raub noted: "Ahmem Zaidan (HZ1HZ) went on the carpet in the ruling assembly to ask the sheiks (with constant raids from Yemenis across their southwest border) to allow radio in the land. An organized group of monitors would be in their interest. Thereby was formed the Dhahran Experimental Radio Association (DERA)."

"W1TYQ (W8NUV) in Dhahran was issued HZ3TYQ. Earlier, while Raub was living in Connecticut and flying from Idlewild to Dhahran, he, Rundy, Brian and Jacob mounted the first operation to the Kuwait Neutral Zone as 9K3TL/NZ. Many high shooters missed them. Vic made two trips to the Kuwaiti Zone as HZ3TYQ/8Z5. W6AM worked him on CW on the first trip. On the second trip Don was out of town and did not allow a second operator to go up to his station to work him."

Don's philosophy toward DXing was best described in a poem written by Theodore Roosevelt. "The credit belongs to the man who is actually in the arena . . . . Who in the end knows the triumph of high achievement." The Kuwait Neutral Zone was one of the three countries Don missed working on phone. The other two were CR8 (Diu/Damao) and FI8 (French Indochina).

One of Don's operators in the fifties, W6VSS (K6UA), put together a big contest/DX station in Fallbrook in the early 60s. Dale Hoppe had bought the property from Cam Pierce (W6QY/K6RU) and put up large Yagis on each band, 40 through 10 meters. On 80 and 160 Dale put up directional wire antennas and prepared to operate multi-operator multi-transmitter in the upcoming DX contests.

Don rarely worked contests any more. He would participate in most and would make a page or two of contacts, but he had won so many over the years that he no longer had the ambition to stay up all night, which was usually required for a serious effort. In 1964 a guest operator, Don Miller, W9WNV, came up to the ranch to work the CQWW CW contest. Don left the station all set up and was with his family that weekend.

During the contest W6AM operated on 40 meters only. A total of 468 contacts in 87 countries and 37 zones worked — all 40 were heard. Afterward Miller reflected: "Right-hand fuse in fuse box number one blew, and believe it or not it took me an hour to find out why I had no high voltage! Also, the H.V. relay in the 40-meter final sticks occasionally. The rhombics gave us a very slight edge over W6VSS on both receiving and transmitting in all directions. We beat W6VSS by about 100 QSOs, 10 countries, and two zones on 40 meters. He had two operators and two rigs on 40!" Dale also had a California kilowatt, his 40-meter amplifier had 750Ts (twice the oomph of Herb Becker's "corn fed" kilowatt of three decades earlier).

With all of the relays and wiring, Don's station was a high maintenance item. If it could be kept on the air all 48 hours in a contest, it could attain a respectable score. As it was, with an hour of prime time missed, the high CW score in North America was achieved by Miller and stood unbroken for some years. Another remarkable aspect of this achievement was the fact that the rhombics didn't perform as well as before, being now mounted in an interlaced configuration. When each antenna had occupied its own piece of real estate its performance had been superior.

For several years Don sponsored one of the CQ CW contest trophies for the multi-operator, multi-Transmitter station that scored highest in the United States In 1964 Buzz Reeves, K2GL, and his team of operators won the W6AM trophy. A former operator of Don's station, John Norback, W6KFV, was among them. Six years earlier the operation of K2GL's station in the ARRL DX contest was featured in the June 30, 1958 issue of *Sports Illustrated.*

A multi-operator team had come up to the ranch for the phone contest a month earlier. Larry Brockman, WA6EPQ, Bob Brockman, WA6IPY, and Jerry Hagen, WA6GLD, put the station on the air that weekend. It was their first attempt at operating from the rhombic farm and they finished a respectable third in the country.

The author's first visit to the rhombic farm was at this time. Jim Lighthipe, K6KCO, was the leader of an electronics-oriented Explorer scout post. Don had an annual open house each June that drew about 100 visitors during the afternoon. Several times a year he would also accommodate special functions, such as a club meeting. Jim arranged a tour of Don's station on May 9, 1965.

There were about a dozen 17-year-olds who had just a couple of years background in high school electronics. When they arrived at the rhombic farm on top of Rolling Hills, there were almost too many sights to absorb at once. Don's house was located in the middle of low hills, with a maze of antenna wire supported by numerous telephone poles of various sizes spread across the property. Inside the building his operating position was stacked with Collins equipment and behind him a line of six-foot-high kilowatt amplifiers, one for each amateur band. It was an impressive layout and that's when the author first heard the term "hundred-foot-long ham shack."

Don gave the youngsters the grand tour and explained the antenna system and how the rhombics were installed. There were eight rhombic antennas situated on the 25-acre piece of property. One more was in the process of being moved from its old location down the hill, but hadn't been re-installed yet. The rhombic lengths varied from a 500-foot "shortie" to nearly 1,100 feet from end to end. All of the land adjacent to his was still vacant; the nearest structure in sight was on a distant hill over half a mile away.

Inside the ham station the author remembers Don tuning a Collins 51J-4 receiver and remarking, *"Hmm, an ET3 (Ethopia). I'd better pick him up."* With that, Don turned the rhombic switch to a North African antenna (as the relays chattered on the back wall), gave the station a short call, and established contact.

When the scouts left at the end of the day, his rhombic farm was all they talked about on the way home. For weeks afterward the main topic of discussion was his station. At one of the meetings a couple of months later there a guest was announced, WN6QNV. Alex Wallace was Don Jr.'s oldest son and the newest amateur in the Wallace family.

On May 12 the author took the novice test at Scott Radio Supply at 344 E. Fourth St. Harold and Evelyn Scott, W6NZP, had established the store in 1929. One of their employees, Bill Whitney, WA6WBB, gave the five-word-per-minute code test and the exam, with Bob Lawson, K6MQG observing. Two weeks later the call sign WN6PNB arrived.

The author worked on the code daily and had it up to 13 wpm by mid-summer. He received his general class amateur radio license three months after the novice, WB6PNB, and DX soon became the primary interest. A crystal-controlled transmitter that ran 75 watts had been built while still a novice, soon followed by a kilowatt amplifier that used a pair of 813s. DX was still not easy to work, being rock-bound at 7010 KC. Soon the early mornings and late afternoons were spent tuning the regions around 7010 and 14020 KC CW, looking for DX stations. On these occasions the unmistakable signal of W6AM was often heard.

Since Don was located at the crest of Rolling Hills, 15 miles southwest of north Long Beach, whenever W6AM transmitted he would pin the signal strength meter of the author's receiver, a Hammarlund Super Pro. When Don worked a station, everyone always tuned down to his frequency to see who it was. Most of the time it was rare DX. However, the station he was in contact with was usually too weak for most others to copy. On the few occasions that the station was strong enough for the moderate stations to hear, the pileup of other stations calling the DX station was so big that the little guys usually didn't have a chance.

In the early sixties Don began to use Collins equipment. As long as he was an RME representative he had used their products almost exclusively. A few years after he retired Don found that Collins equipment had many superior features. Slowly but surely the RME equipment was replaced by Collins.

The Mobile Kilowatt

Don had first put a mobile CW rig in his car shortly after the war. Each time he purchased another car he would have to "rework it" to make it compatible with a transmitter. This was especially true in the mid 1950's when he put a kilowatt amplifier for his mobile rig in the trunk of his car. He shielded the ignition wiring and had the hood, tail pipe, and engine bonded to the frame to minimize any stray noise that the car generated. An extra heavy-duty battery was placed in the trunk, which was connected to the one under the hood. A 70-amp alternator kept them charged.

He enjoyed the mobile kilowatt station, operating it almost exclusively on CW. Virtually every station he worked remarked never having worked a mobile station on CW before, let alone one running a kilowatt. Even on the drive between the Long Beach and Rolling Hills stations, he would make a couple of contacts.

His mobile kilowatt on CW provided many an entertaining moment for him. Jack West, W6VD, recalled, "I was operating mobile CW near Kansas City and worked W6AM/M. He was driving on the Harbor Freeway in Los Angeles using a Swan 400 and a Vibroplex bug. Ace, K6RA/M, broke in. He was using a Swan 500 with a Vibroplex and I was using a Swan 260 and a Vibroplex. We were all mobile-in-motion and had to leapfrog around the band as the Swan transceivers did not have an RIT (receiver incremental tuning, independent of the transmit frequency), but we had a good three-way contact which lasted about 25 minutes."

Don and Bertha went on a round-the-world trip in 1965. They left in late February and flew to Paris, France. Don visited a couple of local amateurs there, went to a DX club meeting, and then he and Bertha continued on to Egypt. At each stop along the way they looked up some of resident hams he had worked previously and Don would usually operate at their station during the visit.

After Aswan, Egypt, they visited Amman, Jordan, then Petra, and then Jerusalem in Palestine. At each stop a number of amateurs always showed Don and Bertha around. They toured the Holy Land. Their journey also took them to Beirut, Damascus, and Teheran. When Don got on the air from Iran, the 20-meter band was open to Russia as well as long path to California. *"When I called CQ,"* he said, *"about a hundred Russians called me. They were so loud that I couldn't hear any fellows from the West Coast who were calling me. They were covered up. So I wrote down all of the Russian call signs that I could copy in the pileup, I sent them all at once, and then sent, 'You're all 599, QRZ W6.' Not quite as many Russians called me that time and I did the same thing again. After a little while no more Russians called and I was able to work some of the fellows back in California."* There were 39 W6s in Don's log from that morning's operation in Iran.

After visits to Isfehau and Shiraz, they left Iran and continued on to India. There they traveled in Bombay, Arangabad, Udaipur, Jaepier, Calcutta, and Agra. Don always brought along a DX call book on his overseas trips, so they visited with amateurs in each city they stopped. After two weeks touring India, they proceeded to Katmandu, Nepal and visited Father Moran, 9N1MM. At each stop Don would use one of the local stations to work a few operators on the West Coast and whoever else called him. *"Some operators get short-tempered when they give a directional CQ and the whole world answers them. When that happened, I found that the best thing to do was work the loud stations as quickly as possible, then I could pick out the weak ones that I wanted to talk to."*

After their visit with Father Moran and sightseeing around Katmandu, Don and Bertha continued on to Bangkok. Thailand was on the banned country list for United States stations, but Don could work Europeans without any conflict. In Cambodia, however, amateur radio was not authorized. When they arrived Taipei, Tim Chen, BV2A, greeted them. He had been licensed as C3YW 16 years earlier, before the mainland had fallen to the Communists. After a few days in Taiwan Don and Bertha flew to Hong Kong and met VS6CT. Phil had known Don for a number of years and gave him the grand tour.

Their next destination was Tokyo. Kan, JA1BK, picked them up atthe airport and had arranged a big dinner with a number of the local DXers in attendance. *"I was sitting at this long table in the seat of honor and I turned to the gentleman next to me. I introduced Bertha and myself and said that I was a retired manufacturers' representative. The fellow introduced himself. He turned out to be the former prime minister of Japan!"*

Hawaii was their next stop. Nose, KH6IJ (formerly

Don operating his mobile 1-kW CW rig.

K6CGK), met them and Don and Bertha stayed with him. Don and Nose first worked each other 30 years earlier. After nearly two months on their trip, Don and Bertha returned to Long Beach.

The author was on the air almost daily in 1966 and 1967. Don sent one of his country's needed lists, with instructions to call him if any were heard. There were several on the list that didn't seem too hard to work. Don replied, "Ever since I missed CR8-Diu, I thought that it would be a good idea to have three cards from each country. That way, if one of the countries split into two, I have a better chance of having a QSL from both of the new ones."

One morning a country was on that Don needed. The author gave him a call, but Don didn't answer his phone. He had worked the new one anyway and later found that other DXers had tried to call him. A short time later Don installed a loud telephone bell in the ranch operating room with a light in front of the operating position that flashed when the phone rang.

The entire complexion of the DXCC changed in the early and middle '60s. In previous years stations that were on the air from remote parts of the world had basic equipment and simple antennas. With few exceptions, only those stations around the world with a strong signal would manage to get through and establish contact with the rare DX stations. Then they would spend 10 or 15 minutes exchanging information about their equipment and antennas. That is why the Honor Roll usually consisted of only 10 stations or so. Those with less dominant stations couldn't get through to the rare stations and so their country totals were significantly less. About the only exceptions had been operators like Bill Hunton, W6ODD, and Bob Roberts, G2RO, whose work took them to the far corners of the world. They were experienced operators and, even with moderate equipment and antennas they could scratch out a contact with some of the weaker stations calling.

In the late '50s and early '60s great numbers of amateurs became interested in DXing. Accordingly, there were adventuresome types who traveled to remote places around the world specifically to contact other amateurs. These expeditions, or DXpeditions as they be came known, made it possible for those with more moderate stations to contact rare locations. As the big guns at the top of the DXCC had contacted these locations previously, and their scores didn't change, it began to get a little crowded at the top.

In the haste to put rare and unusual locations on the air, details such as the location of a country's border, or disembarking from a ship before radio operation commenced, were often overlooked.

At this time a feud was brewing between the ARRL staff and one particular operator. He had established himself as one of the premier operators in the world and soon took to visiting rare countries for a few days to contact as many people as possible.

There had been some abuses by other operators prior to this. At least once, an operator was found not to be located where he had claimed to be. Some years earlier a station in the Sudan indicated that he traveled to a number of rare countries in the region, but was eventually found to be in Khartoum the entire time. Accordingly, the ARRL removed those country credits from stations that had unknowingly added them to their DXCC totals. In the years since, there were a number of other portable stations suspected of not being completely on the up-and-up.

At one rare location in the travels of the person in question, he refused to work a number of stations, including Charlie Mellen, W1FH. Don Wallace heard about it the next day and contacted the operator (who had been a visitor at the rhombic farm) and delivered a stern lecture. *"The DXCC chase included some of the most respected amateurs in the country,"* Don reminded him, *"and that was not the proper way to treat them."* The operator then relented, but the damage was done.

In what was a sad moment for the world of DX, it was demonstrated that this person was not necessarily located where he claimed to be during his foreign travels and radio activity. This was the straw that broke the camel's back and, at that time, the ARRL implemented new rules regarding DXCC qualification. In the future an operator must show a license from the responsible authority in the country where he was located.

This was to have a profound impact on DX activity and the DXCC listings. There were any number of countries around the world that turned their back on radio activity occurring at their doorstep. Often, the authorities there didn't necessarily object to radio activity, but those people in a powerful government position didn't want to take the responsibility of being the first to authorize radio operation in their country. If something went wrong or the government was put in an embarrassing position, they would be held responsible.

Consequently, the clandestine stations that were operating from many of the third world or remote countries no longer counted for the DXCC listing. Those operators who had not previously worked an undercover station from a rare spot would have to wait for an authorized operation. In several cases 25 years have passed and they are still waiting.

Although still scarce to the average amateur, Burma had had moderate radio activity in the years after World War II. W6AM's first QSO with that country dated back to 1928, VU2LZ. In the DX contests of the 1950s it was not unusual for Don to work two or three stations there back to back. Burma entered a period of political turmoil in 1962, and their new government banned amateur radio. The last operation by an operator native to the area is thought to be by XZ2TH that year.

In 1967 Don and Bertha traveled to St. Pierre, a French possession off the northeast coast of Canada. Don looked up an amateur he had worked previously, who helped him get a license, FP8DW. Don was on the air with his host's station from October 5 to 10.

Don held his open house annually and did all that he could to encourage newcomers to the hobby. Steve Massey (N7AHN) recalled: "I was first licensed as WB6SSO in 1966 at the age of 12. WB6PKA and myself used to go up to Don's and use his station for the afternoon. He would give us a ride to the club meeting, working Europeans on 20-meter CW while he drove down the freeway."

Scott Martin (W7SW) had been licensed as WN7LDZ

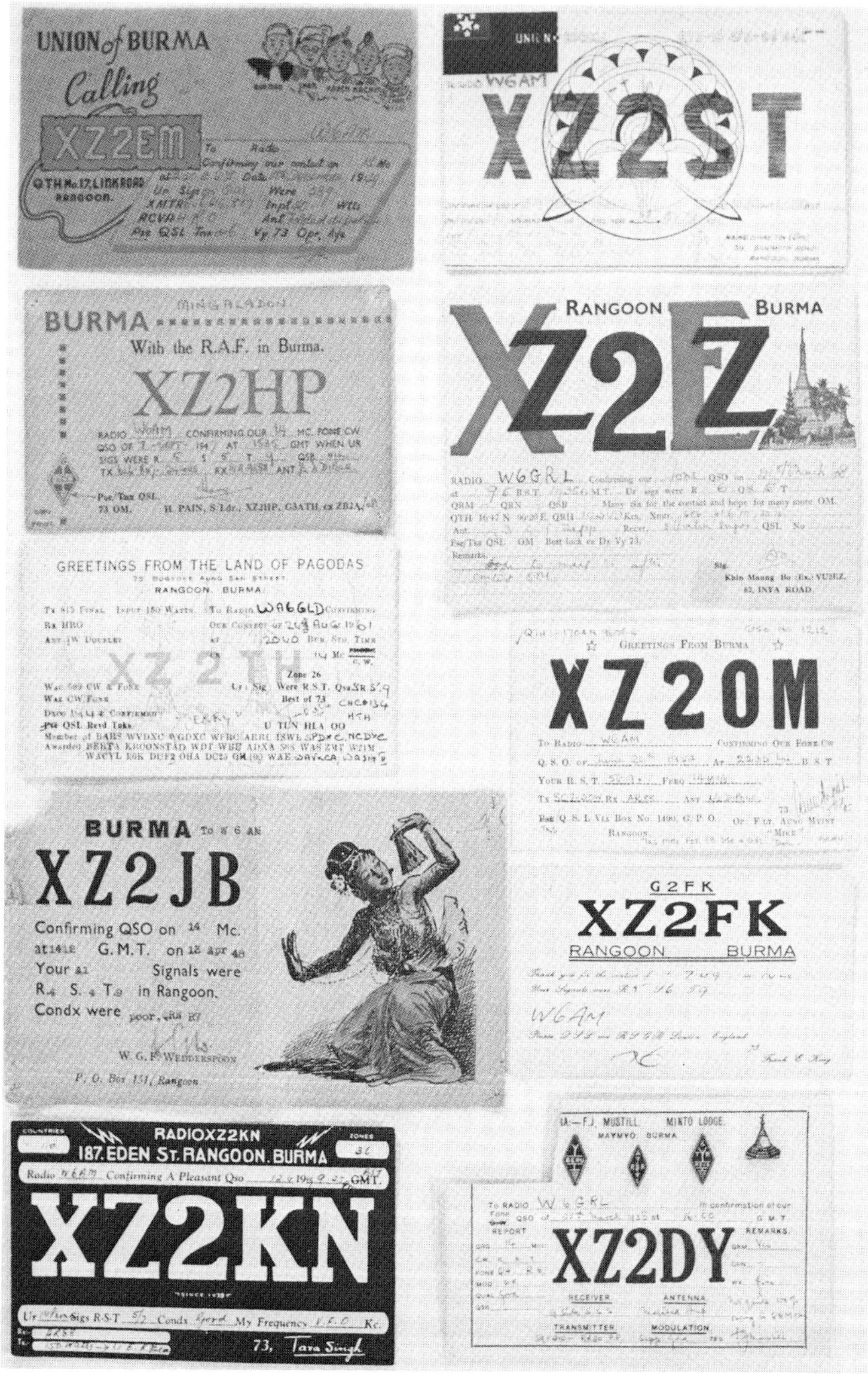

QSLs from Burma.

in Portland. He recalled: "I called CQ and Don called me as W6AM/mobile. Was I ever excited, having seen his picture and read about him in a DX guide book. He was surprised that I knew so much about him, his DX record and his antennas! He was very kind to send slowly and listen to my erratic CW. He was driving and sending CW with a hand key. I have since traveled around the world and made many thousands of contacts but I have never made another CW contact with a mobile station. My QSO with Don was as exciting as my contact with Owen Garriott in the space shuttle almost 20 years later."

Another youngster, Marty Hart, WB6NWW (N6WW) reflected: "I had discovered the wonderful world of DXing and soon learned that W6AM was the master. I rode with him to the Fresno DX convention in 1968. He worked Europeans with his mobile CW rig all the way up and back. I remember DXing from Israel later and Don's signal was the only one that made it through the noise when the band was supposed to be closed to the West Coast."

Lin Hamilton (NJ6Y) went to high school in Torrance and wrote Don a letter asking to visit the rhombic farm. Lin recalled: "I was WN6PAV and Don answered my letter with an invitation to his open house. There were quite a few visitors from abroad and he gave everyone a tour. He told us his radio history and that of the station. When I was married in Palos Verdes at the glass chapel I invited him to join. Don attended the wedding and the reception, driving in an El Dorado that had his mobile CW station in it."

Don always worked a few stations from the mobile whenever he was in the car, even if he was just going to the store. One day W6AM/M was on the freeway and WB2OZW (K2OZ) answered Don's CQ. Paul Adkins remembered: "We were operating CW about 25 words per minute and Don mentioned that he was doing 60 miles-per-hour. I asked him if he was driving his car or was someone else. He replied, '*Who else would be driving*?' He was quite a guy."

From his earliest days Don Wallace had always championed the cause of the beginner in amateur radio. His philosophy was *"Don't make it too hard to get in — you can learn the rest after you're in."* He often spoke before FCC hearings about keeping the entrance requirements into amateur radio as simple as possible. Once someone had a foot in the door, then an incentive licensing system would provide the motivation for them to continue to learn the hobby and upgrade their license.

The ARRL Field Day contest was rapidly becoming the most popular radio activity to be found. Most clubs across the country would gather their membership for an outing of simulated emergency conditions for the

The American Radio Relay League, Inc.
COMMUNICATIONS DEPARTMENT — WEST HARTFORD, CONNECTICUT, U. S. A.
CERTIFICATE OF PROFICIENCY IN USE OF CONTINENTAL CODE
Club Award
This certifies that DON WALLACE (name), W6AM (call), who has participated in the Code Proficiency Program of SOUTHWESTERN DIVISION ARRL CONVENTION 1959 (name of sponsoring radio club) has this date demonstrated his ability to RECEIVE (send, receive) Continental Morse Code at 44 words per minute. The text used in this test consisted of PLAIN LANGUAGE -PENCIL COPY (plain language or code groups) averaging five characters per word or group and sent by AUTOMATIC (manual, automatic) means.
If competition was held, statement of competitive standing appears here: XXX
Competitors ____ (number competing)
25 July 1959
Award Date
Certificate # 9
Club Officers and Witnesses

Over the past half-century Don won dozens of code contests.

weekend. In 1967 there were over 15,000 participants throughout the United States, nearly half again as many as the DX contests brought out. When Ellen White, W1YYM, printed the results, she included a picture of Don operating his mobile CW rig.

About that time, in a *QST* article, Nose, KH6IJ, advised that in CW contests it was wise to learn how to send while holding a pen. He wrote, "Can you imagine setting down and picking up a pen 2500 times over the weekend?" Exactly what many operators had done in the recent CQWW CW contest! Most contesters learned Nose's method and now say that it seems odd to send CW without holding a pen in the same hand.

Even though Don had finally been beaten in a code copying contest in 1959, he was still pretty good at it. Ray Furlong, W6QIL remarked, "At the ARRL convention in Anaheim in the late 1960s I entered the code contest. They gave you a sheet of text with missing words and kept turning up the speed as you tried to fill in the missing blanks. At 50 wpm there were only a few of us left in the competition. I got a cup for third place at 55 wpm, Rod Johnson, W6MUR received second at 60 wpm, and W6AM was first at 65 wpm."

The *HMS Queen Mary* came to Long Beach in 1968. It had been commissioned by the British in 1935 and had served as a troop ship in World War II. During its

The QSL commemorating the last voyage of the *Queen Mary*.

last voyage, from Great Britain to Long Beach, an amateur station operated aboard as GB5QM. For those who contacted the Queen Mary during its last trip, a commemorative QSL could be obtained from Ray Furlong, W6QIL.

Fred Laun, K3ZO, who traveled far and wide in his United States diplomatic assignments, passed through Los Angeles in 1968. "Dave Morgan, K6DDO, took me to see Don Wallace. After showing us the fabulous station, Don took me to lunch at the bottom of the hill at Marineland. I continued to my assignment in Thailand and after a year I was assigned HS5ABD. In January 1970 I fired up on 80 meters and W6AM was the second station that I worked."

By the summer of 1966 the author had obtained a VFO and had the 100 countries confirmed to qualify for the DXCC award. He joined the Southern California DX Club and W6AM was at the first SCDXC meeting he attended. After the meeting Don showed off the mobile radio in his car. He had a Swan transceiver with two VFOs at the driver's position and a kilowatt amplifier in the trunk. He drove around the block and worked a station in Asiatic Russia on CW. Don sent with a speed key that he had next to him on the car seat. Considering that the car was in the middle of high-rise buildings in downtown Los Angeles, it was an amazing feat.

Those who looked at the DXCC Honor Roll about this time would notice that Don was at the top of the listing of countries worked. Even though he was heard to work a lot of rare DX, it still didn't dawn on many just how dominant a signal the rhombic antennas put out.

The next year the author's DX efforts were put on hold. Draft quotas had to be met. He spent a year in the United States Army electronics schools and was then sent to Viet Nam. He was stationed at Pleiku, in the central highlands, as the supervisor of an electronic maintenance (ELM) shop. As it turned out, he was the only person in Pleiku who had any experience with HF radio gear. Others in the unit worked on wire and carrier equipment and also VHF radios.

The local Military Affiliate Radio (MARS) Station heard there was someone who could fix HF radios, specifically the Collins equipment used to run phone patches back to the States. Once the necessary repair work on the KWM-2A transceivers was concluded, they were given a thorough checkout. Of course, that included putting them on 20 meters in the evening after the 12-hour shift was over. Even considering that the location was in a combat zone, it was a young ham's dream come true.

The author selected the call sign of KR6NB when the band was open to the States. Okinawa (KR6) was about

In 1970 Don achieved 5BDXCC.

the closest American possession to Viet Nam. Due to the combat situation, amateur radio operation was not authorized in Viet Nam. When the band wasn't open stateside, WB6PNB/XV5 was used. There were numerous other clandestine operators on the air.

Naturally, there were some drawbacks. Just because one was proficient with radios, it wasn't all gravy. The KWM2As were also used as the backup communications for the remote VHF sites. On four separate occasions during the year the author was air-lifted by helicopter into one or another of these sites during a fire fight in order to get their communications operating. It's certainly remarkable how fast and low one can crawl while holding a KWM-2A.

One evening the author was listening on a Collins R-390 receiver. It was sunrise back on the West Coast of the United States and a station from Pakistan was on 20 meters. The pileup of stations on him was huge, as operators up and down the West Coast were calling. There were familiar call signs, the one that always beat most everyone out. The loudest signal among them was W6AM. What a potent signal he put out with those rhombic antennas. The author worked Don from time to time and he was always the loudest signal on the band.

About once a week a station in Long Beach would be contacted and the author would talk to his parents. One evening Wayne Gingerich, W6EUF, was running a phone patch to them and representatives of the Army

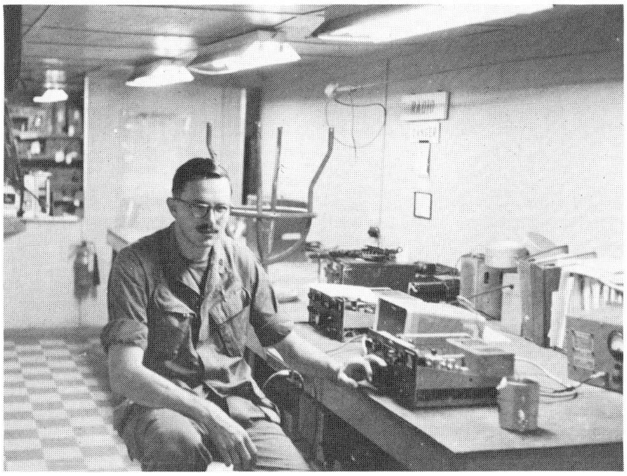

In 1970 the author continued the long tradition of clandestine operation in Indochina; WB6PNB/XV5.

Security Agency (ASA) dropped by his station location at the ELM shop. The ASA was located just down the street from the ELM. It was an awkward moment, as clandestine radio operation in Viet Nam was a court martial offense. As it turned out, they had come to ask a favor. The ASA had been unable to establish communications with advance units in Cambodia and they had been listening to the international contacts occurring just down the street. The author was asked to coordinate the communications for the invasion of Cambodia.

Returning stateside in late 1970, the author told Don what the real location of KR6NB was. At that time he said: *"That used to be French Indochina, I worked FI8ZZ and several other stations there 20 years ago."* About then a look at the latest ARRL DXCC Honor Roll in the December issue of QST revealed that W6AM was at the top of both the CW and phone listing. Don was tied with Ross Hansch, W9BG, on CW with 351 countries and with W8GZ and W2BXA an phone with 349."

Don remarked, *"Sometimes I am not on top of the DXCC in the annual listing, just tied for number one or two. I don't have to worry about it as it has happened several times in the last 15 years and I have always outlived the guy who has passed me up. I am outliving everybody, seems like."*

Don never lost his taste for higher learning. *"Since I've graduated from the University of Minnesota I have taken approximately 110 college extension courses. I don't know anybody who has taken any more than I have, but my wife liked to go to those and we always went to a couple a year. One of them was a science course. It was at UCLA, 12 weeks long. We went over there every week for three hours. You could go as a student for credit, or you could go as an observer and just listen, and we chose to go as observers. However, at the end of the 12 weeks, the instructor said, 'Here are the examinations, come on up front, pick them up, take them home, fill them out and mail them back in a week and we will send you a grade.' I thought, well, for the heck of it, I would pick those up.*

"We never made a note the whole time, my wife and I. We didn't do anything but just enjoy these science talks, a different subject every week. They were fabulous, right up-to-date. So we took the two exams home and that night as I was going to sleep, I read it and I thought, heck, I could answer this thing. So, I got up in the morning and answered it and told the wife. She said, 'I will try it.' Then we compared the two. This was an open book exam. You could look up or do anything you wanted to. You could consult with anybody. There were several hundred questions. Twelve of our answers didn't agree out of the several hundred and, you know my wife would not budge an inch on any one of those answers and neither would I. So I mailed them in and we both got a 'B.' We didn't do anything but sit there and enjoy it. I thought that was pretty good. Here we were, old buzzards in our 70s, who took an undergraduate examination without any work at all.

"Some of the things that lecturer gave were fascinating. I was interested in the fact that on radio night he spent the first five minutes on everything that took place before I went to college. He spent another five minutes on everything that took place up to two years before the lecture. Covered it very thoroughly. Then he spent the next two hours and 40 minutes on things that happened in the last two years. The fellow was really up-to-date. A lot of it was really new to me. I couldn't believe he was so skillful that he covered virtually every new field of radio electronics at that time. That was the bulk of his talk. So I immediately had a lot of respect for him because he did that on all of the programs."

The 5-Band DXCC (5BDXCC) award was announced by the ARRL in 1969. The goal for amateurs was to work and confirm 100 countries on 80, 40, 20, 15, and 10 meters. Don began working stations across the bands, concentrating on 10 meters. He remembered how in past sunspot cycles, sometimes the propagation on that band would be very poor. In that particular cycle, which was near its peak, 10-meter propagation was very good. Those stations that had concentrated on 80 meters were the first to achieve the 5BDXCC. J. Eshleman, W4QCW, was the first to qualify for the award. Windy Windom, W8GZ, gave him a close race for #1, but was hospitalized with a stroke when his last cards came in, and received award #4. On December 7, 1970 W6AM received 5BDXCC award number 54, the sixth to be issued in California.

With his station and rhombic antennas, Don could have been one of the first to achieve the award, but between traveling and his schedule of golf and other activities at the Virginia Country Club, amateur radio was allotted only so much time.

Even though the rhombic antennas were interlaced and didn't work as well as they used to, Don still had a big signal, especially on the lower frequency bands. Jack West, W6VD, recalled, "In the early 1970s I was trying to finish working 100 countries on 80 meters for the wind up of 5BDXCC. One evening on the low end of 80 I heard W6AM work a CT1 in Portugal and Don gave him a 579. I frantically fine-tuned the receiver, cut in a preamp, switched between a 66-foot vertical and a half-wave horizontal dipole, and couldn't even hear a whisper from the CT1. A short time later I chatted with Don at the Fresno DX convention and asked about the CT1 on 80 meters. Don gave credit to his Johnson kilowatt matchbox, explaining that the inductor was wound using a copper bar that provided a high Q and extremely low loss to received signals. He neglected to mention that the matchbox was connected to open-wire feeders going to a 1,000-foot-long rhombic antenna pointed short path to Europe and terminated for unidirectional operation. Now that was HUMILITY!"

In 1971 one of Don's friends from college, Lester Robson, 9KU(K0PEF), passed away. A good friend of his, Dale Russell, W0SMV, took care of selling off his equipment. Dale noticed an antique wavemeter among the items. The wavemeter, from the 1921 era, was used to determine the transmit frequency of stations on the air. He asked Mrs. Robson if he could arrange for it to be displayed with other early items and she agreed. At the rhombic farm Don maintained a corner of the room with a number of early items, such as a spark receiver and spark key. There the wave meter was placed.

In the late 1960s Don was again faced with illness, not his but Bertha's. She had Parkinson's disease. It caused a loss of muscle control and was degenerative. By 1970 she was no longer able to maneuver the stairs in their two-story house in Long Beach. Don had the ranch remodeled at this time. A bedroom, bathroom, and kitchen were added at the south end of the structure, which was completely carpeted. The exterior of the building had shutters added to the windows, a garage built on the north end, and a sloped roof put on with shake shingles. Once the renovation was complete, Don and Bertha moved in. Their house in Long Beach was left intact and Don would stop by a couple of times a week to look in.

Bertha's condition continued to worsen. There was no known cure for her affliction and in 1971 she passed away. Over the years she was the unsung hero. Many wives who were less understanding would have strongly objected to the amount of radio activity in which Don was involved. When one looks at Don's many achievements, Bertha's support, her understanding, and the sacrifices she made in order to accommodate his hobby, shone through.

Bertha Wallace was licensed in 1928, and had the callsign W6MA for 43 years.

Ham Radio's Ambassador

Don sold four lots on the southwest corner of his ranch to pay inheritance taxes. The lots were located where he had long considered building a house.

He now divided his time nearly equally between the ranch and his home in Long Beach. Spending so much time at the ranch had put his home in jeopardy. It had been burglarized once during that time and vandals had put a hose in a basement window. The basement had filled halfway with water, which damaged paperwork stored on lower shelves. Fortunately, the old receivers and homebrew equipment had been placed on the higher shelves and escaped the water.

At the Fresno DX Convention in 1973, John Troster, W6ISQ, presented Don with a plaque commemorating his 60 years of enthusiasm. The September issue of *World Radio News* carried a picture of Don receiving the award and carried the details of the presentation. Most of the front page summarized Don's achievements in the first 30 years of his radio activity, as most of those present were already familiar with the last 30 years. John wrapped up with the comment: "What we have been saying, Don, is that for the last 60-plus years you have done everything a ham can do. . . . You've been a leader in everything you've done . . . and you've stuck with it. Enthusiastically. Don, we are pleased to honor you for 60 years of leadership . . . and inspiration . . . and enthusiasm."

In 1975 *QST* announced that amateurs who had occupied the number 1 position of the DXCC Honor Roll could receive a plaque from the ARRL. It was made of walnut and cost $25. Don wrote the ARRL and asked the plaques be qualified for be sent, along with a bill. Bob White, the DXCC administer, replied, "I found 60 times that you are shown in the #1 spot (since 1956). That comes out to $1,500." Don and Bob conferred on the matter and decided that they would have just two plaques made, one for CW and one for the phone.

Harry, WA6UOR, of the Fresno Amateur Radio Club, invited Don to speak to the group and he put on a special program for their March 12, 1976 meeting. They arranged an "Old Timers Night," and the older members of the club brought early licenses, certificates, QSL cards, and radio parts. Don showed slides and films of "DX and how to do it" and "Rhombic antennas, How and Why." He also helped them get started on a local chapter of the QCWA.

Don's Honor roll #1 phone plaque.

A number of people asked Don how he had come to enjoy such a pleasant life in his late '70s. He responded with the following:

"Get eight hours of sleep per night. Play tennis for one-and-a-half hours, twice per week. Play two or three games of golf each week. Dance two to five times per month. Drink beer regularly, one bottle each Wednesday after golf. Only have one drink at a party. Sip it for the length of the party and have half left at the end. No other time. Don't smoke. I smoked pipes and cigars from the age of 25 to 50. I never quit, I just don't smoke any more.

"Since I graduated from college I have taken over 100 college extension courses. When I was 75 I took speed reading.

"I play gin rummy with Virginia Country Club champions and that keeps me alert mentally, or I lose my shirt. I did not play before I retired, as it was a time-waster when there is work to be done. Go to four or five ham conventions annually, and attend all of the technical sessions. Read all of the ham magazines, and have a good ham station. A couple of hours per day spent on a hobby seems correct. Have a usable ham station in your car, to make driving a pleasure instead of a chore.

"Eat a big breakfast—eggs, bacon, toast, milk & grapefruit. Also eat lunch and dinner, but only about half of what the ordinary person eats. Eat things with lots of seeds: tomatoes, cucumbers, watermelons, canteloupe. Each morning take one vitamin C with Albee. Drink lots of water, every hour or so. Taper off after 5 PM. Drink one cup of coffee at breakfast, and a half a cup at lunch. I used to weigh 245 pounds, but I was not heavy looking; now I weigh 185. I reduced my food intake until my stomach was trim and then I no longer craved much food.

"Wear suspenders, inside. Then you don't have to stick out your stomach to hold your belt and pants up. It can then be pulled in to create a flat stomach. A British ham told me this 30 years ago when I complimented him on his flat stomach. Only go to a doctor when absolutely necessary and then only once every 10 years. My doctor is a great friend and we golf together each Wednesday. I stretch my neck for five minutes each morning and evening (due to a football injury in 1919). Do kick exercises for a few minutes each morning.

"Don't wear headphones. My remaining hearing was going until I switched to a speaker over 20 years ago. Earphones have a negligible air cushion. The radio room is soundproofed. If I knew then what I now know, I never would have used headphones; except when using a crystal set." In the early days, receivers did not have an AVC or AGC system, so when a loud signal was tuned in the level out of the headphones was proportionally louder.

"I don't do anything that worries me if I can help it. There are plenty of other things to do. Competition is stimulating, but if I am beat I don't worry about it. The other chap was simply better. Try to smile and be happy all the time. Sit, stand and walk straight: Try to be taller than your height.

"My wife and I were married for almost 50 years and went together for six years before that. I was sure that she would outlive me. I would rather that she had outlived me as we had a wonderful life together. She was smart. She got her general license, W6MA, in two weeks.

"I enjoy climbing poles and have done so several times monthly since I was 12 years old. The exercise, and working on top of a pole for seven hours, is good for your health. It has probably helped extend my active life. When I was 75 I ran into a story in the paper about speed-reading. I told my son I was signed up for a speed-reading course. He said, 'I wonder if you would take Bobby if he wants to go.' I said, 'Sure, if Bobby wants to come that's fine.' That was his son, a 15 year old boy. Bobby said, 'Oh great, sure, I would like to.' So I picked up Bobby every night for the 12 nights of the course and took him down to the speed reading. The first thing was to find out how good you were. I was something like 275 words a minute reading, 85 percent comprehension. Bobby was something like 150, 50 percent comprehension. When we got through I was 1,200 words a minute , so on some tests I go 2,000. But that was the average of your three bests at 90 percent comprehension. Bobby was 4,000 at 90 percent comprehension. That's the young mind for you. To this day, I can read something at 500 or 1,000 words a minutes and know what I read. I don't do it all the time but it did speed all my reading up from maybe 275 to 400. I just can't read slowly any more: It isn't necessary.

"A lot of people wonder how you become a good radio man and all that. It is a lot of little things put together, no one thing. Getting on top of the DXCC, some of these things probably helped me. I have on overall picture and I have the desire and usually the ability to concentrate when I need something, like a new country, and I get it. I always have thought the most imposing list there is in amateur radio, that I

Don repairing a feed line.

know of, is the DXCC list. All the other things seem to be a passing interest."

Don's oldest son, Bill, was flying over Rolling Hills in 1976, and he had Don Crystal, W6ANH, take a picture of the rhombic farm. Bill sent the picture in to *QST* and it appeared in Rod Newkirk's "How's DX" column in the September issue. The caption noted:

> W6AM has farmed your QTH of the month for thirty years, raising a fine crop of rhombics on Palos Verdes Peninsula. Firing in eighteen directions, nine diamonds fill the 24–acre plot on poles as high as 140 feet. Don's ham shack is just as impressive, enough state–of–the–art equipment to keep W6AM battling for top spot on ARRL's DX century Club Honor Roll.

Fred Tebbets of Long Beach was in Guatemala when a 7.5 magnitude earthquake struck there at 4 AM on February 4, 1976. He recalled, "Betty and I were staying in Guatemala city and were sound asleep when the shaking started. I grabbed her and we stood in the doorway to the bedroom. Electricity went out immediately, of course, and in the moonlight I could see the seven-story hotel across the street clearly. After 30 seconds it collapsed into a pile of rubble. I thought ours would also, but it didn't. There was no fresh water for two days and it was the end of the week before we got a plane out."

Don Wallace was among the dozens of amateurs who handled energency traffic for several days in the after–math of the earthquake. Unlike the one centered in Managua, Nicaragua, three years earlier, for which Don also handled traffic, this one devastated nearly the whole country. W6AM had also been on the air handling traffic when the Alaska earthquake occurred on 1964 and he accumulated a total of 17 public service certificates. The ARRL didn't issue them prior to 1933, otherwise he would have had several more.

An aerial picture of the 25–acre rhombic farm in 1976.

In the middle 1970s the property tax in California was rising at an unprecedented rate. Even though Don had spent $200,000 on the ranch to bring it to its present state, and he was considered to be financially comfortable, the $18,000 annual tax rate on his remaining 25 acres was excessive. As a comparason, an engineer just out of school with a BSEE typically received an annual salary of $12,000 per year. A basic rule of retirement was to spend less than the interest on one's investments. If the principal was annually reduced and one lived a long time, it could be financially disasterous. So Don decided to rent out the cottage on his property.

About one third of the distance between the main structure and the eastern edge of the property, and tucked in among the eucalyptus trees, stood the cottage that Don had built in the late '40s. During the '50s it served as the residence for the foreman of Don's 120 acres and the adjacent property that was farmed at the time. Since Don had reduced his property in 1962 the cottage had not been used.

Don decided to install a pole for commercial VHF and UHF antennas and lease the building and pole to a communications company. The tax relief provided would be substantial. He put a 95–foot pole about 50 feet north of the cottage, bringing the total on the property to 61 poles. Then he signed an agreement with Motorola, which leased space on the pole and the cottage for VHF and UHF communications.

A neighbor on the area, who had a competing commercial interest, complained, and Don entered into a lenghty dialogue with the city. In the opinion of the city attorney Don was in violation of the city code. Don proceeded to make a package of material describing his use of the land and past contributions to the community. The information included data on past limited commercial users such as the Drug Enforcement Agency, Hughes Aircraft, and Los Angeles County emergency radio systems to whom no fee was charged during their 20-year tenure. Attachments also contained copies of public service awards from the ARRL in connection with assistance rendered during earthquakes, certificates of merit from Armed Forces Day messages, letters from the Navy Department in connection with Don's information on portable transmitter operation during World War II, articles written about Don describing his contributions to amateur radio, along with aerial views of the property before and after he purchased it. The stack of material was over an inch thick and Don had 75 complete packages made up. These he distributed to mem-

Don operating W6AM.

bers of the City Council, officials at city hall, and anyone else in a position to influence the city's decision.

He had to appear at a total of 14 City Council and planning commission meetings before the final decision was determined. On October 23, 1976 it was announced that the Council had denied a resident's appeal of the planning commission's approval that Don be issued a conditional use permit for the 95-foot pole. Motorola would remain, and Don would have his tax write-off.

Property taxes continued to rise for another two years. Then, in 1978, the California voters passed Proposition 13. This initiative reduced property taxes to earlier levels, and took the pressure off retired people who had land that was appraised to be high in value.

Don's annual open house at the rhombic farm continued to be a big attraction. On October 23, 1977 he hosted the Santa Barbara Amateur Radio Club for the afternoon. Two weeks later Bob Dyruff wrote,

> "I've been trying to find a moment to thank you for the fun time you and Virginia extended to our gang on Sunday. Many rode down with us in our caravan and even more made their way directly. That was a great show you put on, non-stop. Twenty of us stopped for dinner on the way home and the feeling was unanimous that all had a most enjoyable time and were happy to have seen your fabulous installation. We are still trying to figure out how we can duplicate it here!! Hi!!"

As Don shuttled between appointments, the Virginia Country Club, his house in Long Beach and the ranch, he continued to operate his mobile kilowatt station. Mark Faulhaber remarked, "I had one QSO with Don and it was quite remarkable. It was on 20-meter CW, and what was extraordinary was that he was talking to me while he was driving down the Long Beach Freeway. Incredible!" It was business as usual for Don, though out of the ordinary for many stations that he worked.

Don often gave several amateurs a ride when he went to the convention each year. In the mid-1970s the meeting place was changed to Visalia and it has been held there most years since. Keith Jones, W6OB, recalled: "I rode with Don to the Visalia DX convention in 1978 and on the way home we stopped at Delano and visited Voice of America. It was dark and almost raining. We walked out and looked at the Sterba curtain antenna beamed at Asia. We were all talking and then I looked around for Don. He was halfway up the 325-foot tower. Yes, he made it to the top. After he came down there was a giant arc from the Sterba to ground, 50-kw of RF. They had to break the carrier to get rid of the arc. Don always liked to climb high towers. At heart Don was a 17-year old."

In 1978, shortly after he turned 80, Don Wallace was elected to the *CQ* DX Hall of Fame, which honored those amateurs who had made outstanding contributions to the field of DX. In the DX column of the October issue of *CQ,* John Attaway, K4IIF, devoted over a page in summarizing the radio history of Don. He noted: "Possibly Don's greatest contribution to the DX world has been the personal help and encouragement he has given to hundreds of budding DXers in the U.S., and his inspiration to thousands of DX oriented amateurs the world over. One of the most prominent amateurs started down the DX trail by the efforts of Don Wallace is Lloyd Colvin, W6KG. Lloyd writes that he when was only 12 years old he bought his first *QST* and read an article by 6AM exhorting more amateurs to use 20 meters in their DX activities.

> "Since 1922, Don Wallace has appeared at radio clubs in the U.S. and 89 other countries. The countries that he visited include Pakistan, Formosa, China, Andorra, the Bahamas, Chile, Cuba, Bolivia, Macao, Uruguay, Germany, the Phillipines, Iran, France, St. Pierre and Miquelon, England, Scotland, Hungary, Switzerland, Ecuador, Panama, Honduras, Thailand, the Vatican, Italy, Japan, Jordan, Guantanamo Bay, Guam, the Hawaiian Islands, Alaska, Midway Island, Puerto Rico, American Samoa, the Virgin Islands, Wake Island, Canal Zone, Norway, Argentina, Bulgaria, Peru, Lebanon, Austria, Finland, Denmark, New Guinea, Netherlands, Netherlands Antilles, Brazil, Surinam, Sweden, Egypt, Crete, the Dodecanese, Greece, Turkey, Guatemala, Russia, Kaliningradsk, Ukraine, Georgia, Armenia, Kazakh, Moldavia, Canada, Australia, Belize, Hong Kong, India, Mexico, Syria, Nicaragua, Rumania, Yugoslavia, Venezuala, Albania, New Zealand, Paraguay, Manaco, the Fiji Islands, Israel, Cyprus, Nepal, Trinidad and Tobago and the deleted country of Labrador.
>
> W6AM holds every major DX operating award in the world, and has even worked almost 250 countries from his 1–kilowatt mobile rig. Now in his eighth decade going on nine, Don Wallace is still on the top rung of the DXCC Honor Roll and can be heard in all the pileups.

The article also carried a picture of Don at 6AM in 1927, and another that had been taken recently at the ranch.

The November 1978 issue of *World Radio* carried a picture of Don being presented the *CQ* DX Hall of Fame plaque by Lloyd Colvin, W6KG, at the ARRL National Convention. The article carried a summary of Don's DX accomplishments and radio history and noted:

> Don has traveled in 89 countries, operating as an amateur in most of them. He has been an international ambassador for Amateur Radio goodwill.

Besides being longtime friends both Don and Lloyd associated with the YASME Foundation, which was incorporated as a non-profit organization in 1965 and headquartered in California. Don Wallace was named as its honorary president, acknowledging his contribution to and leadership in the amateur community.

YASME started in 1955 when a young Englishman named Danny Weil decided to sail around the world. Dick Spenceley, KV4AA, helped Danny get his amateur license, VP2VB, and Danny took a rig along on the voyage. Over the next eight years he operated from a number of locations around the world. His vessel, YASME, was lost three times over that period and in 1961 the organization assumed its name.

Lloyd and Iris Colvin W6KG and W6QL began to travel for YASME in 1964 (at their own expense). They have operated from scores of countries since, providing many a DXer with a new country.

The *Queen Mary*'s amateur radio station was formally dedicated on April 22, 1979. The ceremonies were held in the former liner's reconstructed radio room with 400 people attending. The first contact was made by Don Wallace. The trustee of the station, Nate Brightman, K6OSC, remembered the event. "We asked Don to be the first to operate the station, in recognition of the fact that he had one of the first amateur stations in Long Beach 70 years earlier. Don called CQ with the W6RO call and a ZL station in New Zealand answered him. The fellow hadn't worked Don in over 20 years, but recognized his voice and said he knew it was W6AM operating, regardless of what call sign he was using. So they reminisced about the old days and that made it a really special occasion."

During the high sunspot activity of the late 1970s a number of intercontinental DX contacts took place on 6 meters. The card Gary Frey, W6XJ, received from one of the first of these QSOs with a Japanese station stated, "Don't submit this QSL for any awards," and proceeded to describe the operator's kilowatt amplifier. Japan was

In 1978 Don was inducted into the *CQ* DX Hall of Fame.

limited to 50 watts on 6 meters. However, when conditions were marginal, a kilowatt can make the difference.

VHF activity had become more popular in recent years. A neighbor of Don's, John Chambers, W6NLZ, made headlines in 1957 when he worked a station in Hawaii, KH6UK, on 144 MHz. John lived in a Palos Verdes community that had restrictions against ouside antennas. He built a second story room on his house that had a large picture window on its western side. John's parabolic dish antenna was located in this room and fixed toward Hawaii.

On August 10, 1979, the "Second Front Page" of the *Daily Breeze* carried a picture of Don and an aricle by Warren Robak: "What a ham! Don Wallace comes in loud and clear worldwide." The article described Don's radio history and his rhombic antenna farm. *"I like antennas like I believe in football players. I believe a good big one is better than a good little one."* Robak also noted that "Wallace still climbs the lower sections of the poles to make repairs to the antennas."

But Don was slowing down. Throughout the 1970s he continued to climb a maximum of two poles a day to work on his antennas. Now, at 81 years of age, he found that when he had climbed to the top of a 140–foot pole he was too tired to get much work done. Don was a product of the days of wooden ships and iron men, but his time had come. He would have to leave the work at the top of the poles to the younger men, perhaps those who were only in their 60's.

The public seemed fascinated with stories about amateur radio. On August 11 the *San Pedro News Pilot* picked up the story and reprinted it with the headline: "82–Year-Old Peninsula Resident Hams It Up All Across the World."

It was at this time that the author began to spend a lot of time at the ranch. Earlier that summer Don was on the air chatting with another amateur. Don mentioned that he was looking for a certain type of Variac. It had to be the same model number as the other two for his three-phase power supply. Apparently one was available, but it was quite expensive. The author told Don that there was a whole power supply with a three-section Variac at the Hughes Aircraft surplus sale. Don came out to the company and met him and took a look

The QSL of the Queen Mary commemorative station.

at it. It was exactly what he needed. A bid was submitted and was successful, and the power supply was transported up to the ranch.

The thought was to test the power supply. If it was no good the Variac would be removed and used with his three-phase power supply that was spread on the floor behind the amplifiers. If it was good the old one would be disconnected and the new one used.

The power supply was in a seven-foot high rack cabinet that was about three feet wide and weighed about a half a ton. A 220-volt three-phase power line was run over to it, the big switch was thrown, and it worked. Don's old power supply, consisting of transformers and assorted components on the floor, dropped about 1,000 volts when the amplifier was keyed. The new one dropped just 50 volts due to all that iron in the power transformer. So the old supply was disconnected and left in place as an emergency backup. Don was still using the same amplifiers that had been installed in the late 40s and early 50s, and the author started going up to the ranch for an evening every couple of weeks to help Don work on them.

Don had brought a Henry 2-K 1-kilowatt amplifier to the ranch in the middle 70s, and had often used it when one of his other amplifiers was on the blink. With the introduction of commercial linears, the Johnson Kilowatt matchbox was found to work best for them when using the rhombics. Each matchbox was modified by hard wiring around its internal relay and the output coax connector was disconnected, as it would arc over. Just the open-wire feed line RF output was used. Also, the 10-meter position needed a heavy strap added at the internal switch. Otherwise the original strap would melt when transmitting with a kilowatt on that band. An external SWR bridge (Swan SWR-1) was used to tune the matchbox. The internal one in the Johnson matchbox didn't work very well.

As antennas were switched, the matchbox needed to be retuned into each rhombic, even into each end of each one. However, it would tune out the SWR to nothing, indicating a maximum transfer of power from the amplifier to the antenna. When the rig was not in use, the antennas were grounded by another ceramic knife switch behind the operating position. This was for safety reasons, in case of a lightning strike.

One problem did crop up from time to time. As the wind blew and the poles moved slightly, the rhombic wires would lose tension and sometimes momentarily touch the wire below it. Of course, the SWR would rise to nearly infinity for that instant and the amplifier would flash over, usually at its loading capacitor. Not that this was due to a flaw in the Henry 2-K. Any amplifier operating at or near full output capacity will flash over in that situation. When Don bought an Alpha 77 DX amplifier in 1980, it too flashed over from time to time when antenna wires touched. It made for numerous times when the amplifier was taken apart and the arc spots on the plates of the loading capacitor were filed down.

Finally, a special order Alpha had a vacuum capacitor installed in place of the open plate capacitor in the load position. This solved the output capacitor problem, as the vacuum capacitor did not arc over when antenna wires touched. They also incorporated a new design into Don's amplifier, an SWR (actually reflected power) sensor. When the reflected power from the antenna exceeded 350 watts, indicating a 3:1 or higher SWR, the amplifier was automatically knocked off of the air and had to be reset. This, no doubt, saved the 8877 ceramic output tube numerous times.

Learning about Don's rhombic system was to be an enlightening experience. He had installed nearly all Collins Equipment by this time, and when tuning for DX, he always used the 200–cycle filter. *"Where rhombic #15 once stood,* " he remarked, *"a tract of homes was built and there are now 11 hams with 1–kilowatt stations in its place. A Collins receiver is the only one that doesn't overload with fellows that close and the 200–cycle filter lets me operate very close to their transmitting frequency without bothering me."*

For someone whose hearing was beginning to fade, Don had an uncanny ability to tell what kind of transmitter was being used by the other station. One day the author tuned onto a signal and Don remarked, *"That fellow is using a Collins transmitter, its got a good clean tone."* Sure enough, the operator soon mentioned that he had a Collins 32S–3. Regarding another station Don said, *"That signal is kind of squeaky, it is probably Japanese–made."* He was absolutely right. The average ham could not tell the difference between the signals, but Don could almost name the manufacturer of a radio or, if it was home built, what kind of output circuit it used, just by listening to it for a moment.

Sometimes an amplifier would have an intermittent problem that would be very difficult to locate. Don would say, *"Well, let's crank up the power and look for smoke."* He did just that and, when he located the part that was intermittently arcing, he also found that the PA tube, a 450TH, had been overstressed. A $75 tube had been zapped and Don remarked, *"That's just fine, now we know what we have to fix."*

Them there was a lighter side. Occasionally the author would get a little ambitious and, while working behind the amplifiiers, would try to verify that power was off and then misread the wiring or switching network. Full power was still on. *"Oh, you mustn't do that!"* said Don, as the author reached over to to short across two terminals and was left with the smoldering remains of a wire. Don must have committed these same miscalculations decades earlier.

Once the author asked Don if he wanted a couple of othe fellows brought along to help fix problems around the shack. Don felt that there were too many distractions around for much work to get accomplished by visitors. Don noted, *"With one fellow you get a certain amount of work accomplished, with two fellows half that amount, and with three—no work. Let's stick to just the two of us."*

Don had a lot of fun using the kilowatt mobile rig in his car, but other communications services were sometimes susceptible to it. Bart Bartlett, W6OWP, had renewed acquaintances with Don at a convention at the Disneyland Hotel. Bart recalled: "I was in my room and the TV suddenly started 'keying.' It was Don on the air in his mobile! After W6AM/M got a few blocks away the picture returned to normal."

On page 76 in the September 1979 issue of *QST* a "mystery letter" appeared which was submitted by Dave Hardacker, W7TO. He spoke of a QSO with a kilowatt mobile station at 25–30 WPM that was en route to a golf course and he challenged the readers to identify him. The responses in the November issue of QST included:

> Actually I do not see a mystery at all. Everyone knows of Don Wallace, W6AM. He is as much an institution as Arthur Fiedlore. Terry A. Posey, N4KT, Crawfordville, Florida.
>
> The question is too easy. There is only one Don Wallace, W6AM, on his way to Virginia Country Club. WA6YAO.
>
> I had the pleasure of working W6AM's mobile substitute for an antenna farm. 'W7TO is right: Don's mobile CW moves along fast also! N2XJ.
>
> Some years back, with the indicator on my keyer set at 30 WPM, I was answered by W6AM chugging along at yesterday's 60 miles per hour. Truly, his contribution to ham radio has inspired many through the years, myself included. —W6KXI.
>
> One kilowatt mobile. High–speed CW while roaring down the highway. DX in the remote parts of the world. It has to be the old DXer himself, W6AM."—N6JM.
>
> "There is no doubt at all that this is W6AM. Have worked him many times with his kW mobile."-AG4S.
>
> "The world has only produced one Don C. Wallace, W6AM, and that letter only scratched the surface. Someone should write a book about W6AM. I know firsthand, because I used to compete with him in the DX pileups."—W8NBK.
>
> "Don's success at one-handed driving is a sure sign of a misspent youth."—W1AM.
>
> "The mystery letter personality is W6AM, who has been involved in more [extraordinary] operating endeavors that anyone I know. I only wish that I had half the spunk of that old-timer. Hope he goes on forever."—WB2EZG.

W6AM operating mobile in 1977.

Charles Hillinger, a *Los Angeles Times* reporter who was a neighbor of Don's, came by one day and chatted with him. Then he wrote a story that appeared in the Sunday, February 17, 1980 edition: "W6AM, Radio Champ Collects 365 Countries." Four days later the *Times* also published the article "Veteran Ham Going Strong at 81 Years." The story traced Don's radio history and described his rhombic antenna farm. It wound up with "Clicking the key and talking to radio hams everywhere in the world is just as exciting to me today as it was when I was a kid. The world is really our cup of tea." The article was very popular and found its way to the wire services. Commencing in mid-March, it began to appear in other cities across the country and Canada. On March 14 the *Dallas Times Herald* had "Ham a Champ on the Radio"; on the 16th the *Sunday Oregonian* had: "Grand ol' Ham Rules Airwaves"; the *Sunday Patriot News* in Harrisburg, Pennsylvania had "81-Year-Old Ham Credits Success to Antennas"; the *St. Petersburg Times* had "Octogenarian radio ham chats world-wide." The list continued. Clippings were sent from those cities and Albuquerque, New Mexico; Sarasota, Florida; Milwaukee, Wisconsin; Tucson, Arizona; Corpus Christie, Texas; Huntsville, Alabama; Norfolk, Virginia; Stamford, Connecticut; Lorain, Ohio; and Winnipeg, Canada. It probably appeared in every major city across the U.S. and Canada, and even some Australians told Don that they had seen it in their local newspapers.

Bob Clowes, K4MB, in Clearwater, Florida sent along a copy of the article with the note:

> This newspaper item certainly brought back a lot of fond memories of amateur radio activities in the 1920s. When I was 8RL in New Kensington, Pennsylvania I remember having a QSL card from '6AM' in 1927 and I prized it very much. Congratulations to you , sir, for the exceptional successes you have had in DX activities and the help you have been to others.

James Russell, W8BU, in Fairview Park, Ohio sent a clipping of the article and reminisced:

> I am a retired lawyer, having been at it for 50 years, and now ham radio is what is keeping me alive. Got my first license in Cleveland in 1912, 8AL, at the age of 13. We worked in 1923 when you were 9ZT in Minneapolis.

Glen Tapping, VE7CQV wrote:

> Your story appeared on the second front page of the *Winnipeg Free Press*. Your amateur radio experience is quite fantastic and a real inspiration to all ham radio enthusiasts. Your comment that 'Clicking the key . . .' is exactly what ham radio is truly all about. I found the article so great that I was simply compelled to write you and offer my very best wishes to you.

Although many people often dreamed of pursuing the top spot of the DXCC, most did not have the antenna farm or the single-minded purpose necessary in joining the competition. Those who entered even the lower ranks of the chase came to know the exhilaration of chasing a rare station and contacting him. Jerry Hagen, N6AV, once remarked, "Working a new country is better than fooling around.' " Although many enjoy the fun and excitement of chasing DX, some think that Jerry has overstated his case. In any event, it is fortunate that his wife didn't overhear his comment.

In the late 1970s a new kind of interference began to appear in the ham bands. A high-powered pulse of RF, which repeated several times a second, was frequently noted. The net effect to stations in the U.S. was to cause a receiver's AGC to cut back its gain and make it very difficult to copy all but the strongest signals. This pulse

Don at a Morse Telegraph Society meeting in 1980 sending American Morse.

was typically the type used with an over–the–horizon radar and its source was thought to be in the western Soviet Union. Within two years the design of new amateur radio transceivers had a noise blanker designed specifically to reduce this kind of interference, dubbed the "woodpecker." However, only a small pecentage of the amateurs had one of the newer radios. Curiously enough, after an initial flurry of activity, the use of the woodpecker diminished and didn't usually appear during contest weekends. It was often speculated that whoever was programming the transmitting site was an amateur, or had friends who were amateurs, and wanted to minimize disruption of DX contests.

The 1981 DX convention was held in Visalia in mid–April. At the urging of his friends, Doc Stuart, W6GRL, attended for the first time in many years. The author had the good fortune of listening to Don Wallace and Doc reminisce about operating radio in the spark era and about the many experiences they shared in the good old days. That was Doc Stuart's last convention. Don Wallace attended his funeral the next year on January 10, 1982.

The American Radio Relay League, Inc.

DX CENTURY CLUB

This Certifies that Don C. Wallace, W6AM

HAS THIS DAY SUBMITTED EVIDENCE SATISFACTORY TO THE AMERICAN RADIO RELAY LEAGUE THAT HIS STATION HAS CONDUCTED TWO-WAY COMMUNICATION WITH OTHER AMATEUR STATIONS IN AT LEAST ONE HUNDRED DIFFERENT COUNTRIES SINCE NOVEMBER 15, 1945. THIS CERTIFICATE RECOGNIZES HIS OUTSTANDING PERFORMANCE AND ATTESTS HIS MEMBERSHIP IN THE DX CENTURY CLUB.

WEST HARTFORD, CONNECTICUT

June 10, 1948 No. 210

Printed in U.S.A.

Don's DXCC certificate.

The Consummate Radio Man

In 1980 W6AM and a number of other amateurs visited China. They toured extensively, and among other places, stopped in at the Nanking radio factory. Don kept a daily journal during the trip and reported that China's return to the air was imminent. Their experiences were published in the January 1981 issue of *QST*: "China's Ham Radio Slightly Closer" by John Felber, K2BPR, Jean Chittenden, WA2BGE, Irv Emig, W6GC, and Don C. Wallace.

Amateur radio operation had not been permitted in China since 1949, except for a brief period during the late '50s and '60s by BY1PK. In 1982 China opened her doors to amateur radio operation after a quarter-century of silence. The first club stations to be formed were located at the universities. A Canadian amateur of Chinese descent, Tom Wong, VE7BC, traveled to China and received authorization to reactivate BY1PK on March 29, 1982. He set up BY1PK with modern equipment and appeared on 20 meters shortly thereafter, ending the silence.

The author recalled one particular evening at the rhombic farm. Tuning across 20 meters, he found Tom Wong, booming in from Beijing (formerly Peking), with a thunderous pileup. At that time, the rhombic in China was 1,000 feet long and 140 feet high. Don said, *"Go ahead and give him a call."* Wong said he had to QRT at 0400 GMT, and could make just one more contact. The author gave his call, N6AW, once, and settled back, waiting for the frequency to clear. A minute or so later the pileup thinned out and BY1PK came back ". . . you're 5 and 9, November Six America Washington, go ahead." Don was fairly bursting with pride. *"Well, everything seems to be working just fine,"* he said. The next amateur radio operation from Tibet, now a province of China, by an operator native to the area, was five years later, in 1987. China had annexed Tibet in 1974, effectively ending their border dispute.

In 1987 BT0NMN in the Xizang (Tibet) province of China appeared on the air. The operation was part of a communications support team for an assault on Mount Everest. The next year BT0LS and BT0ZML would be active, providing similar communications for the China, Japan, Nepal friendship expedition to Mount Everest.

An operator from the BY4AA club station in Shanghai, Mr. Zhou, operated the BT0LS station from Lhasa. Another operator from the BY4SZ club station in Shanghai, Mr. Kang, operated the BT0ZML station located at a base camp on Mount Everest. Both stations were active for two months, while the team was climbing Mount Everest.

Don Wallace with a tour group in China.

The "How's DX" column of the August 1981 issue of *QST* had an article by Ellen White, W1YL/4, "W6AM, A DX Legend." Ellen's story had a picture of Don at the operating position, and another working up on a pole. Her article recounted Don's 71-year radio history, and his DXCC position the past quarter century. She also noted that he rarely had any pet peeves. Don said, *"In all endeavors there is 1 percent trying to disrupt what the other 99 percent are doing. They're normal like sand traps, hills, lakes and bunkers on a golf course. Without them the game might be dull."* Ellen wrote, "Don's lifelong dreams of a rhombic farm have come true and he notes that *'the reality is even better that the dream.'* "

Don attended the awards dinner of the Radio Club of America (founded in 1909) on November 20, 1981 and was inducted as a Fellow. Two months later he received a letter from Fred Link, W2ALU, the organization's president.

> It was a great treat for me to see you at our banquet. I am certain a lot of other people present

enjoyed meeting and being with you too, as you are in the history of amateur radio — a big name.

I am certain you were satisfied with the prestige of the Radio Club membership. The top echelon of our industry are, or were, members of the Club, and that is why it was important to me that Don Wallace be involved. From my earliest days in amateur radio, going back to about 1919, it seems that I had always known a Don Wallace and your contribution to the state of the art is a legend.

I am anxious to get a copy of the picture of the new group of Fellows to you and, hopefully, you will see fit to place it in your 'shack' with some suitable inscription that will make the occasion a matter of historical record. There were a lot of top industry people in that picture. I'm just happy that you were one of them.

In late 1982 Don received a letter from an aspiring DXer, Tony Spiegel, KC8UR, asking for a few tips. Don wrote,

> *To get a DXCC, just mark the countries you work on an ARRL country list. Circle the ones you get a card from and don't work any more in that country. Out here the best time to get on the air is 7 AM for an hour and the next best is 10 PM for an hour (20 meters). Try those and other times, and then specialize in the two hours best for you. As you know, different countries come in at different times. I tune mostly 14001 to 14027 KHz. If you only plan on one DX bulletin, I'd recommend the LI (Long Island) DX Bulletin.* The assistant editor, WA2BGE was on along on the China trip in 1980.

When he returned from the Visalia DX convention in April 1983, Don received a letter from his old friend Kan Mizoguchi, JA1BK. Kan was going to introduce a DXCC booklet in Japan and wanted to have a photograph from each of the DXCC countries, including deleted ones. He needed to photograph cards from C9 (Manchuria), CN2 (Tangier), CR8 (Diu), CR8 (Goa), FN (French India), I1 (Trieste), I5 (Italian Somaliland), PK1 (Java), PK4 (Sumatra), PK5 (Neterlands Borneo) and PK6 (Celebes and the Molucca Islands). Don had the cards ready the day after the convention and Kan came by to photograph them; all except CR8 (Diu). That was the only country he had missed, W6ODD/CR8 in 1948. Kan's DXCC booklet came out a few months later (in Japanese) and he sent Don a complimentary copy

The author continued to visit the ranch a couple of times a month to work on Don's amplifiers and cabling. They had been neglected for a number of years and had developed an assortment of problems. Everything was

Still climbing poles at 83 years!

old, so while one problem was being fixed, another would develop. Also, now that Don had a "stiff" high-voltage power supply, the amplifiers were stressed more.

Each day before he started operating, Don would test each rhombic for continuity. A ceramic knife switch disconnected the rig and connected a test meter to the antennas. It consisted of a DC power source and a meter rotated, it would check each relay system, feed line,

Don at his station in 1980

The RCA Banquet Fellows. (Don is standing, center)

antenna, and termination for the proper connection. Of course, 4.68 (X 100) was the desired reading.

When he had 32 directions to choose from, the antenna select switch had been marked in a clockwise sense, corresponding to consecutive increments on a world azimuthal map. When the antennas were moved to the 25-acre parcel in the early '60s, some of their feed lines were swapped to account for their new locations and Don had never reorganized them. So, one evening the switch was finally rewired. One understanding between both Don and the author was that once a project was started and the station "off the air" that evening, it had to be operational before the author left. Due to unforeseen circumstances, a couple of nights turned into the wee hours of the morning before everything was back in order.

The antenna selection relays were DPST, with a 3 VAC coil. Each pair of coils was connected in series, to

AIEE/IRE Luncheon Old Timers. Don is standing in the back row, third from the right.

run on 6 volts. Don found that, by using DC to the coils, the relays vibrated less and their contacts didn't have to be cleaned as often. There were so many relays in the system that a 6-volt storage battery was connected across the trunk line to serve as a filter. Another pair of small SPST contacts on each relay was used as an interlock. These were in line with the high-voltage control relay. The high voltage didn't come on and one couldn't transmit until the antenna relay had made a positive connection. This prevented "hot-switching' the RF, which destroyed relay arms, as it had during a contest in the '50s.

One evening Don decided to call a ham he had known for some years and looked in his personal phone book. There were several phone numbers next to the amateur's name, all but two of them crossed out. Then Don said, *"I'm not sure which is his current phone number. He just got married again, for the third or fourth time."* Don paused, and then with a twinkle in his eye said, *"He always marries 'em, you know!"*

Later that evening, once everything was back together and working properly, Don fired up the rig on 20 meters. He worked several strong stations on CW and then heard a weak and watery-sounding DL station in Germany calling. The German was obviously excited about contacting Don. Afterwards Don said, *"That was great. That German was probably a new ham and he was excited about working California. That QSO was the most important thing that I did today."*

Whenever Don thought that a particular station should count as a separate country, he went to great lengths to work it and get a QSL, just in case. He said, *"I keep track of the DXCC by working countries ahead of time. For instance, the Sovereign Military Knights of Malta became a country and I had the card. I have about 60 cards from places somebody thinks ought to become countries.*

"The way to get on top of the DXCC is concentrating and having all possible cards in your possession that could be countries. Years ago, the British published a treatise on the fact that ZD9, Gough Island, and Tristan da Cunha, ought to be separate countries. The ARRL has never recognized them as separate countries because they think the islands are too close. They think they are sort of the same government, but they aren't, according to the British. I worked them both and have the cards.

"Now the African homelands have come on. I have been working all of those I can get hold of and getting their cards. They haven't made any of them countries yet, and I don't care if they do or do not. But if they do, I might have the card for the ones the ARRL counts. A lot of little things like that go into DXing." While we

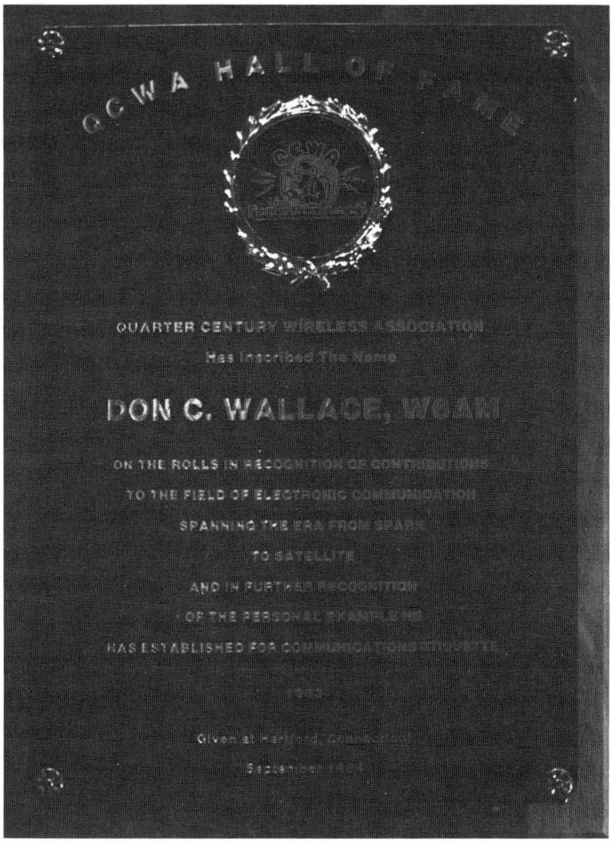

The Quarter Century Wireless Association Hall of Fame plaque.

were working on the equipment Don would often reminisce about the early days of radio or relate some specific DX story.

Soon the station was in good working order, and the author used it in an occasional contest with Don's call sign. He operated in the 1983 CQWW DX contest on 40 meters. It was quite a weekend, and his experiences that weekend were jotted down the following week:

> Contesting at W6AM . . . stepping through a door in time to a half-century earlier. The first impression one gets when they enter the W6AM shack is. . .wire! Crisscrossed next to the ceiling are dozens of pairs of open-wire feed lines. Behind the operating position, layer upon layer of wire interconnects equipment and switches. A matrix of relays on the wall connected by bundles of wire selects the feed lines and terminations to the rhombic antennas. The nine rhombic antennas and feed lines consist of 17 miles of wire. On a post behind the operating position, a bank of nine circuit breakers, with their wiring exposed, controlled power to the equipment. The 220-volt power lines are triple-aught and buried for the 600 feet to the edge of the property.

Since 40 meters was a nighttime band, the plan was to sleep all day until the contest started. The strategy was simple. Get up at 3 AM Friday morning, gather all of the paperwork for the contest, drive up to the rhombic farm at 11 AM and go back to bed. Then get up in the late afternoon, ready for the 4 PM (0000 GMT) beginning of the contest.

Arrival at the ranch was at 11 AM, right on schedule. When Don answered the door he remarked, *"Jan, it's a good thing that you came early. Two antenna wires are touching and intermittently shorting together."* The rhombics were interlaced between 80 and 140 feet high and their separation where they overlapped is less than six feet. It had rained the previous night and a pole had probably settled one-sixteenth inch. Up 140 feet, the top had moved a half-inch. The middle of the antenna it supported dropped two feet, touching the antenna below.

Don went into the work room and gathered the tools that would be needed. Ready to start across the field Don said, *"Oh, you mustn't wear tennis shoes. I have a pair of rubber boots that will fit you."* The author thought to himself, "It's only a little mud." Upon starting out we got less than 50 feet before a giant gob of red clay had formed on each foot, like a bucket of cement. Don had brought along a walking stick. We stopped every few feet and scraped off the clay.

On arriving at the suspect pole we connected a come-along to the top guy wire. After cinching it up a bit the author walked back to the shack to check the antenna, scraping off clay every few feet. A meter at the operating position lets the operator check the 468-ohm impedance of the open-wire feeder to each antenna. After three hours of adjusting the guy wires on two poles, all of the antennas were clear.

The 100 pounds or so of tools were carried back across the field and returned to the work room. The last of the clay was cleaned off and the author sat down at the operating position. It was 3 PM (2300 GMT), just one hour until the contest started. The antenna meter was observed as the antenna switch was turned, giving a quick final check of the antennas. Rhombic #3 seemed to be shorted on one end. Don looked out the window and said, *'There are about 100 quail perched on #3's feed line.'* It was twisted tight. *'They do that when the ground is wet,'* he continued. *'If we take a bamboo pole and reach up 25 feet and tap on the feed line, it will untwist and the quail will return to the brush under rhombic #15.'*

The feed line was cleared and then again observed the antenna meter. It seemed that the meter quivered a bit when antenna #1 or #12 was selected and the wind gusted. Don brought a pair of binoculars out of the work room and looked closely at a pole. *"There it is! A broken insulator at the top of that 30–foot feeder pole,"* he said. Two adjacent feed lines were brushing against each other.

Don got out his safety belt and pole climbing boots while the author got into some overalls. *"Just lean out and do a temporary fix till after the contest,"* he said. A big, fat hawk sitting on the next pole observed the process. While fixing the insulator the lesson learned was to wear a long-sleeved shirt and gloves. Upon climbing down, the time was noted to be 2345 GMT. We sat down in front of the rig, exhausted. "So much for contest strategy," the author thought, as he started digging out slivers.

At last the contest started. The Europeans were Q5 on 40 meters but not very strong. All of a sudden the signals came up to S9 plus. Don said, *"You forgot to throw the receive antenna switch. You were receiving on a three–foot piece of wire between the receiver and the switch."*

The guys down in the Los Angeles Basin were loud. It's a good thing the Collins 75S–3C recerver had a 500–cycle filter. The BFO was adjusted for a zero–beat and then to a tone in the filter's pass band. Don remarked, *"I always use the 200–cycle band pass filter and adjust the BFO for a very low tone, as low as I can copy. You will find that the narrow filter will minimize QRM. Everyone is pretty loud up here."*

Don was to provide a number of interesting insights about operating his station. He said that, when running pileups, if operating conditions permit, the 500–cycle filter could be used. The 200-cycle filter was preferable if there was a lot of interference nearby, but the filter was so sharp that you couldn't hear stations calling you if they were over 200 cycles away from your frequency.

Behind the operating position were a series of six–foot–high rack cabinets, the single–band 1–kW amplifiers. *"I have a piece of one–amp fuse wire in the high–voltage line atter the filter capacitor to protect the power supply,"* Don said. *"It's behind the amplifiers."* The high voltage was bussed to all six finals. The 450TH tubes in the 40–meter rig looked like huge light bulbs. *"Push this switch to select the 40–meter amplifier,"* he continued, as the bulbs began to glow. *"I'm going to dinner at the Virginia Country Club. I hope you do well in the contest. Don't forget to feed the cats."*

The pile of tins on the back porch were for the cats. They were filled from a 25–pound sack of cat food and the 18 cats charged for it. Any field mice around the ranch must have led a precarious existence.

The handle for the Variac was about the size of a steering wheel. Once Don left it was turned up. The RF drive from the 32S–3 transmitter was turned up, but then an arc announced that the one amp fuse wire in the high–voltage line had been blown.

The author squeezed by the BC–610 to get behind the amplifiers and warily observed the three–foot–tall high–voltage capacitor: 277 uF at 5,000 volts. To make sure that it was discharged, a jumper wire between two relays on the wall was borrowed and placed across the capacitor posts. Smoke curled toward the ceiling as the #18 wire vaporized. The fuse wire and jumper were replaced and then it was 0000GMT.

With the rhombics, the DX stations were strong on 40 meters. One tended to tune by them and listen for the weaker signals, which were usually from stateside. VK6HD in western Australia was loudest on the long path, over South America.

During the trip to the refrigerator a short time later, it was noticed that the cats outside had dispersed and there was a skunk at the dish of cat food. A moment later the skunk was brushed aside by a possum, which took its place at the food. Just like a DX pileup: survival of the fittest.

The author had to call one DX station twice. A close eye had to be kept on the S–meter. If the wrong antenna was selected, no matter how loud a station was, the transmitted signal was down 10 dB. As another station was being called the phone rang and, in Don' radio room, the sound almost knocked over the operator. His foot slipped off the foot switch while transmitting, blowing the high–voltage fuse again.

After squeezing past the BC–610 a big screwdriver was slid across the capacitor terminals. The arc destroyed the tip of the screwdriver. This process was a little unnerving. A second piece of fuse wire was paralleled across the first, so it wouldn't blow so often.

A short CQ raised a local station. That station sent, "Hi Don. I recognized the faint 360–cycle tone on your signal. I was an operator for *Press Wireless* before the war. . ." Many operators didn't realize that a contest was in progress and that time was valuable.

Later in the evening, when the operator stood up, a stick protruding from the top of the 40-meter amplifier was bumped and the next transmission resulted in a total blackout. Fortunately, there was enough light from the full moon filtering through the windows for the author to find a flashlight. The line-voltage meter on the modulator indicated there was no voltage coming into the shack. A trip down the driveway revealed the main fuseboxes, 300 feet away, hidden in the shrubs. There were some spare 100-amp replacement fuse inserts handy, and power was restored in a few minutes.

"Once inside the station, the second piece of fuse wire across the capacitor was removed. The next transmission resulted in another blown high-voltage fuse, and a trip behind the BC-610. The shoulder high stick that was bumped turned out to be the 40-meter broadband antenna match adjustment. The drive level was reduced and the antenna match carefully readjusted for maximum RF current, then the drive increased to a normal level.

The station operated properly for a while and around midnight a New Zealand station was contacted. He remarked, "Hi Don, I haven't heard you for quite a while. Our first contact was in 1927 when you were NU6AM. Are you still using the water-cooled rig? The 360-cycle note sounds familiar." The author replied: "Tnx for zone 32, Don says hi. New tx hr. 73." It seems that a number of old-timers liked to say hello and rag-chew a bit when they heard W6AM on.

In the middle of the night the receiver audio seemed to be jumping up and down. The author reached around in back of the 75S-3C receiver to wiggle the antenna input and inadvertently discovered a muffin fan. Fortunately, the damage to the CW hand was only slight.

By 11 PM the Japanese stations were strong. The 200-cycle filter made it easy to carve out a spot at the bottom of the band, at 7001 MHz, and proceed to run them for several hours. During one hour 85 stations were contacted. The signal from the rhombic was so loud that at one point the pileup was overwhelming. Then a light rain started and rain static became noticeable. Various antennas were tried and it was discovered that, by listening on a lower one in the wrong direction, it was quieter and stations could be copied. The upper antennas were acting as an electrostatic shield. However, you had to remember to switch back and transmit on the right antenna.

In the early morning hours one of the rhombic selection relays on the wall developed a bad connection, which caused the amplifier to draw excessive plate current. Again the high-voltage line fuse wire opened up in a dramatic fashion. After it was replaced I noticed that one 450TH final amplifier tube no longer lit up. They had been unusually stressed the previous evening. Once another tube was located and it was changed, the amplifier had to be neutralized. The author perused a stack of material in Don's bookcase, where he discovered a 1935 ARRL Handbook. "Turn off the high-voltage, drive the grid with a low level excitation, and use a rubber-coated stick to adjust the neutralization capacitor. . ." Luckily, there was one right on top of the amplifier and it looked pretty well used. Don must have done this once or twice.

Once the adjustment was complete, the RF drive was slowly turned up and everything seemed OK. Once the relay was cleaned and the station was back on the air at 4 AM, an Irish station was contacted. It being noon in Ireland didn't seem to affect the signal strength, S-9 both ways. Listening to the south, a zone 21 station in Saudi Arabia was found. The rest of the West Coast didn't discover him for some time. A quiet location, and being able to change direction faster than anyone else, gives an edge on the competition.

It was late in the morning, and signals were still loud. A station in the Philippines was still S-7 four hours after sunrise. Don pulled in the driveway a few minutes later. *"Hello, Jan. I thought I would drop by and see how you are doing,"* he said. *"Any problems?"* After being briefed that the station was operating fine and that 40 meters was still open, Don replied, *"You will hear Asia till around 11 AM and Europe is readable at noon. They are pretty tough to raise before 1 PM, however."* So much for the strategy of sleeping in the off–hours.

The author managed an hour's sleep around noon and then slid back into the operating position. Sure enough, the Europeans were in there, not too strong, but workable. Don made a pot of coffee, which lasted through the day and most of the night. The author got his second wind and developed a rhythm of changing rhombics and working stations. Everything

Jan Perkins, N6AW, operating W6AM.

seemed to fall into place, as the same piece of fuse wire survived the next 24 hours. After sunrise on Sunday, eyedrops were needed to read the DX check sheet. When the contest ended at 2400 GMT Sunday, two hours of sleep were necessary before driving was possible. The author left Don's rhombic antenna farm Sunday evening with a different opinion of 40 meters. When the CQWW CW contest results were published, W6AM had taken first place in the U.S.A.

The week after the contest a particularly strong Pacific storm hit Southern California. Dozens of amateurs experienced tower damage. The rhombic farm, which had always been immune to such storms, had one of its 140–foot poles down. Don surveyed the damage a few days later and noticed that 100 feet of his 20–foot–tall oleander shrubs along the eastern border of the property had been almost flattened. Although rare in this area, Don decided that a small twister must have formed.

The damaged corner pole was for one of the higher rhombics, number 4 on central Asia. Its antenna wire was across the antennas below it. The noise from all that wire rubbing together was terrific when the wind blew.

Don purchased two replacement poles, and one 70 feet long and another 80 feet long. He spliced the two of them together with four pieces of the 20–foot–long angle iron that had been salvaged from the damaged pole.

Chip Margelli, K7JA, recalled: "After a twister damaged one of Don's poles and several of his rhombics in 1983 I told him that I would be glad to assist him in gettin them reinstalled. Over a number of weekends Don, my wife Janet, WA7WMB, and I worked feverishly to pull down the damaged wires (which still had their original spring to them, making them a bit like a slinky in the air), make repairs, and hoist the antennas back into position. One long day consisted of four climbs up these poles and it concluded with the final tug–of–war with the antenna wire which (if all went well) would result in the rhombic being in its original position, as good as new.

Don's rhombics were not built casually. A block–and–tackle would not begin to hoist them up. Nor would an ordinary hand winch. A heavy–duty chain hoist was mandatory, as their wires had approximately 2,200 pounds of tension on them.

"So, there I was, 140 feet in the air, holding on for dear life with one hand to this creaking 10–inch diameter pole, straining with everything I had with the other arm to pull on the chain hoist so that the repaired wire would be within the one–inch tolerance Don wanted (on a 1,200–foot peice of wire!). Every tick of the chain hoist caused the pole to creak and bend more, and caused my arm to cramp and lock into position.

"Don, seeing my precarious situation, shouted to me at the top of his lungs, '*Say, Chip, that's great work you're doing up there. You know, that pole hasn't been serviced since 1939!*' And as the next creak went out of the pole I imagined being flung off the top of Palos Verdes by something snapping off, because I knew that the whole thing was originally built by the lowest bidder." Actually, Don was pulling Chip's leg a bit. Although the two 70–foot poles that formed the 140–foot pole that Chip was on were 45 years old, they had been spliced and installed a mere 20 years earlier.

Dave Bell, W6AQ, visited Jordan in November 1983. He said, "I was assigned the call sign JY8AQ, and was operating CW on 20 meter long path one evening. There was quite a pileup and I was working stations at a pretty good clip. Then W6AM called and made a comment that I didn't catch. He was sending a little too fast for me to copy, so I said that there was some interference and please repeat. Don reduced his speed down to under 20 WPM and said, 'Dave, you're doing a good job.' That was very considerate, as he knew that rather than having interference, I probably was not a high-speed CW operator."

Dave was in Jordan to film a sequence with King Hussein in the SAREX Shuttle Amateur Radio Experiment film "Amateur Radio's Newest Frontier." Five years earlier he had filmed portions of "The World of Amateur Radio." An amateur radio operator, Owen Garriotte, W5LFL, was aboard the current space shuttle flight and King Hussein, JY1, was filmed working him on 2 meters.

The past few years Don developed a rather unique (for this area) call sign system. When he was operating at the ranch he used W6AM. When he was mobile he used W6AM/M. When he operated from his home in Long Beach he used W6AM/A ("A" meaning alternate, a common usage in Great Britain). Although he considered Long Beach to be his home, he wanted people to know that if his signal wasn't as strong as they usually heard it, the /A was because he wasn't using a rhombic. It also made it easier to confirm a QSO in the logs when a QSL was received, as Don had a separate log at each location.

In a roundabout way the author discovered that Don enjoyed a special status with the local authorities. One morning a couple of days before a contest the author drove up to the ranch to check out propagation conditions. It was just before sunrise and he had a cup of coffee balanced on his lap. As he turned from Crest Ridge on to High Ridge Road, he didn't quite come to a full stop at the stop sign. The local police pulled him over just a few hundred feet from Don's driveway. The officer examining his license had the sort of look on his face that said he was going to uphold the law and issue a citation for this "California Stop." He asked the author what he was doing out at this early hour and he replied that he was an operator at the radio station across the road. At that point the officer's partner stepped forward and said, "Oh! That's the W Six A portable A fellow. I think just a warning for you will suffice." The first policeman seemed to be speechless and then proceeded to stammer. The author smiled and said, 'Thank you,' and, while they had a heated discussion, drove into the ranch."

Don had never received a traffic violation. In later years his vision wasn't particularly good, especially at night, and he would occasionally get pulled over for weaving in his lane. While chatting with the officer he would show them tickets he had purchased to the policemen's and firemen's ball, receipts from related donations, and kept talking about how lucky the city was to have such a fine police department. Eventually, the officer would just be overwhelmed and Don would get off with a warning.

Some areas of Palos Verdes and Rolling Hills were a bit clannish. During one contest the author went over to a supermarket about half a mile away from the ranch to get some groceries. While cashing a check there the cashier mentioned that they didn't usually cash out-of-area checks. He told her that he had lived in Long Beach for 28 years, just 12 miles away, and had just moved to Cerritos. The cashier replied that Long Beach was out of the area.

In 1983 Don was inducted into the Quarter Century Wireless Association Hall of Fame. For a number of years the QCWA had been doing stories about Don in their annual newsletter and had a write-up that summarized his radio history in the next issue, along with a picture of his being presented the plaque at the annual banquet. A short time later an article by one of the QCWA members, Dr. Hess, W6CK, appeared in the December 1983 issue of *73* magazine. The headline: "Don Wallace: Superham."

In July 1984 Don and a group of friends traveled to Papette, Tahiti, in French Polynesia. He had a French license and operated from there for 10 days as FO0DCW. During that time he made over 1,100 contacts on CW. That was the 100th country from which Don had operated, and the 110th he had visited.

The Olympics were held in Los Angeles in 1984. That summer California stations with a 6 in their call sign could substitute 23 (23rd Olympiad) or 84 (1984 Olympics) in their call sign for a two month period. W23AM and W84AM was on the air daily during this time. Don also operated the Olympic station, NG840, one afternoon, just as he had at W6USA in the Olympic Village 52 years earlier.

Don's cottage was packed with commercial radio equipment during the Olympics. Motorola had installed additional equipment for communications among security personnel, police, fire, medical, and all the other temporary services required for those two months. Round–the–clock personnel were at the cottage to deal with any equipment breakdowns.

Don was at the World's Fair (Louisiana World Exposition) in 1984. Phillip spencer, W5LDH, had met Don as a teenager years earlier, when he operated from British Honduras. He recalled: "Don met with about 25 local hams at my home on May 18. Don told our group many jokes and information about his past expleriences. He operated for a week from the World's Fair next door to the Hilton and contacted his numerous friends throughout the world. There is no doubt that Don was one of the great giants in ham radio and I am proud to say that I met him and enjoyed his tremendous personality."

One afternoon Don called CQ while mobile and was answered by Richard Powell, NF5Z, in Zavalla, Texas. "I told Don that he had an AC hum in his signal and suggested that one of his power supply filters had 'gone west.' (Interesting, out west we refer to a bad compoment as having gone south!) He acknowledged and asked me to QRX while he changed power supplies. Imagine a backup power supply in a kW mobile rig.

"After a few minutes I used the other VFO on my TS–930 to tune around. Sure enough, I found him up a couple of kilohertz, calling CQ NF5Z. He said that, in getting out of his car, he had accidently bumped the tuning knob on the receiver. I gave him a good report and told him that I had to QRT for a sked with my twin brother Bob, N6LNH. It seemed that Bob lived just a short distance from where Don had stopped to make the power supply switch. He called Bob on the landline to explain why I was late for my sked. I shall never forget this pleasant experience with one of the best and most widely respected hams in our radio fraternity."

When asked how he found time to operate so much, Don replied, *"I am quite blessed with two very fine men who are helping me and about 50 others who would like to and have asked me, but I don't let them come up here. Some want to work on aerials, but I have nothing but trouble, and I am a little bit scared. I once let a fel-*

low help me put up an aerial in Long Beach and I had to go up the pole and bring him down. He got stuck up there, froze. So, I am a little hesitant to use other people. The professionals don't get that way because they do it all the time and, if I can't do it myself, I try to hire somebody who is in the business. As far as working on the rig is concerned, I found a fellow who is really fabulous for me. The repairs of the station meet my requirements, which is chasing a new country, keeping the station on the air, and working all DXpeditions, all oddities, all new prefixes, and any exotic stations. Then some nights I get on and call one fellow, then work everybody who answers until the last one quits calling. I figure that's only fair if they want to work me. So Jan Perkins, N6AW, comes down here and he is a transmitter (and receiver) *specialist. He has improved all my transmitting facilities and does a beautiful job, and I appreciate that too. With Jan I am very fortunate.*

"*Jim Spears, AB6R, keeps all of the QSLs up to date. He goes through my log every few weeks and sends out QSLs for all my contacts. Jim also has been taking these piles of QSLs that have been stacked on the tables around here for years and sorts them into the card files. That leaves me time to get on the air and work more fellows.*"

One day in March 1985 the author and several friends were helping Don repair one of his rhombic antennas. Bob Kliman, KB6CIO, and others carried about 75 pounds of tools and equipment out into the field below the station and prepared to splice in a length of #8 copperweld to replace a damaged section of one leg. At each end of each leg of a rhombic antenna there is an insulator with lead inserts connected to a shackle. At the center of the antenna there was a three-foot piece of copperweld wire connecting the two legs together. When a wire eventually breaks, only one leg falls down and needs to be repaired. The shackles were connected to a messenger cable of half-inch galvanized steel cable, which is tied to an eye-bolt at the side pole. The end poles are connected to the Rhombic in the same way.

It was a beautiful day, and Don was showing his guests the method to make the repairs. He firmly grasped a pair of lineman's pliers and cut the copperweld, and then sanded the last several inches of it for a minute. Each end of the two pieces were sanded three separate times. With a nicopress tool Don pressed the fitting on the end of the wire, each press of the tool overlapping each previous clamp mark. (It took four or five). If one inadvertently crimped the wire at some point in the rhombic leg during the repairs, that piece was also cut out and spliced. A crimped wire will rot out (oxidize) and break at that point in a few years. When he was finished, Don sprayed the joint with a clear Krylon spray and said *"The joints usually last 25 years before they rot out. With the Krylon spray it should last at least 40 years."* It was apparent that Don planned to be around to repair the wire the next time it needed servicing.

Don repairing a rhombic in March 1985.

Six weeks from his eighty–seventh birthday, on May 25, 1985, the grand old man of amateur radio passed away. The stroke came while he was playing cards at the Virginia Country Club.

During his life his strength of character and enthusiasm touched us all and helped shape amateur radio. Don Wallace persevered. *"Radio for me 'till I'm 90 years old,"* he once said. From his early days as an experimenter and personal radio operator for President Wilson, through the traffic handling era, to his 30 years on top of the DXCC Honor Roll, he was the consummate radio man. Don Wallace — 86 years of enthusiasm.

End of an Era

Don's last radio contact had been made just a few hours before he passed away, ending radio activity that spanned 75 years, from spark to satellites. Don had been amateur radio's most ardent promoter. During the past three–quarters of a century he had made half a million QSOs and received 100,000 QSLs.

In November 1982 Don had mentioned that he had to do some repairs to one of the rhombics, so one Saturday morning the author went up to the ranch to lend a hand. An insulator had broken on the norhtern end of rhombic #4, pointed at central Asia. Loaded with about 50 pounds of tools, the author climbed up the top of the pole, which was 100 feet high. It took about two hours to repair the problem.

The next week Don left a note in his son's locker at the Virginia Country Club. It read:

> *Don Jr fm dad. The best bet to keep the station together, and/or sell off the equipment would be Jan Perkins, N6AW. He breaks all-time records here when he runs contests, works here a couple of Tuesday evenings per month, is in a transmitter (and receiver) research group at Hughes Aircraft, climbed a 100-foot pole last Saturday, and tries to learn all he can from me, and I learn a lot from him. I think he could dispose of my equipment as well as anybody and he might wish to keep it temporarily as his own station and live here while you negotiate a sale.*

The author was in Lake Tahoe for the weekend when Don passed away and was contacted by Bill Wallace, returning just in time to attend Don's memorial service, which was at the rhombic farm on Wednesday afternoon at 4 PM. The funeral, two hours earlier, had been for family members. About 300 people gathered at the ranch that afternoon, a number of them having flown in to attend the memorial. One at a time, about a dozen amateurs mounted the podium and each related a story about Don Wallace. The author fired up W6AM for its final time on the air, that is, until the FCC reassigns it. Many of the amateurs present took a turn at the operating position and a couple of hundred stations were worked. Chip Margelli, K7JA, worked several DX stations in Africa.

Don's obituary had been carried by local newspapers earlier in the week and it contained a summary of his radio history. On Monday, May 27, Daryl Kelley, the *Times* staff writer, wrote: "Radio Pioneer Don C. Wallace, 86, Dies." and Jack O. Baldwin of the *Long Beach Press Telegram* had: "Country's Oldest Ham, Don Wallace, 86, Dies." The Sunday, June 2 edition of the *Los Angeles Times* also published a picture of Bill Wallace standing at Don's station with the caption: "Death Silences Ham Radio Legend." The article reported:

> The legacy of W6AM will be carried on by Jan Perkins, a Cerritos aerospace engineer Wallace chose to take over his station before his death. 'He will answer the letters that come in from all over the world and will keep talking to people,' said William Wallace. The 25–acre ranch will continue to be a gathering place for hams who want to swap stories, talk about equipment and compete to see who can contact the largest number of hams around the world in a set amount of time.

Don's passing was also noted in the ARRL *Bulletin*, *The DX Bulletin*, the *Westlink Report, World Radio, CQ, QST*, QCWA and SOWP bulletins, the Japanese *S9* magazine, along with many others. The annual ARRL SW Division convention was in August that year and the convention program contained W6AM's radio history and the following salute to Don Wallace:

> It is to this Grand Old Man of Amateur Radio that the 1985 ARRL Southwestern Division Convention is affectionately dedicated.

The author found that operating from the ranch on a daily basis gave him a deeper understanding and appreciation of this almost unparalleled antenna system. One local amateur said that, when he heard the author was working his first long path station each morning, he would go eat breakfast because the rhombic station could work them 20 minutes before he could hear them. One morning, just before sunrise, the author tuned up on 160 meters and called CQ. When the dust had settled, 50 JAs had been worked. When 20 meters sounded dead in the evening, calling CQ usually resulted in a pileup of Europeans calling. When one operates from a dominant station, many of the stations report having just made their first DX QSO.

Four Beverage antennas were strung from a spool of small copperweld, #12, kept in the tool shed. Three were 600 feet long and one was over 1,000 feet. They had to be high enough for the neighboring farmer, Grimes, to pass underneath while riding his tractor, clearance of about 15 feet. Grimes had been farming Don's 25 acres the past 13 years. Even so, at that height the Beverages reduced background noise dramatically on 160 or 80 meters.

The rhombic farm was operated as a multi-operator station with the call N6AW in the CQWW CW contest that fall. Tom Alford, W6KP, tuned up on 40 meters with rhombic #1, pointed at Japan, and was innundated with JAS calling him. He worked them at a furious pace and, after each QSO, the pileup grew larger. Tom made over 1,000 JA QSOs that weekend and probably every one of them sent a QSL card. Following W6AM's example of recent years, all QSLs were answered.

Beginning in 1986 the SCDXC sponsored one of the CQ contest awards as a memorial to W6AM. In 1987 the author began to sponsor an annual trophy in the CQWW contest in Don's memory. The trophy was awarded for the high score in the United States on 7 MHz CW. The winner of it that first year was Jim McCook, W6YA, who ran a close race with Jim Stevenson, KM6B.

After one year of operating the station and competing one last time in the CQ World Wide and ARRL DX contests, while also having a final open house at the ranch, the serious dismantling effort began. Then the realization of the magnitude of the task ahead became apparent. Escrow closed on the property on May 5, 1986. With the help of dozens of amateurs the rhombic antenna farm was dismantled during the last half of that year.

At that time there were ten 140-foot poles, five 100-foot poles, nine 80-foot poles, and 17 feeder poles standing. Seventeen miles of #8 copperweld wire remained in the air (about six tons!). Another four miles of half-inch galvainized guy cable held up the poles (another six tons!).

The 80-foot poles were guyed about five feeet from the top, in three directions with half-inch galvanized cable. They were on the ground (not in a hole), with a metal spike protruding two inches up from the ground underneath. Irv Emig, W6GC, said that Don once told him, *"The more pole that you put in the ground, the less of it that sticks out!"*

The 95-foot poles were guyed at 65 feet and 90 feet, in three directions with half-inch galvanized cable. They also were on top of the ground and were about three feet across at their base. They had been installed 40 years earlier with a block-and-tackle that was about a foot across with three-fourths-inch line (rope). Presumably, a smaller pole was used as a gin pole. Over the years they settled into the ground about two or three inches. One pole had a huge crack in the side of it, near the base, and Don Barnett, N6EKX, was able to stick his arm in all the way to his shoulder.

The 140-foot poles were guyed at 70 feet (in the middle of the splice), and 135 feet, in three directions with half-inch galvanized cable. The base of the pole was set in the ground three feet.

During the dismantling of the ranch the splices were removed from most of the 140-foot poles after they were taken down. A crane from H & S Martins Electrical in Bakersfield proved invaluable. The splices were stacked near the poles and each one had about a 10-degree bend in the middle from being bolted to the poles. About a week later it was noticed that most of them had straightened out. After 20 years, half-inch-thick pieces of steel that weighed 350 pounds still had a memory.

Some interesting aspects of #8 copperweld wire were discovered. In the past, watching Don cut it looked easy. To the uninitiated, it takes a full minute, and a certain amount of jumping up and down, to cut a piece of #8 copperweld with lineman's pliers. After a daily sesssion of doing this for a week, one can cut 30 or 40 pieces of it a day before his hand stops functioning. In another month cutting copperweld wire seems like using a pair of scissors on paper.

As work continued on the ranch, word got around and hams from all over the southland came out to lend

The chief cook and W6AM team mascot.

a hand. One fellow who had known Don for 50 years called from San Francisco and wanted to take a week off work to join us. Each evening after work the team toiled until sunset. On weekends omelettes were fixed at sunrise. We rolled wire and pulled poles all day long, then barbecued steaks were served at sunset for the survivors. The author had never fancied himself a tower person, but before it was over he had been up every pole.

A camera was always carried along during those last months and the final days of the ranch were recorded. A total of 25 rolls of film was taken, along with part of one videocassette. Several others who were helping out also took numerous rolls of film.

Ron Brunner was the safety engineer retained by the Wallace family and, as he observed our various activities, he had veto power. The first day we were out in the field he checked the author's climbing belt (Don's, of many years), and found evidence of impending weakness. It was replaced with another one, 25 years newer, which met his approval. Several years earlier Ron Camp, K6EVR, was on a pole at his house in Temple City when his safety belt broke. He recovered from his injuries and was considered lucky.

There were some interesting developments that occurred during this period. while the team was loading some guy cable on a truck, Chris McCarroll stopped by. His father had farmed the land for many years and still had an open storage area bordering High Ridge Road. He referred to it as his steel pond. McCarroll had 40 or 50 tons of assorted pieces of steel there, plus a couple of old trucks and tractors. Those items that McCarroll wanted would have to be moved in the coming weeks. Dale Rice, K6SQA, asked McCarroll if he needed a hand and did he want to part with his Mack truck that was parked among the scrap iron. McCarroll was using his tractor and didn't need any help, but said that, for a case of cold beer, Rice could have the truck. So Rice rustled up a case of cold beer for McCarroll. Rice came back the next week with another truck to tow the first one home. He spent an afternoon checking oil levels and cleaning electrical connections on the Mack truck. The engine fired right up and he drove it home. Sometimes a case of beer will take you quite a ways.

The phone couldn't be heard when the team was out in the fields some distance from the house, so an answering machine was set to activate after four rings had been installed by the bed. Within 100 feet of the house the phone could be heard, as it was connected to the fire alarm bell that was extremely loud. Inside the building the sound almost knocked one out of one's chair. The author was normally a light sleeper but, several times, after putting in a 12-hour day outside, he woke up the next morning to find a message on the machine. Incredibly, he had slept through four rings of that bell!

Bob Kliman, KB6CIO, was a volunteer fireman in Silverado Canyon. When the dismantling effort started Kliman loaned the author a pair of his leather gloves. By the end of the summer, before they could be returned, holes had been worn in both of the palms. Kliman was astounded, as they were the high-quality type that usually last 20 years. They did see a lot of mileage at the ranch that summer.

Another local institution came to an end that year. Marineland, located three miles southwest of the ranch on a bluff overlooking the Pacific Ocean, had been a fixture on the Palos Verdes Peninsula since the 1950s. In February 1986 it was sold and its doors closed within two weeks. The marine life still remaining was transferred to Sea World, located in San Diego.

The author was unable to locate anyone who could use Don's 4,000 feet of poles at their station. They were donated to the Palos Verdes Horsemen's Society for reinforcing equestrian trails in the area. Cleyton, AD6C, and Dave, W6QHS, were given several miles of copperweld wire for installing vee beams and rhombics at their stations. Most of the #8 copperweld wire at the ranch was 45 years old and still usable. Jim, KM6B, used nearly a mile of it in his 160-meter Beverage antennas.

The big push was to get the outside work finished before the first rain fell that autumn. After a rainfall the fields would turn into a sea of mud for several days and whatever outside work remained would get very messy and would be delayed. Once the outside work was out of the way, the inside projects could continue with leisure.

At this time, while emptying the storage shed outside the main building, a partial case of dynamite was discovered, along with a quantity of blasting caps. The case was marked 1963 and it had probably been used when Don was relocating the poles from the 120 acres to the remaining 25 acres. His nearest neighbors were still two miles away then. Being in the middle of 25 acres, the blasting caps were easy to dispose of. To a passerby, they just sounded like a firecracker. As for the dynamite, the team wasn't sure what to do with it. The author had handled a few Claymore mines in Viet Nam, but unstable dynamite was another matter. Then one person had an idea. He brought his truck up to the ranch with a tub of water in the back of it. The sticks were placed in the water, one at a time, and taken down to the author's boat. A trip to Catalina Island had been planned in a few weeks. That weekend, while halfway across the channel, the sticks were dropped over the side, again one at a time, in 2,500 feet of water.

It was later discovered that Don had some old dynamite destroyed three years earlier. Apparently, the

Moored at Catalina Island.

proper way to dispose of is is by burning. Anyway, it's now in a safe place.

That fall a letter was received from Ian Cummings, KB1SG (ex-WB6ABP), in Dayville, Connecticut. Cummings wrote:

> I lived in Palos Verdes Estates and recall my frequent trips by bicycle up the hill to operate W6AM. I now find myself in Massachusetts and retain a great deal of nostalgia towards Don. I was saddened to hear of the dismantling of the old *Press Wireless* rhombics. I am currently looking to relocate to a part of New England where acreage is affordable, with the ultimate intent of erecting a rhombic farm not unlike that of W6AM (although of perhaps a bit smaller scale). I am writing to you in hopes that you may have access to construction and design details specific to the W6AM antenna farm.

Two weeks were spent taking pictures of a number of aspects of Don's installation that were unique to his switched antenna system, along with an eight-page letter describing Don's rhombic system, which were then sent to Cummings. Judging by the progress reports that have been received, he is well on his way to having a good antenna system installed.

In the fall of 1986 the *Public Broadcasting Service* showed a 30-minute presentation about Don and his rhombic farm on local TV networks. it had been made by Wayne Overbeck, N6NB, the year before, shortly before Don became a silent key.

The visits to the ranch continued to wrap up loose ends until the spring of 1987. A trap vertical antenna was installed outside the building and was used occasionally during this time. The electrical service was pulled by the Southern California Edison Company in early March. The weekend before, during the ARRL DX contest, Jerry Hagen, N6AV, made the last QSO from the former rhombic farm with that vertical antenna. The December 1988 issue of *QST* had a brief article about dismantling the ranch entitled "The W6AM Rhombic Antenna Farm Dismantled."

The Wallace family donated one acre of the property to the city. It was a strip bordering the eastern edge along High Ridge Road, which was developed into a jogging trail. A total of 83 homes were constructed on the remaining 24 acres at the site. In June 1987 the "Hundred Foot Long Ham Shack" was dismantled.

Once the ranch projects were completed the author had a lot of free time on his hands and decided to write a book about Don. Raymond Simmons, W3COY, in Allentown, Pennsylvania noted: "I saw you ad concerning Don Wallace, W6AM. I came across some articles in 2 old *QST* magazines and hope they may be of use to you. Ray's books were very helpful. One of the author's assistants at the ranch inadvertently removed the 1925 to 1936 period of Don's *QST* and they were never recovered. Otherwise, Don had a complete set of *QST* in binders going back to their first issue, December 1915.

A short time later a note was received from Vern weiss, WA9VLK.

> For years a dear friend of mine, Russ Davenport, W9HPR, and I admired the work and the station of Don Wallace, W6AM. Mr. Wallace always served as a "shining light" as Russ and I often referred to his manyfold achievements. Yesterday W9HPR suffered a stroke. I want to put together a get well gift that will lift his spirits during his recovery period. What a thrill Russ would get if I could mount one of Don's insulators on a handsome plaque with an engraved plate saying "Insulator from 80-meter Rhombic, W6AM, Rolling Hills, California." and another engraved with "To W9HPR, Ham of the Year." If at all possible, could you send me two insulators with some type of documentation signed by yourself indicating that the insulators were removed from actual W6AM service and, if possible, specifics as to what band/firing direction of the Rhombic from which the insulators were taken.

Once in a while s heart-warming note such as Vern's would be received, which made the project of dismantling the rhombic farm and disseminating the hardware worth the effort many times over.

In 1988 the Central California DX Club sponsored a plaque in the AARL DX contest for the world high phone score on 14 MHz, as a memorial to W6AM. The winner was Econ. Alfredo Caviedes Vega, HC1HC. Vega wrote:

> In 1985 W6AM sponsored a DX contest plaque, which HC1HC won. I hope to achieve a plaque on each of the other bands to display with my first two from W6AM.

Don's equipment was in demand with those who knew him. John Weiner III, WB8IPG, purchased a Collins 75S-3A receiver and 32S-3 transmitter, along with a W6AM Shortwave Manual, which he still uses in his station.

Vern Weiss wrote:

> We all admired W6AM and his enthusiasm and achievements. His logbooks must have been voluminous. Had he not been taken from us I am certain two classes of ham operators would exist today: those who have worked W6AM and those who will work W6AM.

In 1986 the SCDXC began to sponsor an annual W6AM memorial award, which was to be given to the most active Dxer in the club. The first winner was announced at the 1987 Visalia DX convention: Ray Harper, N6DKP. His picture appeared in the October 1987 issue of *CQ*.

In 1988 the AARL was soliciting donations for the restoration and expansion of the Maxim Memorial station, W1AW. The author made a $1,000 donation to the fund in Don's name and inscribed on the permanent plaque at W1AW is the inscription "In memory of Don C. Wallace, W6AM."

Don's Long Beach home was sold the year after he passed away and escrow closed on August 12, 1986. The last of the equipment had long since been removed. The team had marveled at the four-foot squirrel cage fan in the basement that for 50 years had pulled the heat out of the radio room. Don Jr., Bill, Alex, Jonie, and Anne Wallace, along with the author met there that evening for a last tribute to what was known for over a half-century as "The shrine of amateur radio in Long Beach."

When the ranch was being dismantled the author receiver a letter from Don Wallace Jr.

> As Trustee of the Don C. Wallace family trust, I hereby appoint you as trustee of the Don C. Wallace amateur radio call letters of W6AM. On behalf of the Don C. Wallace children, Betty Jean Wallace Green and Wm. H. Wallace, we also urge that you take all of the necessary steps to attempt to procure the approval by the FCC to permit continuation of the use of W6AM in connection with such activities as would be appropriate in honoring the lifetime of service with distinction to amateur radio by my father.
>
> Prior to his death, Don. C. Wallace appointed you as the person who should actively take charge of perpetuating the historical features of the Wallace Radio Ranch station and of his station at 4214 Counry Club Drive, Long Beach. Since his death in May of 1985, you have accumulated, with the consent of our family, many amateur items which represent the chronological advancement enjoyed by amateur radio, particularly under the leadership of people like dad.
>
> As Trustee of the W6AM Call, we hereby authorize you to do any and all things which you believe are necessary to assist in obtaining the perpetuation of this call in commemoration of the history of amateur radio.

Bill Wallace and the author.

In mid-October, the author received a letter from the vice-president of the SCDXC, Steve Locks, W6FRZ. It read:

> On October 9, 1986 the Southern California DX Club met and at this monthly meeting a motion was made, seconded and voted unanimously to ask you as the trustee of the Don Wallace, W6AM call to permit the SCDX Club to use W6AM as our club call sign.

Although a number of other clubs were also interested in Don's call as a memorial station, similar to W1AW, the SCDXC was the organization with which he was most closely identified.

Jay Halliday, W6EJJ, first vice-president of the ARRL wrote Bill Wallace and noted:

> I hope that we can work together to develop a suitable memorial to your father. I also feel that the League can be of help in insuring that the W6AM call

sign is reassigned in accordance with your family's wishes.

A short time later I received a letter from Bill Wallace.

> As custodian of the Don Wallace memorabilia and as his biographer, the Wallace family hereby authorizes you to petition the FCC for perpetuating the W6AM call sign by extending its use to the Southern California DX Club.

Accordingly, on behalf of the Wallace family and the SCDXC, and with the support of the ARRL, the author requested the FCC to issue W6AM to the SCDXC as the Don Wallace memorial call sign with the author as trustee.

The FCC has yet to act on my petition. The W6RO memorial station at the Queen Mary was one of the last club station licenses issued. In the meantime, the author's request remains on file with the chief of the Personal Radio Bureau of the FCC, in the event that they decide to once again issue club call signs.

While sorting Don's famous back room and the basement of his Long Beach house, the author discovered his Hoover Cup receiver, assorted spark equipment, dozens of telegraph keys, and numerous homebrew rigs. He saved everything. Many of these items are being prepared for a museum display. A rhombic antenna will be constructed on some acreage northeast of Los Angeles and Don's restored equipment will be installed there, all operational. The property is far enough away from other residents to permit Don's early "Rock Crushers" to be lit off without disturbing anyone.

One afternoon in the autumn of 1986 several hams had come up to the ranch and helped take down the rhombic breakout frame. While coiling up the wire and stacking the redwood a tremendous racket started up. It was late September. There were thousands of birds flying by and the noise from all of them chirping was deafening. We puzzled over this for a few minutes and Don Barnett said: "The antenna wire's gone! They have been roosting on it, when they migrated through here for the past 48 years, and now it's gone." And with that it was final. We knew an era had ended.

Appendix 1

The Morse Code and the Continental Code

Continental (International Morse)		Morse (Land Telegraph)	
A	dit dah	dit dah	A
B	dah dit dit dit	dah dit dit dit	B
C	dah dit dah dit	dit dit dit	C
D	dah dit dit	dat dit dit	D
E	dit	dit	E
F	dit dit dah dit	dit dah dit	F
G	dah dah dit	dah dah dit	G
H	dit dit dit dit	dit dit dit dit	H
I	dit dit	dit dit	I
J	dit dah dah dah	dah dit dah dit	J
K	dah dit dah	dah dit dah	K
L	dit dah dit dit	daaah (long)	L
M	dah dah	dah dah	M
N	dah dit	dah dit	N
O	dah dah dah	dit dit	O
P	dit dah dah dit	dit dit dit dit dit	P
Q	dah dah dit dah	dit dit dah dit	Q
R	dit dah dit	dit dit dit	R
S	dit dit dit	dit dit dit	S
T	dah	dah	T
U	dit dit dah	dit dit dah	U
V	dit dit dit dah	dit dit dit dah	V
W	dit dah dah	dit dah dah	W
X	dah dit dit dah	dit dah dit dit	X
Y	dah dit dah dah	dit dit dit dit	Y
Z	dah dah dit dit	dit dit dit dit	Z
1	dit dah dah dah dah	dit dah dah dit	1
2	dit dit dah dah dah	dit dit dah dit dit	2
3	dit dit dit dah dah	dit dit dit dah dit	3
4	dit dit dit dit dah	dit dit dit dit dah	4
5	dit dit dit dit dit	dah dah dah	5
6	dah dit dit dit dit	dit dit dit dit dit dit	6
7	dah dah dit dit dit	dah dah dit dit	7
8	dah dah dah dit dit	dah dit dit dit dit	8
9	dah dah dah dah dit	dah dit dit dah	9
10	dah dah dah dah dah	daaaaaah (very long)	10

The Continental (International Morse) code had eleven letters and nine numbers that were different than the Morse code. Also, six of the Morse code characters used a peculiar spacing between dits, and two had a dah that were of different lengths.

International Morse code characters in 1925.

DEPARTMENT OF COMMERCE
BUREAU OF NAVIGATION
RADIO SERVICE

INTERNATIONAL MORSE CODE AND CONVENTIONAL SIGNALS

TO BE USED FOR ALL GENERAL PUBLIC SERVICE RADIO COMMUNICATION

1. A dash is equal to three dots.
2. The space between parts of the same letter is equal to one dot.
3. The space between two letters is equal to three dots.
4. The space between two words is equal to five dots.

A	. —
B	— . . .
C	— . — .
D	— . .
E	.
F	. . — .
G	— — .
H	. . . .
I	. .
J	. — — —
K	— . —
L	. — . .
M	— —
N	— .
O	— — —
P	. — — .
Q	— — . —
R	. — .
S	. . .
T	—
U	. . —
V	. . . —
W	. — —
X	— . . —
Y	— . — —
Z	— — . .
Ä (German)	. — . —
Á or Å (Spanish-Scandinavian)	. — — . —
CH (German-Spanish)	— — — —
É (French)	. . — . .
Ñ (Spanish)	— — . — —
Ö (German)	— — — .
Ü (German)	. . — —
1	. — — — —
2	. . — — —
3	. . . — —
4	. . . . —
5	. . . . .
6	— . . . .
7	— — . . .
8	— — — . .
9	— — — — .
0	— — — — —

Period	. . . . . .
Semicolon	— . — . — .
Comma	. — . — . —
Colon	— — — . . .
Interrogation	. . — — . .
Exclamation point	— — . . — —
Apostrophe	. — — — — .
Hyphen	— . . . . —
Bar indicating fraction	— . . — .
Parenthesis	— . — — . —
Inverted commas	. — . . — .
Underline	. . — — . —
Double dash	— . . . —
Distress Call	. . . — — — . . .
Attention call to precede every transmission	— . — . —
General inquiry call	— . — . — — . —
From (de)	— . . .
Invitation to transmit (go ahead)	— . —
Warning—high power	— — . . — —
Question (please repeat after.....)—interrupting long messages	. . — — . .
Wait	. — . . .
Break (Bk.) (double dash)	— . . . —
Understand	. . . — .
Error	. . . .
Received (O. K.)	. — .
Position report (to precede all position messages)	— . — .
End of each message (cross)	. — . — .
Transmission finished (end of work) (conclusion of correspondence)	. . . — . —

Many of the punctuation characters have changed over the decades.

Appendix 2

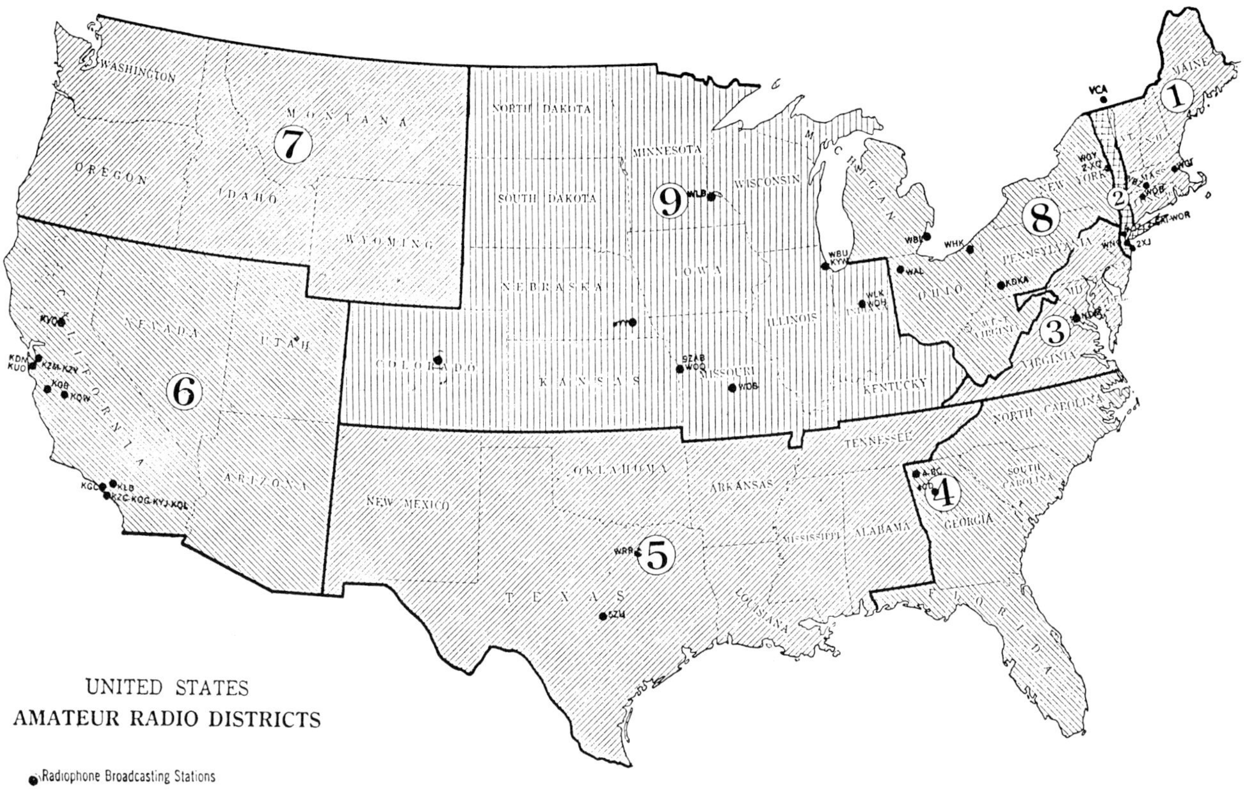

Call district partition - 1912

Call district partition - 1945

Appendix 3

International Prefixes (*Radio*, February 1927)

NEW INTERNATIONAL INTERMEDIATES, EFFECTIVE 0000 G.M.T., FEBRUARY 1, 1927

EUROPE

EA—Austria
EB—Belgium
EC—Czechoslovakia
ED—Denmark and Faroe Ids.
EE—Spain and Andorra
EF—France and Monaco
EG—Great Britain and Northern Ireland
EH—Switzerland
EL—Italy
EJ—Jugo-Slavia
EK—Germany
EL—Norway, Spitzbergen and Franz Josef Land
EM—Sweden
EN—The Netherlands
EO—Irish Free State
EP—Portugal, Madeira Is., and the Azores
EQ—Bulgaria
ER—Rumania
ES—Suomi (Finland)
ET—Poland, Esthonia, Latvia, Courland and Lithuania
EU—U. S. S. R. ("Russia"), including Ukraine
EV—Albania
EW—Hungary
EX—Luxemburg
EY—Greece
EZ—Zone of the Straits

ASIA

AA—Arabia
AB—Afghanistan
AC—China (including Treaty Ports) including Manchuria, Mongolia, and Tibet.
AD—Aden
AE—Siam
AF—French Indo-China
AG—Georgia, Armenia and Azerbaijan
AH—Hedjaz
AI—India (and Baluchistan) and Goa
AJ—Japan and Chosen (Korea)
AK—(Unassigned)
AL—(Unassigned)
AM—Federated Malay States (with Straits Settlements)
AN—Nepal
AO—Oman
AP—Palestine
AQ—Iraq (Mesopotamia)
AR—Syria
AS—Siberia, including "Central Asia"
AT—Turkey
AU—(Unassigned)
AV—(Unassigned)
AW—(Unassigned)
AX—(Unassigned)
AY—Cyprus
AZ—Persia

NORTH AMERICA

NA—Alaska
NB—Bermuda Id.
NC—Canada, Newfoundland and Labrador
ND—Dominican Republic
NE—(Unassigned)
NF—Bahama Ids.
NG—Guatemala
NH—Honduras
NI—Iceland
NJ—Jamaica
NK—(Unassigned)
NL—Lesser Antilles
NM—Mexico
NN—Nicaragua
NO—British Honduras
NP—Porto Rico and Virgin Ids.
NQ—Cuba and Isle of Pines
NR—Costa Rica
NS—Salvador
NT—Haiti
NU—United States of America
NV—(Unassigned)
NW—(Unassigned)
NX—Greenland
NY—Panama
NZ—Canal Zone

SOUTH AMERICA

SA—Argentina
SB—Brazil, Trinidad Id., and St. Paul Id.
SC—Chile
SD—Dutch Guiana
SE—Ecuador and Galapagos Archipelago
SF—French Guiana
SG—Paraguay
SH—British Guiana
SI—(Unassigned)
SJ—(Unassigned)
SK—Falkland Ids. and Falkland Dependencies
SL—Colombia
SM—(Unassigned)
SN—Ascension Id.
SO—Bolivia
SP—Peru
SQ—(Unassigned)
SR—(Unassigned)
SS—(Unassigned)
ST—(Unassigned)
SU—Uruguay
SV—Venezuela and Trinidad
SW—(Unassigned)
SX—(Unassigned)
SY—(Unassigned)
SZ—(Unassigned)

AFRICA

FA—Abyssinia
FB—Madagascar, Reunion Id., Comoro Id., etc.
FC—Belgian Congo, Ruanda, Urundi
FD—Angola and Kabinda
FE—Egypt
FF—French West Africa, including French Sudan, Mauritania, Senegal, French Guinea, Ivory Coast, Upper Volta, Dahomey, Civil Ter. of the Niger, French Togoland, etc.
FG—Gambia
FH—Italian Somaliland
FI—Italian Libya (Tripolitania and Cyrenaica)
FJ—Somaliland Protectorate and Socotra
FK—Kenya, Zanzibar Protectorate, Uganda, Anglo-Egyptian Sudan, and Tanganyika Territory.
FL—Liberia
FM—Tunisia, Algeria, Morocco (including the Spanish Zone), Tangier
FN—Nigeria
FO—Union of South Africa, Northern and Southern Rhodesia, Bechuanaland Protectorate, and Southwest Africa
FP—Portuguese Guinea and Cape Verde Ids.
FQ—French Equatorial Africa and Cameroons
FR—Rio de Oro and adjacent Spanish Zones, Ifni, and Canary Ids.
FS—Sierra Leone
FT—Eritrea
FU—Rio Muni (Spanish Guinea) and Fernando Po
FV—French Somaliland
FW—Gold Coast Colony, Ashanti, Northern Territories and British Togoland
FX—Seychelle Dependencies
FY—(Unassigned)
FZ—Mozambique

OCEANIA

OA—Australia (and Tasmania)
OD—Dutch East Indies*
OE—Melanesia*
OH—Hawaiian Ids.
OI—Micronesia*
OO—Polynesia*
OP—Philippine Ids.
OZ—New Zealand
*To be further partitioned when activity warrants.

SHIP STATIONS

Ship stations with amateur calls will place an X before their usual intermediate. E. g., Australian 3AA at sea, calling U. S. 1AW, would send "1AW NUXOA 3AA". The reply would be "3AA XOANU 1AW".

Appendix 4

ARRL Country list (*QST*, January 1937)

• I. A. R. U. NEWS •

Devoted to the interests and activities of the

INTERNATIONAL AMATEUR RADIO UNION

Headquarters Society: THE AMERICAN RADIO RELAY LEAGUE, West Hartford, Conn.

MEMBER SOCIETIES

American Radio Relay League
Associazione Radiotecnica Italiana
Canadian Section A.R.R.L.
Ceskoslovensti Amatéri Vysilaci
Deutscher Amateur Sende-und-Empfangs Dienst
Experimenterende Danske Radioamatorer
Irish Radio Transmitters Society
日本アマチュア無線聯盟
Liga Colombiana de Radio Aficionados
Liga Mexicana de Radio Experimentadores
Nederlandsche Vereeniging voor Internationaal Radioamateurisme
Nederlandsch-Indische Vereeniging Voor Internationaal Radioamateurisme
New Zealand Association of Radio Transmitters
Norsk Radio Relæ Liga
Oesterreichischer Versuchssenderverband
Polski Zwiasek Krotkofalowcow
Radio Club Venezolano
Radio Society of Great Britain
Rede dos Emissores Portugueses
Reseau Belge
Reseau des Emetteurs Français
South African Radio Relay League
Suomen Radioamatööriliitto r.y.
Sveriges Sandareamatorer
Unión de Radioemisores Españoles
Union Schweiz Kurzwellen Amateur
Wireless Institute of Australia

Countries:

This department is going to overrun its allotted space this month, anyway, so we haven't much time for introductory comment. We suggest you turn to page 40 of the October, 1935, issue of *QST* for the background (it was a piece called "How to Count Countries Worked," you may remember).

Yes, that's the subject. "How to count countries worked. . . ." In that discussion we set forth how ordinary counting methods led one into no end of trouble, and then set forth a more or less simple rule for arriving at a total figure for legitimate countries worked. That is, we thought it was simple. Some of the gang have since registered disagreement. Finally, we slid gracefully (?) out from under the question of actually making a *list* of all the countries of the world by citing the plain fact that one could easily reach a total of six or seven hundred—and there weren't enough pages available in *QST!*

There came a time, though, when Fate caught up to us—when we could no longer elude the problem—when we had not only to actually list the countries of the world for ham purposes but also to boil the list down to a reasonable figure (a reasonable figure being, say, a couple of hundred). The instrument in the hands of Fate which brought about this pretty pass was the new A.R.R.L. Map of the World. On that map, like any good map, countries had to be shown. Not only did they have to be shown on the map, with a different color to each country (no simple problem in itself, with sixty times as many countries as colors!), but there had to be an indexed reference list at the bottom.

The connection between this map list and an "official list of countries, etc." was obvious. It offered a perfect vehicle for the general popularization and appreciation of a suggested standard list—not a perfect list, for no such list could be perfect, since inevitably there will be different persons who will be dissatisfied with different aspects of it —but at least a generally-recognized standard to which everyone could refer and use as a start for further argument.

And so we present to you not an "official list of countries," since "official" presumes general sanction and that can come only with time, but a list that we feel is just about as complete as could be made, fair to the DX man from the standpoint of giving credit where credit is due, and certainly a very definite starting point. Only official prefixes are included, so don't worry about the absence of "K5," "OM," etc.

We would like to help towards the acceptance of a standard of "countries worked." Our contribution follows. Your comment is invited.

Country	*Prefix*
Abyssinia, see Ethiopia	
Aden	
Aegean Islands	
Afghanistan	YA
Alaska	K7
Albania	ZA
Aldabra Islands	
Algeria	FA
Andaman Islands	
Andorra	
Anglo-Egyptian Sudan	ST
Angola	CR6
Arabia, see Saudi Arabia	
Argentina	LU
Ascension Island	ZD8
Australia	VK
Austria	OE
Azores Islands	CT2
Bahama Islands	VP7
Bahrein Islands	VS8
Balearic Islands	EA6
Baluchistan	
Barbados	VP6
Bechuanaland	
Belgian Congo	ON
Belgium	ON
Bermuda Islands	VP9
Bhutan	
Bismarck Archipelago	
Bolivia	CP
Borneo, Netherlands	PK5
Brazil	PY
British Cameroons, see Nigeria	
British Honduras	VP1
British North Borneo	VS4
Brunei	
Bulgaria	LZ
Burma	
Cameroons, Franch	FE8
Canada	VE
Canal Zone	
Canary Islands	EA8
Cape Verde Islands	CR4
Caroline Islands	
Celebes and Molucca Is.	PK6
Ceylon	VS7
Chile	CE
China	XU
Chosen (Korea)	J8
Christmas Island	ZC3
Cocos Islands	ZC2
Colombia	HJ
Comoro Islands	
Cook Islands	ZK1
Corsica	
Costa Rica	TI
Crete	
Cuba	CM-CO
Curacao and Netherlands West Indies	PJ
Cyprus	ZC4
Czechoslovakia	OK
Danzig	YM
Denmark	OZ
Dominican Republic	HI
Eastern Island	
Ecuador	HC
Egypt	SU
Eritrea	
Estonia	ES

Country	Prefix
Ethiopia (Abyssinia)	ET
Faeroes, The	OY
Falkland Islands	VP8
Fanning Island	VR3
Federated Malay States	VS2
Fiji Islands	VR2
Finland	OH
Formosa, see Taiwan	
France	F
French Cameroons, see Cameroons	
French Equatorial Africa	FQ8
French Indochina	FI8
French Oceania	FO8
French West Africa	FF8
Galapagos Islands	
Gambia	ZD3
Germany	D
Gibraltar	ZB2
Gilbert & Ellice Islands and Ocean Island	VR1
Goa	CR8
Gold Coast (and British Togoland)	ZD4
Gough Island	
Great Britain	G
Greece	SV
Greenland	OX
Guadaloupe	FG8
Guam	K6
Guatemala	TG
Guiana, British	VP3
Guiana, Neth. (Surinam)	PZ
Guiana, French, and Inini	FY8
Guinea, Portuguese	CR5
Guinea, Spanish	
Haiti	HH
Hawaiian Islands	K6
Hejaz	HZ
Honduras	HR
Hong Kong	VS6
Hungary	HA
Iceland	TF
Ifni	
India	VU
Inini, see Guiana, French	
Iran (Persia)	EP
Iraq	YI
Ireland, Northern	GI
Irish Free State	EI
Italy	I
Jamaica and Cayman Islands	VP5
Japan	J
Java	PK
Kenya	VQ4
Kerguelen Islands	
Korea, see Chosen	
Laccadive Islands	
Latvia	YL
Leeward Islands	VP2
Liberia	EL
Libya	
Liechtenstein	
Lithuania	LY
Luxembourg	LX
Macau	CR9
Madagascar	FB8
Madeira Islands	CT3
Maldive Islands	VS9
Malta	ZB1
Manchukuo	
Marianas Islands	
Marshall Islands	
Martinique	FM8
Mauritius	VQ8
Mesopotamia, see Iraq	
Mexico	XE
Midway Island	K6
Miquelon & St. Pierre Islands	FP8
Monaco	
Mongolia	
Morocco, French	CN
Morocco, Spanish	EA9
Mozambique	CR7
Nepal	
Netherlands	PA
Netherlands West Indies, see Curacao	
New Caledonia	FK8
Newfoundland and Labrador	VO
New Guinea, Neth.	PK6
New Guinea, Territory of	VK9
New Hebrides, British	YJ
New Hebrides, French	FU8
New Zealand	ZL
Nicaragua	YN
Nicobar Islands	
Nigeria (British Cameroons)	ZD2
Niue	ZK2
Non-Federated Malay States	VS3
North Borneo, see British North Borneo	
Norway	LA
Nyasaland	ZD6
Ocean Island, see Gilbert & Ellice Islands	
Oman	
Palau (Pelew) Islands	
Palestine	ZC6
Panama	HP
Papua Territory	VK4
Paraguay	ZP
Persia, see Iran	
Peru	OA
Philippine Islands	KA
Phoenix Islands	
Pitcairn Island	VR6
Poland	SP
Portugal	CT
Portuguese India, see Goa	
Principe & Sao Thome Islands	
Puerto Rico and Virgin Islands	K4
Reunion Island	FR8
Rhodesia, Northern	VQ2
Rhodesia, Southern	ZE
Rio de Oro	
Roumania	YR
St. Helena	ZD7
Salvador	YS
Sardinia	
Samoa, U. S.	K6
Samoa, Western	ZM
Sandwich Islands	
Sarawak	VS5
Saudi Arabia	
Seychelles	VQ9
Siam	HS
Sierra Leone	ZD1
Socotra	
Solomon Islands	VR4
Somaliland, British	VQ6
Somaliland, French	FL8
Somaliland, Italian	
South Georgia	VP8
South Orkney Islands	VP8
South Shetland Islands	VP8
Southwest Africa, see Union of South Africa	
Soviet Union	U
European States	U1-7
Asiatic States	U8-9-0
Spain	EA
Spitzbergen, see Svalbard	
Straits Settlements	VS1
Sumatra	PK4
Surinam, see Guiana, Neth.	
Svalbard (Spitzbergen)	
Sweden	SM
Switzerland	HB
Syria	
Taiwan (Formosa)	J9
Tanganyika Territory	VQ3
Tangier Zone	
Tannu Tuva	
Tasmania	VK7
Tibet	
Timor, Portuguese	CR10
Togoland, British, see Gold Coast	
Togoland, French	FD8
Tokelau (Union) Islands	
Tonga (Friendly) Islands	VR5
Transjordan	ZC1
Trinidad and Tobago	VP4
Tristan de Cunha	ZU9
Tunisia	FT4
Turkey	TA
Uganda	VQ5
Union Islands, see Tokelau Islands	
Union of South Africa	ZS-ZT-ZU
United States	W (N)
Uruguay	CX
Venezuela	YV
Virgin Islands, see Puerto Rico	
Wake Island	K6
Windward Islands (Grenada)	VP2
Wrangel Island	
Yemen	
Yugoslavia	YT-YU
Zanzibar	VK1

There were a total of 231 countries on the first ARRL DXCC country list. Only official prefixes were listed. Unofficial prefixes are listed in parentheses.

Byron Goodman put out an updated DXCC list in the January, 1938 issue of QST. The previous list had been discussed with a number of prominent DXers and Herb Becker, W6QD. After a final check with several geographical authorities, the following countries were added:

Cayman Islands	VP5	Southwest Africa	ZS3
Chagos Islands	VQ8	Soviet Union	
Channel Islands	G	White Russian	U2
Cocos Island	TI	Ukranian SSr	U5
French India	FN	Transcaucasian SFSR	U6
Franz Josef Land		Uzbek SSr	U8
Isle of Man	G	Turkoman SSR	U8
Jan Mayen Land	OY	Turks and Caicos Islands	VP5
Jarvis Island	K6	Virgin Islands	K4
Kuwait		Wales	GW
Scotland	GM		

Twenty countries were added to the 1937 DXCC list and one was removed, the Bismarck Archipelago, bringing the total number of countries to 250.

The last prewar ARRL DXCC country list appeared in the January, 1939 issue of QST. The additions consisted of Pacific Islands

Baker, Howland, and Am. Phoenix Is.	KF6
Clipperton Island	
Johnston Island	KE6

Austria was removed from the list, as it had been annexed by Germany the previous March. The British and French New Hebrides were combined into one country, and the Sandwich Islands were removed. The total number of DXCC countries was still 250.

Radio country list (*Radio*, January 1937)

COUNTRIES OF THE WORLD AND PREFIXES

Country	Prefix
Abyssinia, see Ethiopia	
Aden	
Aegean Island	
Afghanistan	YA
Alaska	K7
Albania	ZA
Aldabra Islands	
Algeria	FA
Andaman Islands	
Andorra	
Anglo-Egyptian Sudan	ST
Angola	CR6
Arabia, see Saudi Arabia	
Argentina	LU
Ascension Island	ZD8
Australia	VK
Austria	OE
Azores Islands	CT2
Bahama Islands	VP7
Bahrein Islands	VS8
Balearic Islands	EA6
Baluchistan	
Barbados	VP6
Bechuanaland	
Belgian Congo	ON
Belgium	ON
Bermuda Islands	VP9
Bhutan	
Bismarck Archipelago	
Bolivia	CP
Borneo, Netherlands	PK5
Brazil	PY
British Camerons, see Nigeria	
British Honduras	VP1
British North Borneo	VS4
Brunei	
Bulgaria	LZ
Burma	VU
Cameroons, French	FE8
Canada	VE
Canal Zone	NY, K5
Canary Islands	EA8
Cape Verde Islands	CR4
Caroline Islands	
Celebes and Molucca Islands	PK6
Ceylon	VS7
Chile	CE
China	XU
Chosen (Korea)	J8
Christmas Island	ZC3
Cocos Islands	ZC2
Colombia	HJ
Comoro Islands	
Cook Islands	ZKI
Corsica	
Costa Rica	TI
Crete	
Cuba	CM, CO
Curacao and Netherlands West Indies	PJ
Cyprus	ZC4
Czechoslovakia	OK
Danzig	YM
Denmark	OZ
Dominican Republic	HI
Easter Island	
Ecuador	HC
Egypt	SU
Eritrea	
Estonia	ES
Ethiopia (Abyssinia)	ET
Faeroes, The	OY
Falkland Islands	VP8
Fanning Island	VR3
Federated Malay States	VS2
Fiji Islands	VR2
Finland	OH
Formosa, see Taiwan	
France	F
French Cameroons, see Cameroons	
French Equatorial Africa	FQ8
French Indochina	FI8
French Oceania	FO8
French West Africa	FF8
Galapagos Islands	
Gambia	ZD3
Germany	D
Gibraltar	ZB2
Gilbert & Ellice Islands, and Ocean Island	VR1
Goa	CR8
Gold Coast (and British Togoland)	ZD4
Gough Island	
Great Britain	G
Greece	SV
Greenland	OX
Guadeloupe	FG8
Guam	OM, K6
Guatemala	TG
Guiana, British	VP3
Guiana, Neth. (Surinam)	PZ
Guiana, French, and Inini	FY8
Guinea, Portuguese	CR5
Guinea, Spanish	
Haiti	HH
Hawaiian Islands	K6
Hejaz	HZ
Honduras	HR
Hong Kong	VS6
Hungary	HA
Iceland	TF
Ifni	
India	VU
Iran (Persia)	EP
Iraq	YI
Ireland, Northern	GI
Irish Free State	EI
Italy	I
Jamaica and Cayman Islands	VP5
Japan	J
Java	PK
Kenya	VQ4
Kerguelen Islands	
Korea, see Chosen	L8
Laccadive Islands	
Latvia	YL
Leeward Islands	VP3
Liberia	EL
Libya	
Liechtenstein	
Lithuania	LY
Luxembourg	LT
Macau	CR9
Madagascar	FB8
Madeira Islands	CT3
Maldive Islands	VS9
Malta	ZB1
Manchukuo	MX
Marianas Islands	
Marshall Islands	
Martinique	FM8
Mauritius	VQ8
Mesopotamia, see Iraq	
Mexico	XE
Midway Island	K6
Miquelon & St. Pierre Islands	FP8
Monaco	
Mongolia	
Morocco, French	CN
Morocco, Spanish	EA9
Mozambique	CR7
Nepal	
Netherlands	PA
Netherlands West Indies, see Curacao	
New Caledonia	FK8
Newfoundland and Labrador	VO
New Guinea, Neth.	PK6
New Guinea, Territory of	VK9
New Hebrides, British	YJ
New Hebrides, French	FU8
New Zealand	ZL
Nicaragua	YN
Nicobar Islands	
Nigeria (British Cameroons)	ZD2
Niue	ZK2
Non-Federated Malay States	VS3
North Borneo, see British North Borneo	VS4
Norway	LA
Nyasaland	ZD6
Ocean Island, see Gilbert & Ellice Islands	VR1
Oman	
Palau (Pelew) Islands	
Palestine	ZC6
Panama	HP
Papua Territory	VK4
Paraguay	ZP
Persia, see Iran	EP
Peru	OA
Philippine Islands	KA
Phoenix Islands	
Pitcairn Island	VR6
Poland	SP
Portugal	CT
Portuguese India, see Goa	CR8
Principe & Sao Thome Islands	
Puerto Rico and Virgin Islands	K4
Reunion Island	FR8
Rhodesia, Northern	VQ2
Rhodesia, Southern	ZE
Rio de Oro	
Romania	YR
St. Helena	ZD7
Salvador	YS
Sardinia	
Samoa, U. S.	K6
Samoa, Western	ZM
Sandwich Islands	
Sarawak	VS5
Saudi Arabia	
Seychelles	VQ9
Siam	HS
Sierra Leone	ZD1
Socotra	
Solomon Islands	VR4
Somaliland, British	VQ6
Somaliland, French	FL8
Somaliland, Italian	
South Georgia	VP8
South Orkney Islands	VP8
South Shetland Islands	VP8
Southwest Africa, see Union of South Africa	
Soviet Union	U
European States	U1-7
Asiatic States	U8-9-0
Spain	EA
Spitzbergen, see Svalbard	
Straits Settlements	VS1
Sumatra	PK4
Surinam, see Guiana, Neth.	PZ
Svalbard, (Spitzbergen)	
Sweden	SM
Switzerland	HB
Syria	
Taiwan (Formosa)	J9
Tanganyika Territory	VQ3
Tangier Zone	
Tannu Tuva	
Tasmania	VK7
Tibet	
Timor, Portuguese	CR10
Togaland, British, see Gold Coast	
Togoland, French	FD8
Tokelau (Union) Islands	
Tonga (Friendly) Islands	VR5
Transjordan	ZC1
Trinidad and Tobago	VP4
Tristan da Cunha	ZU9
Tunisia	FT4
Turkey	TA
Uganda	VQ5
Union Islands, see Tokelau Islands	
Union of South Africa	ZS-ZT-ZU
United States	(N) W
Uruguay	CX
Venezuela	YV
Virgin Islands, see Puerto Rico	K4
Wake Island	K6
Windward Islands	
Wrangel Island	
Yemen	
Yugoslavia	YT-YU
Zanzibar	ZK1

There were a total of 234 countries on the first country list of the world by *Radio* magazine.

Herb Becker put out an updated country list in the January, 1938 issue of *Radio.* The previous list had been reviewed by John Hunter, G2ZQ, H.A.M. Whyte, G6WY, and By Goodman, W1JPE. The following countries were added:

Chagos Islands	VQ8	Soviet Union	
Channel Islands	G	White Russian SSR	U2
French India	FN	Ukranian SSR	U5
Franz Josef Land		Transcaucian SFSR	U6
Isle of Man	G	Uzbek SSR	U8
Jan Mayen Island	OY	Turkoman SSR	U8
Kuweit		Turks and Caicos Islands	VP5
Scotland	GM	Virgin Islands	K4
Southwest Africa	ZS3	Wales	GW

Seventeen countries were added to the 1937 Radio country list and one was removed, the Bismarck Archipelago, bringing the total number of countries to 250.

Herb Becker put out an updated country list in the January, 1939 issue of *Radio*. The following countries were added:

Baker, Howland and Am. Phoenix Is	KF6
Johnston Island	KE6

Austria was removed from the country list, as it had been annexed by Germany the previous March. The total numer of countries was now 251.

The last prewar country list by Radio appeared in the January, 1940 issue. The following countries were added:

Clipperton Island	
Darien (Kanto-shu)	J8P

Darien was Japanese territory located at the southern tip of Manchuria. The British and French New Hebrides were combined into one country, and the Sandwich Islands were removed. The total number of countries was still 251.

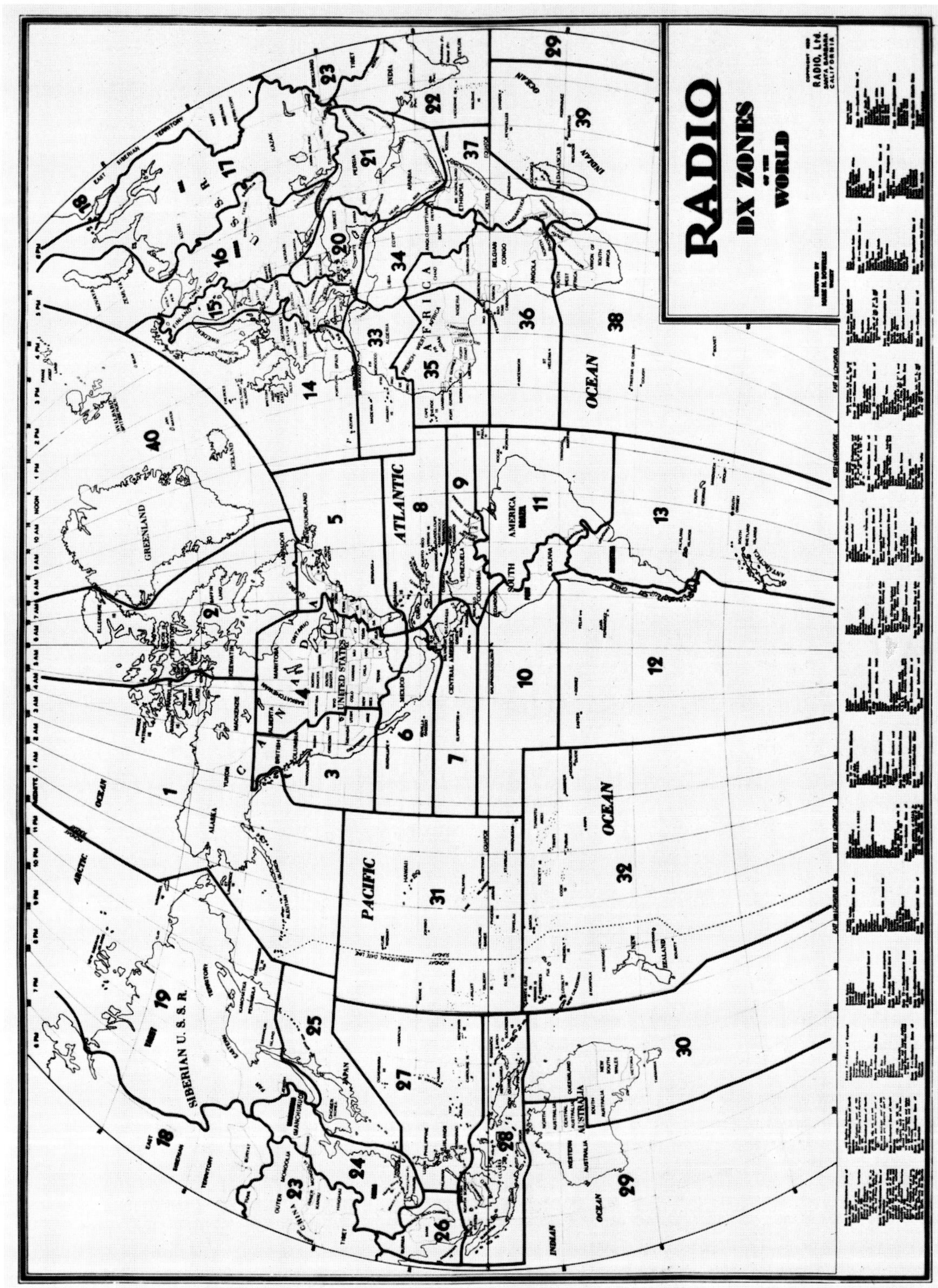

R/9 zone map of the world.

Appendix 7

Last prewar "WAZ" Honor Roll listing (*Radio*, February 1941)

CW and PHONE	Z	C
ON4AU	40	158
G2ZQ	40	147
J5CC	40	130
W8CRA	39	156
W2BHW	39	156
W8BTI	39	154
W1BUX	39	153
W2GTZ	39	153
W2HHF	39	152
G6WY	39	151
W6GRL	39	151
W6CXW	39	150
W2GT	39	150
W9TJ	39	149
W4CBY	39	145
W6CUH	39	143
W6KIP	39	143
W8OSL	39	143
W9KG	39	141
W6ADP	39	140
W6BAX	39	140
W3AG	39	140
W8OQF	39	139
W8LEC	39	136
W6QD	39	135
W9TB	39	134
W2ZA	39	134
VK2EO	39	133
G5BD	39	133
W2GVZ	39	132
W4CYU	39	132
W3EVT	39	131
W8PQQ	39	131
W5KC	39	130
W2GWE	39	129
W6KRI	39	129
W1ADM	39	128
VE4RO	39	126
W6VB	39	125
W7BB	39	123
W6HX	39	123
G5BJ	39	120
W8JSU	39	120
W2IYO	39	119
W2CYS	39	117
G2LB	39	115
W4IO	39	115
W7DL	39	115
W3BEN	39	115
W2GNQ	39	113
W6FZL	39	112
ON4HS	39	111
ON4FE	39	110
W1AQT	39	110
W6FZY	39	109
W6SN	39	100
W9NRB	39	98
W6GPB	39	94
XE1BT	39	90
K6AKP	39	78
W1CH	38	150
W2GW	38	143
W5VV	38	144
W3HZH	38	139
W3EMM	38	139
W5BB	38	138
W8BKP	38	138
ZL1HY	38	138
W3EPV	38	136
W9GDH	38	134
W3HXP	38	133
W4FVR	38	130
W9FS	38	130
W3EAV	38	130
W8JMP	38	127
W2GRG	38	127
ON4EY	38	126
W8ZY	38	125
W3GAU	38	125
W1BXC	38	125
W3EVW	38	124
W3GHD	38	121
W8AU	38	120
W8LYQ	38	120
W8DFH	38	119
W9PST	38	119
W8QXT	38	119
W8JIN	38	119
W3FQP	38	119
W8DWV	38	118
W1GDY	38	118
W2BMX	38	118
W1BGC	38	117
W6AM	38	117
LU7AZ	38	116
W3DDM	38	116
W9UQT	38	116
W9ELX	38	116
W8MTY	38	114
W9KA	38	114
W8LFE	38	113
G6CL	38	112
W8HWE	38	112
G2QT	38	112
W8EUY	38	112
W9CWW	38	112
W2BXA	38	111
W6GRX	38	111
LY1J	38	110
W1AB	38	110
W6HZT	38	110
W4MR	38	108
W8KWI	38	108
W8BOX	38	106
W9ADN	38	106
W8OE	38	106
W6NLZ	38	106
W9PK	38	105
W8GBF	38	105
W3KT	38	105
ON4UU	38	104
G2IO	38	103
W8BWB	38	98
J2KG	38	95
G6XL	38	95
ON4FQ	38	92
W9VDQ	38	79
SU1WM	37	138
W2BJ	37	134
W6GAL	37	131
W8KKG	37	127
W7AMX	37	125
J2JJ	37	123
W2IOP	37	122
W1RY	37	120
W6MVK	37	118
G6NF	37	115
W9RCQ	37	114
W3TR	37	113
ON4FT	37	112
W9RBI	37	112
W6MEK	37	112
W6ADT	37	111
W1IED	37	111
G2MI	37	110
W7AYO	37	110
W8DOD	37	110
W6ITH	37	109
VE2EE	37	108
W4DMB	37	108
W5ENE	37	107
W9PTC	37	103
W3FJU	37	103
W9GBJ	37	103
G6GH	37	102
W3AYS	37	102
VK2DA	37	101
W6FKZ	37	101
W6JBO	37	101
W8KPB	37	100
W4DMB	37	100
W9AJA	37	99
W4EQK	37	99
ON4VU	37	99
W3RT	37	99
W3EXB	37	98
ZL2CI	37	97
W6DLY	37	97
W6MHH	37	95
W6MCG	37	92
G2UX	37	91
W2BSR	37	90
W6MUS	37	89
W9UBB	37	77
W8AQT	36	120
K4FCV	36	112
W8AAJ	36	107
W3GGE	36	106
W6BAM	36	106
W9AFN	36	105
W5PJ	36	105
W8QDU	36	105
W5ASG	36	104
SP1AR	36	103
W8LDR	36	101
W6NNR	36	100
W6KWA	36	99
W8LZK	36	99
G6BJ	36	99
VE1DR	36	98
W9VES	36	98
W9GKS	36	97
W8AAT	36	96
G6YR	36	94
W2IZO	36	94
VE5AAD	36	92
W4ADA	36	90
W1APU	36	91
W9LBB	36	90
W8JAH	36	89
OK2HX	36	86
VK2NS	36	84
W6TI	36	80
W7DSZ	36	73
W2GXH	36	71
W1WV	35	119
W8OXO	35	113
W6GHU	35	103
W4QN	35	103
W9PGS	35	103
W6HJT	35	103
W9VKF	39	101
K6NYD	35	100
W8CLM	35	99
W8OUK	35	99
W8CJJ	35	98
W2WC	35	98
OK1AW	35	96
W9EF	35	94
G6QX	35	94
W8NV	35	94
W3DRD	35	93
W6AQJ	35	92
VE5ZM	35	92
W6ONQ	35	92
LU3DH	35	89
W9GNU	35	88
W9ERU	35	88
K6CGK	35	88
W9VDX	35	86
W6KQK	35	85
ON4NC	35	82
GI6TK	35	80
W4ELQ	35	80
W8QIZ	35	78
W6GK	34	105
W6HEW	34	103
K7FST	34	102
W8CED	34	102
W8BSF	34	100
W1APA	34	98
W2BZB	34	99
VK2AS	34	94
W8HGA	34	93
W3EYY	34	91
W9MQQ	34	89
W2FLG	34	89
W6TE	34	88
G6WB	34	88
W6CVW	34	88
VK2OQ	34	87
G5VU	34	85
W9BCV	34	83
W6PNO	34	83
ZS1CN	34	82
VK2TF	34	81
W6MJR	34	81
ON4SS	34	80
W6HIP	34	76
VK2TI	34	75
W7AVL	34	75
W8JK	34	75
ZL2VM	34	72
W6LHN	34	71
VK2AGJ	34	70
VK2EG	34	70
VE5MZ	34	69
VK2VN	34	63
W9QOE	34	56
F8XT	33	112
W8ACY	33	106
W3DAJ	33	97
W6KEV	33	96
W8BWC	33	93
W6KUT	33	90
W6CEM	33	88
W3LE	38	128
F8UE	38	103
W6OCH	36	107
W6ITH	36	103
W3FJU	36	87
VE1CR	36	81
W1ADM	35	101
W9NLP	35	95
W9TIZ	35	93
KA1ME	35	79
F8VC	35	78
W4CYU	34	100
ON4HS	34	92
W9ELX	34	92
W6EJC	34	84
W7BVO	34	80
W4DAA	34	71
W2IXY	33	105
W6NNR	33	92
GM2UU	33	84
F8XT	33	70
W3FAM	33	68
W6MLG	32	97
W8LFE	32	91
W2IKB	32	90
W4DRZ	32	89
W9BEU	32	88
W9QI	32	86
W1HKK	32	85
W8QXT	32	85
G5BY	32	85
VK4JP	32	85
W4DSY	32	84
W6OI	32	83
W9TB	32	82
W6IKQ	32	80
VE1DR	32	59
W1AKY	31	93
W3EMM	31	88
W8LAC	31	85
G6BW	31	83
G3DO	31	78
W1KJJ	31	78
W6FTU	31	77
G8MX	31	73
W8RL	31	71
W9UYB	31	71
W6AM	31	67
F8KI	31	58
W9ZTO	31	53
W4EEE	30	86
W2GW	30	86
W1JCX	30	83
W8AAJ	30	82
W2IUV	30	79
W2AOG	30	77
W9BCV	30	68
W6MZD	30	52
G6DT	29	83
W4BMR	29	80
K6NYD	29	78
CO2WM	29	78
W9RBI	29	71
W6NLS	29	64
W6GCT	29	62
W6NRW	29	60
W2DYR	28	80
W2GRG	28	74
W6PDB	28	65
W8NV	28	65
W7EKA	28	63
VE2EE	28	62
W4DRZ	28	62
W1BLO	28	62
VK2AGU	28	61
W6MPS	28	60
W3EWN	27	93
W2HCE	27	76
W5CXH	27	52
G5ZJ	26	77
W5ASG	26	62
W5VV	26	61
W4EQK	26	61
W8QDU	26	61
W9NMH	26	61
W5DNV	26	60
VK2OQ	26	56
W4TS	26	51
VE4SS	26	50
W6FKK	26	47
W7AMQ	26	47
K6LKN	26	46
G6CL	26	46

Postwar ARRL DXCC Country List

Official List for ARRL DX Contest and the Postwar DXCC (*QST*, February 1947)

Country	Prefix
Aden and Socotra Island	VS9
Afghanistan	YA
Alaska	KL7
Albania	ZA
Aldabra Islands	
Algeria	FA
Andaman Ids. and Nicobar Ids	
Andorra	PX
Anglo-Egyptian Sudan	ST
Angola	CR6
Argentina	LU
Ascension Island	ZD8
Australia (including Tasmania)	VK
Austria	OE
Azores Islands	CT2
Bahama Islands	VP7
Bahrein Island	VU7
Baker Island, Howland Island and Am. Phoenix Islands	KB6
Balearic Islands	EA6
Barbados	VP6
Basutoland	ZS4
Bechuanaland	
Belgian Congo	OQ
Belgium	ON
Bermuda Islands	VP9
Bhutan	
Bolivia	CP
Bonin Islands and Volcano Islands (e.g., Iwo Jima)	
Borneo, British North	VS4
Borneo, Netherlands	PK5
Brazil	PY
British Honduras	VP1
Brunei	VS5
Bulgaria	LZ
Burma	XZ
Cameroons, French	FE8
Canada	VE
Canal Zone	KZ5
Canary Islands	EA8
Cape Verde Islands	CR4
Caroline Islands	
Cayman Islands	VP5
Celebes and Molucca Islands	PK6
Ceylon	VS7
Chagos Islands	VQ8
Channel Islands	GC
Chile	CE
China	XU, C
Christmas Island	ZC3
Clipperton Island	
Cocos Island	TI
Cocos Islands	ZC2
Colombia	HK
Comoro Islands	
Cook Islands	ZK1
Corsica	
Costa Rica	TI
Crete	SV
Cuba	CM-CO
Cyprus	ZC4
Czechoslovakia	OK
Denmark	OZ
Dodecanese Islands (e.g., Rhodes)	SV5
Dominican Republic	HI
Easter Island	
Ecuador	HC
Egypt	SU
Eire (Irish Free State)	EI
England	G
Eritrea	I6
Ethiopia	ET
Faeroes, The	OY
Falkland Islands	VP8
Fanning Island (Christmas Island)	VR3
Fiji Islands	VR2
Finland	OH
Formosa (Taiwan)	
France	F
French Equatorial Africa	FQ8
French India	FN
French Indo-China	FI8
French Oceania (e.g., Tahiti)	FO8
French West Africa	FF8
Fridtjof Nansen Land (Franz Josef Land)	
Galapagos Islands	
Gambia	ZD3
Germany	D
Gibraltar	ZB2
Gilbert & Ellice Islands and Ocean Island	VR1
Goa (Portuguese India)	CR8
Gold Coast (and British Togoland)	ZD4
Greece	SV
Greenland	OX
Guadeloupe	FG8
Guantanamo Bay	NY4
Guatemala	TG
Guiana, British	VP3
Guiana, Netherlands (Surinam)	PZ
Guiana, French, and Inini	FY8
Guinea, Portuguese	CR5
Guinea, Spanish	
Haiti	HH
Hawaiian Islands	KH6

Country	Prefix
Honduras	HR
Hong Kong	VS6
Hungary	HA
Iceland	TF
Ifni	
India	VU
Iran	EP-EQ
Iraq	YI
Ireland, Northern	GI
Italy	I
Jamaica	VP5
Jan Mayen Island	
Japan	J
Jarvis Island, Palmyra group (Christmas Island)	KP6
Java	PK
Johnston Island	KJ6
Kenya	VQ4
Kerguelen Islands	
Korea	
Kuwait	
Laccadive Islands	VU4
Leeward Islands	VP2
Liberia	EL
Libya	(LI)
Liechtenstein	HE1
Little America	KC4
Luxembourg	LX
Macau	CR9
Madagascar	FB8
Madeira Islands	CT3
Malaya	VS1, VS2
Maldive Islands	
Malta	ZB1
Manchuria	
Marianas Islands (Guam)	KG6
Marshall Islands	
Martinique	FM8
Mauritius	VQ8
Mexico	XE
Midway Island	KM6
Miquelon and St. Pierre Islands	FP8
Monaco	
Mongolia	
Morocco, French	CN
Morocco, Spanish	EA9
Mozambique	CR7
Nepal	
Netherlands	PA
Netherlands West Indies	PJ
New Caledonia	FK8
Newfoundland and Labrador	VO
New Guinea, Netherlands	PK6
New Guinea, Territory of	VK9
New Hebrides	FU8, YJ
New Zealand	ZL
Nicaragua	YN
Nigeria	ZD2
Niue	ZK2
Norway	LA
Nyasaland	ZD6
Oman	
Palau (Pelew) Islands	
Palestine	ZC6
Panama	HP
Papua Territory	VK4
Paraguay	ZP
Peru	OA
Philippine Islands	KA
Phoenix Islands (British)	
Pitcairn Island	VR6
Poland	SP
Portugal	CT
Principe and Sao Thomo Islands	
Puerto Rico	KP4
Reunion Island	FR8
Rhodesia, Northern	VQ2
Rhodesia, Southern	ZE
Rio de Oro	
Roumania	YR
Ryukyu Islands (e.g., Okinawa)	
St. Helena	ZD7
Salvador	YS
Samoa, American	KS6
Samoa, Western	ZM
Sarawak	VS5
Sardinia	
Saudi Arabia (Hedjaz and Nejd)	HZ
Scotland	GM
Seychelles	VQ9
Siam	HS
Sierra Leone	ZD1
Sikkim	(AC3)
Solomon Islands	VR4
Somaliland, British	VQ6
Somaliland, French	FL8
Somaliland, Italian	
South Georgia	VP8
South Orkney Islands	VP8
South Sandwich Islands	VP8
South Shetland Islands	VP8
Southwest Africa	ZS3
Soviet Union:	
European Russian Socialist Federated Soviet Republic	UA1-3-4-6
Asiatic Russian S.F.S.R.	UA9-0
Ukraine	UB5
White Russian Soviet Socialist Republic	UC5
Azerbaijan	UD6
Georgia	UF6
Armenia	UG6
Turkoman	UH8
Uzbek	UI8
Tadzhik	UJ8
Kazakh	UL7
Kirghiz	UM8
Karelo-Finnish Republic	UN1
Moldavia	UO5
Lithuania	UP
Latvia	UQ
Estonia	UR
Spain	EA
Sumatra	PK4
Svalbard (Spitzbergen)	
Swan Island	KS4
Swaziland	
Sweden	SM
Switzerland	HB
Syria	(AR)
Tanganyika Territory	VQ3
Tangier Zone	EK
Tannu Tuva	
Tibet	AC4
Timor, Portuguese	CR10
Togoland, French	FD8
Tokelau (Union) Islands	
Tonga (Friendly) Islands	VR5
Trans-Jordan	ZC1
Trieste	
Trinidad and Tobago	VP4
Tristan da Cunha and Gough Island	ZD9
Tunisia	FT4
Turkey	TA
Turks and Caicos Islands	VP5
Uganda	VQ5
Union of South Africa	ZS
United States of America	W, K
Uruguay	CX
Venezuela	YV
Virgin Islands	KV4
Wake Island	KW6
Wales	GW
Windward Islands	VP2
Wrangel Islands	
Yemen	
Yugoslavia	YT-YU
Zanzibar	VQ1

The postwar ARRL country list was made up by G2MI, W6QD and an ARRL Headquarters group consisting of five-men. Twenty countries were added to the prewar DXCC list:

Country	Prefix
Austria.	OE
Basutoland	
Bonin & Volcano Islands	
Dodecanese	SV5
Guantanamo Bay	NY4
Little America	KC4
Ryukyu Islands.	
Sikkim	AC3
South Sandwich Islands	VP8
Soviet Union	
Azerbaijan	UD6
Georgia	UF6
Armenia.	UG6
Tadzhik	UJ8
Kazakh	UL7
Kirghiz.	UM8
Karelo-Finnish Republic	UN1
Moldavia	UO5
Swan Island.	KS4
Swaziland	
Trieste	

A total of fourteen countries were combined into seven countries; Aden and Socotra, Andaman and Nicobar Islands, Australia and Tasmania, Federated Malay States and the Straights Settlements, Marianas Islands and Guam, Saudi Arabia and Hejaz.

Six countries were removed from the prewar list; Aegean Islands, Baluchistan, Danzig, Isle of Man, Non-Federated Malay States, and the Transcaucasian SFSR. These totals combined for a net gain of seven on the DXCC list, for a total of 257 countries.

The country list was in a state of constant change in the coming years, as political boundaries became better defined. The following additions were noted in *QST* :

Date	Country	Prefix
Jul 47	Isle of Man	GD
Mar 48	Lebannon	AR8
	Pakistan	AP
	San Marino	M1
Sep 48	Little America (KC4)	deleted
	Antarctica	
Oct 48	Marion Island	ZS2
Feb 49	Norfolk Island	VK9
	Vatican	HV1
May 49	Heard Island	VK1
Jun 49	Israel (14 May 1948)	4X4
Jul 49	Macquarie Island	VK1
Sep 49	New Foundland & Labrador (VO) 1 Apr 49	deleted
Apr 51	Saar (8 Nov 47)	9S4
	Amsterdam & St Paul (separate from Kerguelen Island)	FB8Z
May 52	Singapore (1 Apr 46) (VS2 counts separately for Malaya)	VS1
Oct 52	Qatar	MP4Q
Mar 53	Sultanate of Oman (Muscat split, now Trucial Oman)	MP4B
Jul 54	San Andres & Providencia	HK0
Nov 54	Navassa Island	
Jan 55	French India (FN8) (1 Nov 54)	deleted
Jun 55	Wallis & Futuna Island	FW8
	Tannu Tuva	deleted
Sep 55	Kermadec	ZL
Nov 55	French Indochina (FI8) 19 Jul 55	deleted
	Laos (20 Jul 55)	XW8
	Cambodia (20 Jul 55)	3W8
	Viet Nam (20 Jul 55)	FI8
Mar 56	Tromelin.	FB8
Jul 56	Nauru Island	VK9
Aug 56	Revilla Gigedo Islands	XF4
Sep 56	Aves Island	YV0
Jan 57	Aland Island	OH0
Apr 57	Saar (9S4) (1 Apr 57)	deleted
	Trieste (I1) (1 Apr 57)	deleted
May 57	Ghana (5 Mar 57)	ZD1
	Gold Coast (ZD4) (4 Mar 57)	deleted
Dec 57	Rodriguez Island	VQ8
Mar 58	Fernando da Noronha.	PY0
May 58	Lord Howe Island	VK2
Jun 58	Windward & Leeward Islands redefined	
	Anguilla	VP2E
	Antigua & Barbuda	VP2A
	British Virgin Islands	VP2V
	Dominica	VP2D
	Grenada & Dependencies	VP2G
	Monserrat	VP2M
	St. Kitts & Nevis	VP2K

Date	Country	Prefix
	St. Lucia	VP2L
	St. Vincent & Dependencies	VP2S
	Caymans combined with	
	Jamaica (VP5)	
	Above effective 1 June 58	
Aug 58	Chatham Is	ZL
Oct 58	Trindade & Vaz	PY0
Jan 59	Rep of Guinea.(1 Oct 58)	7G1
	(formerly p/o F-W Afr)	
	Manihiki	ZK1
Jun 59	Seranna Bank & Rondor Clay	KS4
Jan 60	Cargados Carajos (St. Brandon)	VQ8
	Willis Island	VK4
Apr 60	Aland & Campbell.	ZL
Sep 60	Marcus Is	KG6I
	Mali Federation. (20 Jun 60)	
	incl Senegal & Sudanese Reps	
	(formerly p/o French West Afr)	
	Mauritania (20 Jun 60)	5T
	(formerly p/o French West Afr)	
	Ruanda-Urundi (1 Jul 60)	9U5
	(formerly p/o Belgian Congo)	
	Somalia Rep (1 July 60)	6O
	(comprises British Somaliland and Italian Somaliland).	
	British Somaliland (VQ6)	deleted
	effective 1 July 60	
	Italian Somaliland (MD4, I5)	deleted
	(1 July 60)	
	Karelo-Finnish Republic	
	(UN1)	deleted
	effective 1 July 60.	
	Tangier (EK) (1 July 60)	deleted
	Wrangle Island (UA0)	deleted
	Cayman Is. separate from	
	Jamaica	
Oct 60	French West Africa (FF8)	deleted
	effective 7 August 60.	
	Dahomey (1 Aug 60)	TY
	Niger Rep (3 Aug 60	5U
	Voltaic Rep (5 Aug 60)	XT
	Ivory Coast (7 Aug 60)	TU
	French Equatorial Africa	
	(FQ8)	deleted
	effective 17 Aug 60.	
	Chad (11 Aug 60)	TT8
	Central African Rep.	TL8
	(13 Aug 60)	
	Congo (15 Aug 60)	TN8
	Gabon (17 Aug 60)	TR8
Jan 61	Kaliningradsk	UA2
	Mali Federation	deleted
	Mali Republic 20 Jun 60)	TZ
	Senegal Republic (20 June 60)	6W
Mar 61	Bajo Nuevo	HK0
May 61	Kure Island	KH6
	East Pakistan	AP
Jun 61	Malpelo	HK0
Aug 61	Damao	CR8
	Diu	CR8
Sep 61	Damao & Diu same country	
	Kuwait/Saudi Arabia	
	Neutral Zone	8Z5
Dec 61	Kamaran Island	VS9K
Apr 62	Goa (CR8) (31 Dec 61	deleted
	Diu, Damao (CR8) (31 Dec 61)	deleted
Sep 62	Ruanda-Urundi (9U5)	deleted
	(30 Jun 62)	
	Ruanda (1 July 62)	9X5
	Urundi (1 July 62)	9U5
	Guam	KG6
Mar 63	Bouvet Island	3Y
	Eritrea (ET2) (14 Nov 62).	deleted
	Channel Islands split	
	Jersey	GC
	Guerensey & Dependencies	GC
Apr 63	Juan de Nova (25 June 60)	FR7
Jun 63	Gloriso (25 June 60)	FR7
Jul 63	Netherlands New Guinea	
	(JZO) (1 May 63)	deleted
	Java (PK1, 2, 3) (1 May 63)	deleted
	Sumatra (PK4) (1 May 63)	deleted
	Netherlands Boroneo (PK5)	deleted
	(1 May 63)	
	Celebes & Molucca (PK6)	deleted
	(1 May 63)	
	Indonesia (1 May 63)	PK
Dec 63	Singapore (VS1)	deleted
	Malaya (VS2) (15 Sep 63)	deleted
	Sarawak (VS4) (15 Sep 63)	deleted
	British North Boroneo (ZC5)	deleted
	(15 Sep 63)	
	Manchuria (C9) (16 Sep 63)	deleted
	Singapore (VS1) & Malaya	
	(9M2)	
	Sarawak (VS4) & North	
	Borneo (ZC5)	
Jan 64	Kuria Muria Islands	VS9H
Feb 64	ITU Headquarters	4U1
Apr 64	Crozet	FB8W
Aug 64	Saudi Arabia/Iraq Neutral Zone	8Z4
Jun 65	San Felix	CE0X
Oct 65	Singapore reinstated (8 Aug 65)	9V1
Nov 65	St. Peter & St. Paul Rocks	PY0
Jan 66	Spratley	

	Ebon Atoll	HC8
	Cormoran Reef	TI9
Jul 66	Desroches	VQ9
	Maria Theresa	FO8
	Minerva Reefs	1M4
Dec 66	Ebon Atoll Invalid	
	Cormoran Reef Invalid	
Mar 67	Farquahr	VQ9
Jul 68	Kuria Muria Islands (VS9H (30 Nov 67)	deleted
	Palestine (ZC6) (1 Jul 68)	deleted
Sep 68	Blenheim Reef	
	Geyser Reef.	
Aug 69	Ifni (EA9) (13 May 69)	deleted
Mar 70	Market	OJ0
Apr 71	Kuwait/Saudi Arabia Neutral Zone (9K3/8Z5) (15 Dec 69)	deleted
Aug 71	Abu Ail & Jabel At Tair	
Oct 71	Annobon Island	3C0
May 72	Ryukyu Is. (KR6) (15 May 72)	deleted
Sep 72	Maria Theresa annuled	
	Minerva Reef (1M4) (15 Jul 72)	deleted
Oct 72	Mellish Reef	VK9
Feb 73	Mt. Athos.	SV
	Swan Island (KS4) (1 Sep 72)	deleted
Nov 73	Germany (DL & DM) (17 Sep 73)	deleted
	Federal Republic of Germany	DL
	German Democratic Republic.	DM
Jun 74	Tibet (AC4) (June 1, 74)	deleted
	Zanzibar (VQ1) (June 1, 74)	deleted
Sep 75	Sikkim (AC3) (30 Apr 75)	deleted
	Blenheim (30 Jun 75)	deleted
Dec 75	Papua New Guinea	P2
	Tuvalu. formerly p/o Gilbert & Ellice Is.	VR8
	Papua Territory (VK9) (15 Sep 75)	deleted
	Territory of New Guinea (VK9) (15 Sep 75)	deleted
Jan 76	Sable Island	VX9
	St Paul Island	VY0
May 76	Okino Torshima.(30 May 76)	7J1
Nov 76	Portugese Timor (CR10) (14 Sep 76)	deleted
Feb 77	Comoro Islands	D6A
	Mayotte	FH8
	Comoros (FH8) (5 Jul 75)	deleted
	Aldabra (VQ9) (28 Jun 76)	deleted
	Desroches (VQ9) (28 Jun 76	deleted
	Farquhar (VQ9) (28 Jun 76)	deleted
Jan 78	Geyser Reef (1 Mar 78)	deleted
Jul 78	Rio de Oro (EA9)	deleted
Oct 78	UN Headquarters, New York (Feb 4, 78)	4U1
	Southern Sudan (May 7, 72)	ST0
Aug 79	Desecheo (1 Mar 79)	KP4
Dec	Canal Zone (KZ5) (1 Oct 79)	deleted
Sep 80	Okino Torishima (7J1) (Dec 1, 80)	deleted
Nov 81	Sovereign Military Order of Malta	1A0
May 82	Kamaran (Mar 11, 82)	deleted
Dec 82	Serrana Bank (KS4) (17 Sep 81)	deleted
	Baja Nuevo (HK0) (17 Sep 81)	deleted
	Saudi Arabia/Iraq Neutral zone (8Z4) (17 Sep 81)	deleted
Mar 85	Cyprus military bases separate from 5B4	ZC4

Some countries that were deleted over the years do not appear on the deleted country list (See Appendix 11). Apparently no QSOs were made from Wrangle Island, Tannu Tuva, or the Mali Federation while they were on the active country list.

At the time Don Wallace became a silent key there were 315 countries on the ARRL country list and 52 deleted-countries. The submission date for ZC4 was June 1, 1985, one week after he passed away. On CW Don had QSLs from all 315 countries on the current list, and all but one on the deleted country list; CR8- Diu, Damao (366/367). On phone Don had QSLs from all 315 countries on the current list, and and all but three on the deleted country list; CR8-Diu, Damao, FI8-French Indochina, and 8Z5-Kuwait/Saudi Arabia Neutral Zone (364/367).

Appendix 9

Postwar DXCC List (*QST* August, 1948)

DX CENTURY CLUB AWARDS

DXCC Certificates based on postwar contacts with 100-or-more countries have been issued to the amateurs listed below. The countries-worked totals indicated have been certified by examination of written evidence under the award rules as published in March 1947 *QST*.

W1FH........(1)..189
W6VFR.......(3)..181
G2PL........(90)..179
W8HGW......(7)..175
W4BPD......(4)..171
G6ZO........(10)..171
W2BXA.......(6)..170
W3BES.......(16)..170
W3GAU......(15)..169
W2HHF......(14)..165
W6SAI.......(17)..164
W8RDZ......(20)..164
W1TW.......(11)..162
W1CH........(5)..160
W1IAS........(8)..160
W2QKS.....(25)..160
W6GAL.....(18)..157
W2AGW......(29)..157
W9ANT......(32)..157
W3JNN.....(48)..157
W6GHU......(23)..156
W5ASG.......(26)..154
W6QJU.......(70)..153
WØYXO......(60)..151
ZS2X........(28)..150
W2GWE......(37)..150
W4CYU......(53)..150
PAØUN.......(30)..149
W4AIT.......(35)..149
W8NBK......(46)..149
W7FZA.......(85)..148
W6DI.......(164)..148
ZL1HY.......(12)..147
W1AXA.......(9)..146
W2GUM......(39)..146
W1ME.......(44)..146
W3RCQ......(67)..146
W2CYS........(2)..145
W1ADM......(22)..144
W7AMX......(42)..144
GW3ZV......(84)..144
W6SN........(72)..143
W3IYE.......(94)..143
W3KT........(57)..142
HB9CE.......(19)..141
W6TT........(43)..141
WØNUC......(50)..141
W2DS.......(122)..140
LU6DJX...(135)..140
W2AQW.....(184)..139
W6FSO......(138)..138
W3GHD......(65)..137
WØNTA......(79)..136
W6MEK.....(105)..136
W5KC........(36)..135
VE7ZM......(161)..135
W2COK......(40)..134
W6WKU.....(113)..134
W3EPV.....(56)..133
HB9CX.....(13)..132
W8LEC.....(24)..131
W9RBI.......(62)..131
VE3QD.......(91)..131
W4PN........(45)..130
W8FJN.......(49)..130
WØGKS......(71)..130
W6LER....(109)..130
W3JTC......(112)..130
PY1AJ......(124)..130
W2CZO......(144)..130
W3EVW......(51)..129
W4MR.......(54)..129
W4BRB......(38)..128
G2AJ........(100)..128
W2HZY.....(139)..128
LA7Y........(59)..127
NY4CM......(75)..127
W3FGB....(148)..127
ON4JW......(156)..127
W2NSZ......(157)..127
W5FNA......(61)..126
W9KOK......(73)..126
W1JYH.....(82)..126
W2BRV.....(114)..126
W7DL.......(118)..126
W3DRD......(63)..125
PAØJQ.......(86)..125
W2ALO.....(103)..125
W2MEL......(88)..124
OK1FF......(130)..124
ZL1BY......(169)..124
W8FJL.......(83)..123
W3KQF.....(117)..123
W9NDA......(93)..122
W2LJR.....(125)..122
W1DX......(138)..122
G5OO.......(145)..122
W3BXE......(64)..121
W4DKA.......(76)..121
W1BIH......(111)..121
W6GFE.....(115)..121
W3CPV.....(116)..121
CE3AG.....(123)..121
W2AFU.....(128)..121
W2PWP.....(153)..121
W8JIN.......(27)..120
G8KP........(66)..120
W6ANN.....(120)..120
W3DKT.....(131)..120
W1ENE.....(133)..120
W5LGG.....(152)..120
W9UIG.....(182)..120
W6RBQ.....(188)..120
W2CWE.....(195)..120
W4JXM......(92)..119
W2IOP.......(34)..118
W6KUT.....(159)..118
W1CLZ.....(171)..118
W3DPA......(47)..117
W4FIJ......(110)..117
W1ZL.......(186)..117
W7BD........(33)..116
W3TIF.......(68)..116
W6OMC.....(127)..116
W2QHH.....(134)..116
OK1LM.....(154)..116
W6RDR.....(200)..116
W6MHH.....(98)..115
W6BAM.....(119)..115
W6AM......(210)..115
W3LNE......(58)..114
W8CVU......(69)..114
W2SAI.......(97)..114
HB9DO.....(151)..113
W2RGV.....(155)..113
W4OM.......(89)..112
W7GUI......(189)..112
W1FJN......(196)..112
I1KN.......(208)..112
W9ERU......(80)..111
W6PB.......(132)..111
W1KFV.....(175)..111
W5BGP.....(201)..111
W7BE.......(204)..111
W2ITD......(205)..111
W6TI........(77)..110
W8FGX......(78)..110
I1IR.........(96)..110
W6BPD......(99)..110
W5CPI......(108)..110
W2GNQ.....(141)..110
G8IL.......(142)..110
KP4KD.....(167)..110
W1LOP......(183)..110
ZS6DW......(55)..109
W6CEM.....(176)..109
W6PUZ......(185)..108
W1WK......(197)..108
ZL2QM......(166)..107
W8SDR.....(207)..107
W4FPK.....(147)..106
G8TD.......(202)..106
W9DUY......(52)..105
ON4MS.....(174)..105
W6KRI.......(81)..104
G3FJ........(104)..104
W2TQC.....(136)..104
WØSQO.....(162)..104
W6NNV.....(180)..104
W6EBG.....(181)..104
W6LDJ......(192)..104
J2AHI......(193)..104
W2UFT.....(206)..104
W6WB......(106)..103
W7DXZ.....(129)..103
W8ERA....(158)..103
VK2DI......(168)..103
W8UAS......(173)..103
W1BLO.....(178)..103
CM2SW.....(190)..103
WØAIW.....(191)..103
W8LYQ......(31)..102
EI9J........(177)..102
PAØGN......(179)..102
W1JLT......(194)..102
W6AMA......(41)..101
W8HRV.......(74)..101
W2QKE......(95)..101
W6SC.......(102)..101
W4JV.......(107)..101
W9TWC.....(121)..101
G4JZ........(140)..101
W9FJB......(149)..101
W6POZ.....(160)..101
W3OP.......(163)..101
W1NMP.....(165)..101
WØDAE.....(198)..101
W3ZN.......(203)..101
W6MJB......(21)..100
W1BUX......(87)..100
W6DUC.....(126)..100
W3FJU......(137)..100
W5ENE.....(143)..100
W4KIT......(146)..100
W6VBY.....(150)..100
I1IV........(170)..100
W2BZS......(172)..100
W8WZ......(187)..100
W6TZD.....(199)..100
W1HRI......(209)..100

RADIOTELEPHONE

W1FH........(1)..152
W6DI........(14)..135
W1JCX........(2)..133
W4CYU.......(5)..132
W1HKK.......(3)..124
G2PL.........(7)..124
W6VFR......(24)..118
W2BXA.......(6)..117
W1NWO.......(9)..114
VQ4ERR.....(10)..114
W1MCW.....(18)..112
W2AFQ......(16)..111
W1FJN.......(21)..111
W2ZW........(4)..110
W1ADM......(22)..110
XE1AC.......(26)..108
TI2OA.......(12)..106
SU1HF.......(13)..105
G3YM........(11)..104
G6WX.......(15)..101
W8BF........(17)..101
W2RGV......(20)..101
ZL1HY.......(25)..101
W9NDA.......(8)..100
W4EWY......(19)..100
W5BGP......(23)..100

The complete listing as of June 15, 1948, of those that have qualified for the postwar DXCC award. Figures in parentheses following each call indicate the order in which the awards were issued.

WAZ Honor Roll (CQ November 1949)

CW & PHONE

WAZ

Call	Countries
W6VFR	229
W2BXA	224
W6EBG	224
W6ENV	224
W3BES	221
W3GHD	221
W6ADP	218
W6GRL	217
G2PL	216
W6PFD	216
W6ITA	214
W6MEK	213
W8BHW	211
G6ZO	211
W6SN	211
W6SAI	210
WØYXO	210
W2AQW	208
W8HGW	208
W3LOE	207
W4BPD	206
W6FSJ	205
VK3BZ	205
W6TT	204
VE7ZM	204
W6MX	202
W3EVW	202
W7GUI	202
W7AMX	201
W6MJB	201
W9KOK	200
W4CYU	199
W6DI	199
ZL1HY	199
W6OMC	198
W6SYG	198
W2IOP	197
W6MVQ	197
PY1AJ	196
W9VW	195
LU6DJX	195
W6NNV	195
W6GAL	193
VE7HC	193
W6AVM	192
W6RM	192
W6PKO	192
W6ZCY	191
W5KC	191
ZS2X	191
VK2DI	191
W6DZZ	191
W6HX	190
W6AM	190
WØNTA	188
VE4RO	186
W6AMA	186
ZL2GX	185
W2CZO	185
W6PB	185
W6SA	184
W6UCX	184
W6RBQ	183
W6AOA	181
W6PQT	181
W6KRI	181
W6SRU	181
W6SC	180
VK2ACX	180
W6RW	179
W6TI	173
CE3AG	178
W7DL	177
WØUOX	177
CX1FY	176
W1AB	175
W6PCS	174
W6WKU	174
W7FZA	174
W6TS	174
W6TZD	173
G5YV	172
OK1LM	172
LA7Y	171
W6CIS	170
W6BAM	170
KH6BA	169
W6DLY	169
W5AFX	169
ON4JW	169
W6RLN	168
W6ANN	167
W6GDJ	167
W6UHA	167
W6EFM	167
VK3CN	167
W6PZ	166
W6DUC	166
G3DO	166
W6UZX	166
W8SDR	166
KH6MI	166
W6EPZ	162
W6KUT	161
KH6IJ	161
W6IBD	161
W6PUY	160
W7BD	157
W7BE	156
W6BVM	155
W6BAX	155
G3AAM	154
W6KEV	153
W6BPD	152
WØOUH	152
VK2QL	151
W6LRU	150
I1IR	150
W6LEE	150
W6FHE	150
W6PH	150
OK1FF	148
W7DXZ	146
W6AYZ	146
W9NRB	145
W6MUC	145
W6QD	145
W6LER	145
W6LN	145
ON4TA	144
W6LDD	144
OK1CX	144
JA2KG	143
W6WWQ	142
W6CEM	136
G3AZ	133
W6TEU	133
W6RDR	133
W6MHB	130
W6YZU	129
W7GBW	127
G8IP	127
G5BJ	126
PK6HA	124
W6NRQ	123
W6MLY	123
W6BIL	115
W7KWA	98

39 Zones

Call	Countries
WØNUC	211
W2PEO	208
W2HHF	203
W8NBK	203
W9IU	202
W9ANT	202
W4AIT	201
W2HZY	200
W2NSZ	199
W9RBI	199
W2GWE	195
W6OEG	195
W3KT	194
PY1DH	194
F8BS	194
W1JYH	191
W3JNN	191
W3DPA	191
W8JTC	191
W1BIH	191
W3EPV	191
W1ENE	190
W2CWE	190
W3DRD	183
W4GG	183
W8RDZ	184
W4INL	183
W3OCU	181
W5ASG	177
W1ZL	176
I1KN	173
W3DKT	173
W6IFW	172
WØSQO	171
W9LNM	170
W2EMW	170
W1NMP	169
W3JKO	169
VO6EP	168
W2CYS	167
OK1VW	167
KH6MI	166
W8LEC	166
W2CNT	166
W4DKA	165
WØDU	165
W9MXX	164
W9VND	164
W6EAK	163
W3KDP	162
W4BRB	162
G5DQ	160
W9LM	159
WØGKS	158
W4OM	158
WØAIW	157
G8KP	156
W2WZ	156
W9YNB	155
WØEYR	153
G6QR	152
W2RDK	152
G2AJ	151
W9FKC	150
W4VE	150
SM5WI	148
DL2KW	147
G2WW	147
W2COK	146
W2GUR	146
W2MEL	145
W6BUD	145
W6JZP	141
W9DUY	140
G6BQ	140
G3FJ	139
W2BJ	138
W8VLK	137
W6EHV	136
OK1AW	135
W6ID	135
W6BZE	134
W9TQL	135
OK1CX	133
KH6PY	132
G2VD	132
W8WWU	132
G5RV	132
WØOUH	131
VK4RC	131
W6LGD	131
W6RLQ	130
G2FSR	130
W5CPI	130
VR5PL	124
G5VU	124
G3AAK	122
GM3CSM	121
G8RL	120
G5WM	120
W7BTH	120
W6MUF	118
DL3DU	118
G6BS	117
W6NRZ	117
G3QD	116
W7ETK	115
G3TK	114
W7HXG	112
W7ASG	110
W7GXA	105
KG6AL	104
W6LEV	103
W7ENW	101
W6WJX	101
W6AX	93

38 Zones

Call	Countries
VE3QD	190
W2PUD	180
KP4KD	168
CM2SW	167
W8KPL	166
W3IYE	161
W8FJN	160
W6CTL	148
W3LVJ	145
W8CVU	141
W2WZ	138
TF3EA	137
W9FKH	135
W4FPK	131
G8IL	131
G5CI	130
W2PQJ	130
W3ZN	129
G6LX	126
W9MZP	126
GW3AX	123
W9TB	122
GW4CX	120
W6ATU	117
W7EYS	107
G3ZI	107
W6VAT	107
C1CH	84

37 Zones

Call	Countries
W2HMJ	173
W1KFV	168
W2ZA	160
W2RGV	149
W4TWO	146
W8EYE	142
W3WU	142
W4ML	135
GM2UU	133
W2AYJ	133
W9TQL	129
WØAZT	129
W4DIA	129
W9ABA	128
W3FYS	119
G4CP	117
VE1EA	116
WØFET	115
G4AR	108
W8HSW	104
W7PK	104
G5MR	100
W2BLS	99
G3AAE	99
KL7PJ	98
W2SGK	95
KL7KV	88

36 Zones

Call	Countries
W4LVV	147
W4ILA	147
W2GVZ	138
OA4AK	128
HC2JR	125
W3AYS	124
W2WC	124
WØRBA	122
SVIRX	119
MD5AK	118
DL1DA	116
W2BF	115
G2CNN	114
ZS2AT	114
G2AKQ	112
W5CD	108
W2JA	102
G2AO	100
W5BK	99
G6WX	95
GM2AAT	75

35 Zones

Call	Countries
W4RBQ	138
W8ZMC	137
W4DHZ	132
W1BFT	130
VE3AAZ	130
W9WCE	130
W9CKP	122
W9RQM	119
CO6AJ	119
W9FNR	112
W9LI	118
VE3ACS	117
W6ZZ	114
W8AVB	113
VE1PQ	111
FE8AB	107
W2HAZ	106
VE5JV	103
WØGBJ	101
WØFWW	99
W6ETJ	95

34 Zones

Call	Countries
W4IYT	127
W1MRP	104
W8JM	89
W9WEN	83
W8PCS	80

33 Zones

Call	Countries
W4QN	110
W2SEI	100
OE1FF	91
W8QUS	85
KH6VP	83

32 Zones

Call	Countries
VE3AGC	107
W2OEC	105
G6QX	102
VE1NE	98

30 Zones

Call	Countries
OK1WY	103
ZL1QW	82

PHONE ONLY

39 Zones

Call	Countries
W6DI	187
W6VFR	164
HB9DS	145
W7HTB	143
VE7ZM	140

38 Zones

Call	Countries
W2BXA	171
W4CYU	170

37 Zones

Call	Countries
XE1AC	176
W1JCX	170
W8REU	163
W9RBI	162
G2PL	154
W6PXH	150
W6WNH	150
W8BF	146
G3DO	144
W3JNN	136
W1HKK	136
W6KQY	135
W6TT	133
G6LX	124
G2AJ	121
F8VC	115

36 Zones

Call	Countries
W1NWO	163
W3LTU	156
W1MCW	152
W7MBX	144
W9HB	139
W4ESP	136
W2D	135
W9BZB	130
W4INL	129
W1FJN	128
GM3UU	127
G6BW	127
VE3BNQ	108
G5YV	106
G6WX	105
W6AM	104
W3DHM	96
W6SA	92
F8DC	87

35 Zones

Call	Countries
VK3BZ	151
W4HA	136
W6PCK	129
W6CHV	128
G8QX	123
W2GHV	118
G3FU	115
W9CKP	114
W5LWV	108
W4OM	106
W3PA	105

34 Zones

Call	Countries
W8BIQ	120
W2RGV	118
W8ZMC	118
W6UZX	113
W9MIR	113
WØEYR	112
W4LZM	109
W1BPH	105
W8UIG	100
W4IWO	99
W8QBF	92

33 Zones

Call	Countries
W9RNX	126
W5ASG	119
HC2JR	115
W2ZW	115
W5ALA	112
WØANF	112
W4LZM	109
WØHX	107
WØANE	106
VE3ZM	100
W2PQJ	100

32 Zones

Call	Countries
W9MIR	113
WØEYR	112
W9WCE	111
W8BFQ	101
WØSQO	95
W1BFB	94
WØAIW	93
W2HY	85
W2SVK	84

31 Zones

Call	Countries
W4BA	106
W9CZC	100
WØPUE	95
W1MRP	89

30 Zones

Call	Countries
WØUYC	88
OA4AK	86
W7JUO	80

Appendix 11

Postwar Deleted Country List (*ARRL*, 1985)

Credit for any of these countries can be given if the date of contact with the country in question agrees with the date(s) shown in the corresponding footnote.

(AC3[1])[2] ... Sikkim
(AC4[1])[3] ... Tibet
C9[4] ... Manchuria
CN2[5] ... Tangier
CR8[6] ... Damao, Diu
CR8[6] ... Goa
CR8[34] ... Port. Timor
DA,DJ,DK,DL,DM[7] ... Germany
EA9[8] ... Ifni
EA9[38] ... Rio de Oro
ET2[9] ... Eritrea
FF8[10] ... Fr. West Africa
FH[35] ... Comoros
FI8[11] ... Fr. Indo-China
FN[12] ... French India
FQ8[13] ... Fr. Equatorial Africa
HKØ[42] ... Bajo Nuevo
I1[14] ... Trieste
I5[15] ... Italian Somaliland
JD/7J[40] ... Okino Tori-shima
JZØ[16] ... Netherlands New Guinea
KP3,KS4,HKØ[42] ... Serrana Bank & Roncador Cay
KR6,8,JR6,KA6,[17] ... Ryukyu Islands
KS4[18] ... Swan Islands
KZ[39] ... Canal Zone
P2,VK9[19] ... Papua Territory
P2,VK9[19] ... Terr. New Guinea
PK1,2,3[20] ... Java
PK4[20] ... Sumatra
PK5[20] ... Netherlands Borneo
PK6[20] ... Celebe & Molucca Isls.
UN1[21] ... Karelo-Finnish Rep.
VO[22] ... Newfoundland, Labrador
VQ1,5H1[23] ... Zanzibar
VQ6[24] ... British Somaliland
VQ9[36] ... Aldabra
VQ9[36] ... Desroches
VQ9[36] ... Farquhar
VS4[25] ... Sarawak
VS9H[26] ... Kuria Muria Is.
ZC5[25] ... British North Borneo
ZC6[27] ... Palestine
ZD4[28] ... Gold Coast, Togoland
(1M[1])[29] ... Minerva Reefs
70/VS9K ... Kamaran Is.
8Z4[43] ... Saudi Arabia/Iraq Neutral Zone
9K3,8Z5[30] ... Kuwait/Saudi Arabia Neutral Zone
9M2[25] ... Malaya
9S4[31] ... Saar
9U5[32] ... Ruanda-Urundi
[33] ... Blenheim Reef
[37] ... Geyser Reef

1. Unofficial prefix.
2. (AC3) Only contacts made before May 1, 1975, will count for this country. Contacts made May 1, 1975, and after, count as India.
3. (AC4) Only contacts made before June 1, 1974, will count for this country.
4. (C9) Only contacts made before September 16, 1963, will count for this country.
5. (CN2) Only contacts made before July 1, 1960, will count for this country.
6. (CR8) Only contacts made before January 1, 1962, will count for this country.
7. (DA, DJ, DK, DL, DM) Only contacts made before September 17, 1973, will count for this country.
8. (EA9) Only contacts made May 13, 1969, and before, count for this country.
9. (ET2) Only contacts made November 14, 1962, and before, will count for this country.
10. (FF8) Only contacts made August 6, 1960, and before, will count for this country.
11. (FI8) Only contacts made before December 21, 1950, will count for this country.
12. (FN) Only contacts made before November 1, 1954, will count for this country.
13. (FQ8) Only contacts made August 16, 1960, and before, will count for this country.
14. (I1) Only contacts made before April 1, 1957, will count for this country. Contacts made April 1, 1957, and after, count as Italy.
15. (I5) Only contacts made June 30, 1960, and before, will count for this country.
16. (JZØ) Only contacts made before May 1, 1963, will count for this country.
17. (KR6, JR6, KA6) Only contacts made before May 15, 1972, will count for this country. Contacts made May 15, 1972, and after, count as Japan.
18. (KS4) Only contacts made before September 1, 1972, will count for this country. Contacts made September 1, 1972, and after, count as Honduras.
19. (P2, VK9) Only contacts made before September 16, 1975, will count for this country.
20. (PK1, 2, 3, 4, 5, 6) Only contacts made before May 1, 1963, will count for this country.
21. (UN1) Only contacts made June 30, 1960, and before, will count for this country. Contacts made July 1, 1960, and after, count as European Russian S.F.S.R.
22. (VO) Only contacts made before April 1, 1949, will count for this country.
23. (VQ1, 5H1) Only contacts made before June 1, 1974, will count for this country.
24. (VQ6) Only contacts made June 30, 1960, and before, will count for this country.
25. (VS4, ZC5, 9M2) Only contacts made September 15, 1963, and before, will count for this country.
26. (VS9H) Only contacts made before November 30, 1967, will count for this country. Contacts made December 1, 1967, and after, count as Oman.
27. (ZC6) Only contacts made July 1, 1968, and before, will count for this country.
28. (ZD4) Only contacts made March 5, 1957, and before, will count for this country.
29. (1M) Only contacts made July 15, 1972, and before, will count for this country. Contacts made July 16, 1972, and after, count as Tonga.
30. (9K3, 8Z5) Only contacts made before December 15, 1969, will count for this country.
31. (9S4) Only contacts made before April 1, 1957, and after, count as Germany.
32. (9U5) Only contacts made between July 1, 1960, and July 1, 1962, will count for this country.
33. (Blenheim Reef) Only contacts made between May 4, 1967, and July 1, 1975, will count for this country. Contacts made July 1, 1975, and after, count as Chagos.
34. (CR8) Only contacts made before September 15, 1976, will count for this country.
35. (FH) Only contacts made before July 6, 1975, count toward this country.
36. (VQ9) Only contacts made before June 29, 1976, count for these countries.
37. (Geyser Reef) Only contacts made between May 4, 1967 and March 1, 1978, will count for this country.
38. (EA9) Only contacts made before January 8, 1976 will count for this country.
39. (KZ) Only contacts made before October 1, 1979 will count for this country.
40. (JD/7J) Only contacts made between May 30, 1976 and November 30, 1980, will count for this country. Contacts made December 1, 1980, and after, count as Ogasawara.
41. (70/VS9K) Only contacts made before March 11, 1982, will count for this country.
42. (HKØ) (KP3,KS4,HKØ) Only contacts made before September 17, 1981, will count for these countries.
43. (8Z4) Only contacts made before December 26, 1981, will count for this country.

The deleted countries of Sikkim (Don's QSL from AC3SQ is a telegram), Tibet, Manchuria, Tangier, Goa, Portugese Timor, Germany, and Ifni.

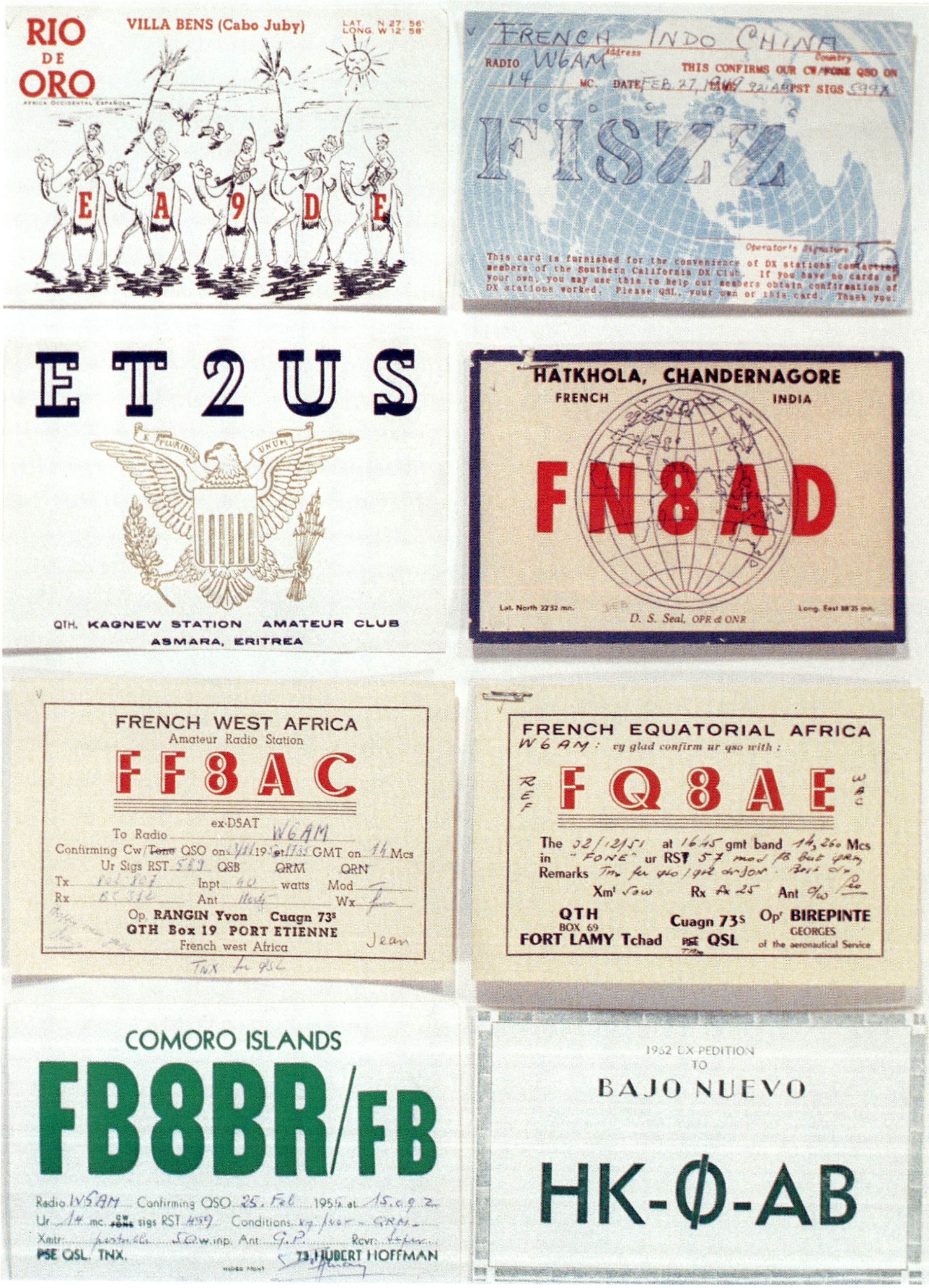

The deleted countries of Rio de Oro, Eritrea, French West Africa, Comoros, French Indochina, French India, French Equatorial Africa, and Baja Nuevo.

The deleted countries of Trieste, Italian Somaliland, Okino Torishima, Netherlands New Guinea, Serrana Bank, Ryukyu Islands, Swan Islands, and Canal Zone.

The deleted countries of Papua territory, Territory of New Guinea, Java, Sumatra, Netherlands Borneo, Celebe & Molucca Islands, Karelo-Finnish Republic, and Newfoundland & Labrador.

The deleted countries of Zanzibar, British Somaliland, Aldabra, Desroches, Farquhar, Sarawak, Kuria Muria Islands, and British North Borneo.

The deleted countries of Palestine, Gold Coast & Togoland, Kamaran, Saudi Arabia/Iraq Neutral Zone, Kuwait/-Saudi Arabia Neutral Zone, Minerva Reefs, Malaya, and Saar

The deleted countries of Ruanda-Urundi, Blenheim Reef, and Geyser Reef.

Don's DXCC certificate.

Appendix 13

Banned Country List

The Federal Communications Commission Public Notice, December 21, 1950, stated that communications between amateur radio stations licensed by the FCC and foreign amateur radio stations are permissable subject to the limitations of section 1 of article 42 of the Radio Regulations Annexed to the International Telecommunications Convention in Atlantic City in 1947.

Section 1 of this article provides as follows: Radio communications between amateur stations of different countries shall be forbidden if the administration of one of the countries concerned has notified that it objects to such radio communications.

According to information obtained by the commission from the Department of State certain foreign countries object to the exchange, internationally, of amateur radio communications and others impose specific limitations upon such communications. The names of the countries forbidding exchange, internationally, of amateur communications and countries imposing restrictions on such exchange, together with the terms of the restrictions, are hereinafter set forth.

Administrations which forbid radio communications between their amateur stations and amateur stations in other countries:

Indonesia
Japan (excluding amateur stations of Allied Occupation Forces as authorized by the Supreme Commander, Allied Powers)

Administrations which forbid all amateur radio operation:

Indochina
Iran
Lebanon
Netherlands Antilles
Thailand

The following administrations have placed the special restrictions noted on amateur radio communications:

Australia (Commonwealth of): Amateur stations in Australia are authorized to conduct radio communications for purely experimental purposes with amateur stations in other countries and the administrations of which permit such radio communication.

Austria: The reception of foreign amateur station transmissions is permitted, but transmissions by Austrian amateur stations are strictly forbidden by the Allied control authorities in Austria.

Accordingly, United States amateur licensees are warned that international communications are limited by treaty as indicated above. The foregoing does not in any way modify and should not be confused with the provisions of Section 2 of Article 42 of the International Radio Regulations (Atlantic City, 1947) which prohibits the use of amateur stations for transmitting international communications on behalf of third parties except when permitted by special arrangements between the countries concerned.

This notice supercedes and cancels Public Notices of October 12, 1949 (mimeo No. 41636) and November 4, 1949 (Mimeo No. 42642).

From time to time QST would update the banned country list. The clarifications are as follows:

Jan 5	PK7 & some PK6 okay to work. Indonesia still on banned list.
Dec 52	Netherlands Antilles removed from banned list (Mar 11, 52). Lebanon (AR) and Japanese nationals removed from banned list (15 Oct 52). Banned countries include Indonesia, French Indochina, Iran, Thailand and Austria (Except Allied occupation forces).
Aug 53	Korea on banned country list after 1 June 53
Jun 54	Austrian nationals removed from banned list (1 Apr 54).
Nov 55	Indochina partitioned in three countries. Laos removed from banned list. Cambodia and Viet Nam still on banned list.
Apr 56	Thailand removed from banned list (1 Sep 55).
Jun 59	Banned list includes Cambodia, Indonesia, Iran, and Viet Nam.
Feb 65	Banned list includes Cambodia, Indonesia, Thailand and Viet Nam.
Feb 69	Indonesia removed from banned list.
Oct 69	Thailand removed from banned list.
Jun 72	Cambodia removed from banned list.
Jul 74	Viet Nam removed from banned list. No restrictions on U.S. Stations.

Appendix 14

Southern California DX Club Charter Members (1949)

W6AM	Don Wallace (WU, 6OC, 9BU, 9XI, 9ZT/9XAX, 6AM, 6MA, 6ZZA)
W6ADP	Glen Means
W6ANN	Bill Adams (W6BA)
W6AOA	Frank Cuevas
W6AVM	Byron Guthiel (K6TX)
W6BPD	Dick Grove
W6BXL	Charles Bailey
W6CYI	O.J. Mills
W6DLY	Guy Martin
W6DUC	Victor Peterson
W6ENV	Andy Elsner
W6FSJ	Wendell Pierce
W6GFE	Homer Biedebach
W6HX	Ted Gillette
W6HZT	Stan Carr
W6IBD	Warren Davis (K6NA, W6EE)
W6ID	Vaughn Perry
W6ITA	Gordon Marshall (W7VV, W6RR)
W6JZP	D. Jackson
W6KYV	David Kennedy
W6LEE	Bob Turk
W6LS	Leo Sheppard
W6MUB	Arthur Enochson (W6EA)
W6NNV	Norm Wasson
W6NRQ	Carl Louis
W6PFD	Mark Graffis
W6PQT	C.B. McKnight
W6PDB	Ken Moore
W6PUY	E.J. Real (K6OJ)
W6PUZ	Don Roberts
W6QJU	Bob Martin
W6RLN	K.C. Jones
W6RW	R.D. Mace
W6SAI	Bill Orr
W6SRF	R.W. Davis
W6SYG	Frank Newton
W6TZD	E.C. Dvorak
W6UCX	John Linden
W6UHA	Maxine Willis

1920's English & French Stations

Appendix 16

1920's Australian & New Zealand stations

Bibliography

Bucher, E.E.: "The Wireless Experimenters Manual." Wireless Press, Inc., New York, 1920.

Bucher, E.E.: "Practical Wireless Telegraphy," Wireless Press, Inc., New York, 1917.

Tyne, G.F.J.: "The Saga of the Vacuum Tube." Howard W. Sams and Co. Inc., Indianapolis, 1977.

"Encyclopaedia Britannica World Atlas." C.S. Hammond & Company, Inc., New York, 1945.

"Radio Stations of the United States." Department of Commerce, Bureau of Navigation, Washington, D.C., July 1, 1913 & 1916.

"Amateur Radio Stations of the United States." Department of Commerce, Bureau of Navigation, Washington, D.C., June 30, 1920-1931.

"Consolidated Radio Callbook." The Consolidated Radio Callbook Co., Inc., New York, July 1920.

"Citizens Radio Amateur Call Book." Citizens Radio Service Bureau, Chicago, 1928.

"Radio Amateur Callbook Magazine." Radio Amateur Callbook, Inc., Chicago, 1929-1946.

Handy, F.E.: "The Radio Amateurs Handbook." The American Radio Relay League, Inc., Hartford, 1926.

Wallace, D.C.: "Short Wave Manual." Pacific Radio Publishing Co., San Francisco, 1933.

Radio Technical Staff: "The Radio Handbook." Pacific Radio Publishing Co., San Francisco, 1935.

Jones, F.C.: "The Radio Handbook." Pacific Radio Publishing Co., San Francisco, 1936.

DeSoto, C.B.: "200 Meters and Down." The American Radio Relay League, Inc., Hartford, 1936.

Ford, R.W.: "Wind Between the Worlds." David McKay Company, Inc., New York, 1957.

Palmer, Eric. Jr. "Riding the Air Waves with Eric Palmer, Jr." Horace Liveright, Inc. New York, 1930.

Terman, F.E.: "Radio Engineering." McGraw-Hill Book Company, Inc., New York & London, 1932.

Terman, F.E.: "Radio Engineers' Handbook." McGraw-Hill Book Company, Inc., New York & London, 1943.

Harper, A.E.: "Rhombic Antenna Design," D. Van Nostrand Company, Inc., New York, 1941.

Hawkins, J.N.A.: "The Radio Antenna Handbook." Radio, Ltd., Los Angeles, 1936.

Grammer, G. and Goodman, B.: "The ARRL Antenna Book." The American Radio Relay League, Inc., Hartford, 1939.

Orr, W.I. "All About Cubical Quad Antennas." Radio Publications, Inc., Wilton, CT.

"The Hatchet," *USS George Washington,* Feb-Aug 1919.

"QST." Maxim, H.P. and Tuska, C.D.: Hartford, Dec 1915-Sept 1916. The QST publishing Co., Hartford, Oct 1916-Sept 1917. The American Radio Relay League, Inc., Hartford, June 1919-Dec 1988.

"Radio." Pacific Radio Publishing Co. San Francisco, Apr 1924-Dec 1935. Radio, Ltd., Los Angeles, Jan 1936-Feb 1941. Editors & Engineers, Ltd., Santa Barbara, Mar 41-Aug 42.

"R/9." R9 Ltd., Los Angeles, Jan 1934-Dec 1935.

"CQ." Radio Magazines, Inc., NY. Jan 1945-Aug 1. Cowan Publishing Corp., New York, Sept 1950-1960. CQ Publishing Co., New York. CQ Communications, Inc., New York.

Glossary

AC	Alternating Current
AGC	automatic gain control
AM	amplitude modulation
AP	Associated Press
ARALB	Associated Radio Amateurs of Long Beach
ARRL	American Radio Relay League
ASA	Army Security Agency
AVC	automatic voltage control
AWA	Antique Wireless Association
BBC	British Broadcasting Corporation
BCL	broadcast listener
BFO	beat frequency oscillator
BPL	Brass Pounders League
BSEE	Bachelor of Science in Electrical Engineering
CIA	Central Intelligence Agency
CPO	chief petty officer
CPS	cycles per second
CQWW	CQ World Wide
CW	carrier wave
DC	direct current
DERA	Dhahran Experimental Radio Association
DPST	double-throw single-throw
DXCC	DX Century Club
ECO	electron coupled oscillator
ELM	Electronic Maintenance Shop
ERP	effective radiated power
FAA	Federal Aviation Administration
FBI	Federal Bureau of Investigation
FCC	Federal Communications Commission
FM	frequency modulation
FRC	Federal Radio Commission
FSK	frequency shift keying
GE	General Electric
GM	General Motors
GMT	Greenwich Mean Time
HF	high frequency
HMS	Her Majesty's Service
Hz	Hertz
IEEE	Institute of Electrical and Electronic Engineers
INA	International News Association
IRC	international reply coupon
IRE	Institute of Radio Engineers
KC	kilocycle (kilohertz)
KVA	kilo volt-amps
KW	kilowatt
LA	Los Angeles
LF	low frequency
LBYC	Long Beach Yacht Club
LORAN	Long Range Navigation
MARS	Military Affilate Radio System
MC	megacycle (megahertz)
NCDXC	Northern California DX Club
OARA	Oregon Amateur Radio Association
OB	old boy
OM	old man
OSS	Office of Strategic Services
OWI	Office of War Information
PARA	Philippine Island Amateur Radio Association
PE	professional engineer
PI	Philippine Islands
PST	Pacific Standard Time
QCWA	Quarter Century Wireless Association
RAF	Royal Air Force
REF	Reseau des Emetteurs Francais
RCA	Radio Corporation of America
RCAC	RCA Communications
RF	radio frequency
RIT	receive-independent-transmit
RME	Radio Manufacturers Engineers
RPM	revolutions per minute
RST	readability, signal strength & tone
SAREX	Shuttle Amateur Radio Experiment
SCAP	Supreme Commander, Allied Powers
SCDXC	Southern California DX Club
SCM	Section Control Manager
SK	silent key
SMOM	Sovereign Military Knights of Malta
SOWP	Society of Wireless Pioneers
SPST	single-pole single-throw
SS	sweepstakes
SSB	single sideband
SWL	short wave listener
SWR	standing wave ratio
TRF	tuned radio frequency
TV	television
TVI	television interference
TWA	Tibet workers of America
UCLA	University of California at Los Angeles
UHF	ultra high frequency
UMAA	University of Minnesota Alumni Association
UPI	United Press International
USSR	Union of Soviet Socialist Republics
VAC	volts alternating current
VFO	variable frequency oscillator
VHF	very high frequency
VIP	very important person
VJ	Victory in Japan
WAC	Worked All Continents
WAS	Worked All States
WAZ	Worked All Zones
WERS	War Emergency Radio Service
WPM	words per minute
5BDXCC	Five Band DX Century Club

Index